ONE WEEK LOAN

Pharmacology for Chemists

PHARMACOLOGY FOR CHEMISTS

Second Edition

Joseph G. Cannon

OXFORD
UNIVERSITY PRESS
2007

OXFORD
UNIVERSITY PRESS

Oxford University Press, Inc., publishes works that further
Oxford University's objective of excellence
in research, scholarship, and education.

Oxford New York
Auckland Cape Town Dar es Salaam Hong Kong Karachi
Kuala Lumpur Madrid Melbourne Mexico City Nairobi
New Delhi Shanghai Taipei Toronto

With offices in
Argentina Austria Brazil Chile Czech Republic France Greece
Guatemala Hungary Italy Japan Poland Portugal Singapore
South Korea Switzerland Thailand Turkey Ukraine Vietnam

Published by Oxford University Press, Inc.
198 Madison Avenue, New York, New York 10016

www.oup.com

Oxford is a registered trademark of Oxford University Press

Library of Congress Cataloging-in-Publication Data
Cannon, Joseph G., 1926–
  Pharmacology for chemists / Joseph G. Cannon—2nd ed.
    p. cm.
Includes bibliographical references and index.
ISBN 978-0-8412-3927-2
1. Pharmacology. I. Title.
RM300.C36 2006
615′.1024541—dc22   2005057231

9   8   7   6   5   4   3

Printed in the United States of America
on acid-free paper

To Lynne, with continued love

Still with me, still for me

Preface to Second Edition

In the approximately six years since the first edition of this book appeared, great changes and profound advances have occurred in the pharmacological sciences, as must be expected in every area of science. I have constantly and continually revised the didactic material for my American Chemical Society short course (upon which the first edition of this book was based) to attempt accurately to reflect the state of 21st century pharmacology and, from time to time, to look further into the future. In this second edition of the book, my goal has been to effect similar changes in *Pharmacology for Chemists*.

I have attempted to maintain a critical attitude toward the contents of this book, retaining essentially unchanged those portions that I view as still being relevant, appropriate, and contemporary, while adding new information and ideas and deleting or changing narrative material and topics that I concluded to be no longer valid or to be less useful to chemist readers. Many diagrams and illustrations have been changed, added or deleted. Literature reading references at the end of each chapter have been extensively updated. The glossary at the end of the volume has been expanded, and some definitions have been modified, better to reflect contemporary thought.

In accord with my original plan for the first edition, I have produced a volume of modest size, such that chemist readers will be less likely to be intimidated by the depth and magnitude of the discussions. The book is still intended to be an *introduction* to pharmacology, and it is still aimed at chemist readers for their self-study efforts.

My wife, Dr Carolyn Cannon, utilized her computer and organic chemical skills to construct the chemical structures in the book, a computer-based task far beyond my ability. To her, all thanks.

Preface to First Edition

This book derives from a three-day short course, "Pharmacology for Chemists," which I have offered under the auspices of the American Chemical Society for approximately 20 years. The course originated from a concern that many newly graduated Ph.D.-level chemists enter medicinal chemistry/drug design research groups in the pharmaceutical industry, in academia, and elsewhere, well prepared in appropriate areas of chemistry, but with no knowledge of and no appreciation for pharmacology and physiology. Productive medicinal chemical research requires that chemist practitioners be skilled in integrating their chemistry expertise with biology, and that they participate in meaningful dialog with pharmacologists and other biologist collaborators. Chemists must appreciate and understand the biological significance of their chemical efforts.

In considering the nature of an abbreviated set of lectures in pharmacology, I concluded that a course aimed at medicinal chemistry researchers should be different from one designed for the education of health care practitioners, whether physicians, pharmacists, dentists, or nurses. I attempted to identify those aspects of pharmacology that have been useful to me and my research group in our medicinal chemical research: inter alia, physical chemical aspects, molecular-level biological mechanistic phenomena, and metabolism. Most contemporary pharmacology textbooks are clinically oriented, and they feature extensive discussions of drug therapy and specifics of clinical uses of individual drugs, dosage strategies and regimens, and empirical descriptions of side effects. This approach is frequently not optimally beneficial to the chemist reader.

Pharmacology is not a "pure," monolithic science, but rather an *olio* of physiology, anatomy, organic chemistry, biochemistry, and physical chemistry. I concluded that a course format would be most appropriate which would demonstrate to chemists that they can (and should) utilize their chemical knowledge in understanding pharmacological principles. The short course, and now this book, attempt to utilize chemistry as the foundation upon which to construct a working knowledge of pharmacology in conjunction with discussions of elements of physiology and anatomy. I have frequently stated, only partly facetiously, that pharmacology might be considered a branch of organic chemistry.

The line of demarcation between chemical pharmacology and medicinal chemistry is not sharp. I have attempted, however, to remain cognizant that the present effort is intended to be an exposition of *pharmacology* and not of medicinal chemistry.

This book was written for the organic chemist who has a grasp of physical chemistry and of biochemistry but only a modest knowledge of biology. I have expanded the coverage of most topics beyond what was possible in a three-day oral presentation. However, this book is not, nor is it intended to be, a complete "textbook of pharmacology." Space (and other) constraints did not permit inclusion of all categories of drugs. Topics were selected on the basis of three criteria:

1. My (admittedly prejudiced) assessment of the overall importance in contemporary drug therapy and research.
2. How well the drug category lends itself to use and exposition of pharmacological principles. Is the topic a good teaching tool?
3. My competence to write about the topic.

I have consciously attempted to present a work of modest size which will be less likely to intimidate the prospective reader. The discussions are intended to introduce the subjects addressed, with the goals of imparting fundamental information and of stimulating the reader subsequently to expand his/her pharmacological horizons beyond the superficial, occasionally simplistic coverage provided. Each chapter ends with a listing of references to more extensive and more detailed literature. It is my hope that this book will give the reader sufficient and appropriate background to facilitate further, more advanced independent study, even in categories of drugs not addressed in the book. A glossary of commonly employed terms is included for convenience.

The volume comprises three sections: Part I contains discussions of a considerable number and variety of chemical and biological topics and concepts upon which pharmacology is based. Parts II and III address specific categories of drugs and attempt to demonstrate that the concepts and phenomena described in Part I can be exploited to rationalize the experimentally observed pharmacology of the drugs described and perhaps additionally to assist the reader in gaining that most precious of intellectual gifts, *understanding*.

Contents

Part I: Chemical and Biological Bases of Pharmacology

1 Some General Concepts of Pharmacology   3

   1.1   Introduction   3
   1.2   Biological Membranes   3
   1.3   Water: Structure and Pharmacological Significance   8
   1.4   Drug Dissolution   8
   1.5   Penetration of Membranes by Organic Molecules   9
   1.6   Lipophilicity, Hydrophilicity, and Partition Coefficient   11
   1.7   Drug Absorption and Transport   11
   1.8   The Blood–Brain Barrier   16
   1.9   Transport Across the Placental Membrane   18
   1.10 Storage Sites for Drugs and Their Metabolites   18
         Bibliography   19
         Recommended Reading   19

2 Pharmacokinetics   21

   2.1   Introduction   21
   2.2   Bioavailability   21
   2.3   Compartments   22
   2.4   Clearance   23
   2.5   Volume of Distribution   24
   2.6   First-Order Elimination Kinetics   25
   2.7   Biological Half-Life   27
   2.8   Model for Combined Absorption and Elimination   28
   2.9   Nonlinear Pharmacokinetics   28

2.10 Loading Dose   29
      Bibliography   29
      Recommended Reading   30

3 Drug Metabolism   31

3.1   Metabolism of Xenobiotics   31
3.2   Aspects of Functional Microanatomy of the Kidney   31
3.3   Role of the Kidney in Drug Excretion   33
3.4   Extrarenal Routes of Drug and/or Drug Metabolite Excretion   34
3.5   Undesirable Metabolic Consequences   35
3.6   Differences in the Metabolic Fate of Enantiomers   36
3.7   In Vivo Drug Metabolism   37
3.8   Chemical Aspects of Drug Metabolism   40
3.9   Animal Species Differences in Drug Metabolism   50
3.10 Human Genetic Variation in Drug Metabolism   52
3.11 Age and Gender Differences in Drug Metabolism   54
3.12 Prodrugs and Latentiated Drugs   54
      Bibliography   56
      Recommended Reading   56

4 Drug Receptors   57

4.1   Receptor Sites and Drug Binding Sites   57
4.2   Structurally Nonspecific and Structurally Specific Drugs   63
4.3   Agonists and Antagonists: Occupancy Theory of Drug Action   64
4.4   Competitive and Noncompetitive Antagonists   66
4.5   Induced Fit   66
4.6   Partial Agonists and Inverse Agonists   67
4.7   Enzymes as Drug Receptors   67
      Bibliography   72
      Recommended Reading   73

5 Principles of Pharmacological Assays   74

5.1   Affinity (Binding) Assays   74
5.2   Quantification of Biological Responses   76
5.3   Tachyphylaxis: Drug Tolerance   79
5.4   Cumulation   80
5.5   Quantitative Expression of Effective Dose   81
5.6   Establishment of Pharmacological Antagonism as
      Competitive or Noncompetitive   82
5.7   Quantal Assays   83
5.8   Therapeutic Index   85
5.9   Numerical Expression of Dose   87
      Bibliography   88
      Recommended Reading   88

Part II The Peripheral and Central Nervous Systems

6 Some Basic Concepts of the Anatomy and Physiology of the
Nervous System   91

    6.1   Aspects of Functional Anatomy of the Nervous System and Its
Related Effector Organs   91
    6.2   Trans-synaptic Nerve Impulse Transmission   102
    6.3   Introduction to the Autonomic Nervous System   105
        Bibliography   108
        Recommended Reading   108

7 The Noradrenergic and Dopaminergic Nervous Systems   109

    7.1   The Noradrenergic System   109
    7.2   The Dopaminergic System   125
        Bibliography   130
        Recommended Reading   130

8 The Cholinergic System   131

    8.1   Acetylcholine-Inactivating Enzymes   131
    8.2   Anatomy/Physiology of the Cholinergic Nerve Terminal   132
    8.3   Cholinergic Receptors and Receptor Subtypes   134
    8.4   Muscarinic Agonists and Partial Agonists   135
    8.5   Nicotinic Agonists   136
    8.6   Indirect-Acting Cholinergic Agents   138
    8.7   Therapeutic Uses of Cholinergic Receptor Stimulants   141
    8.8   Nicotinic Receptor Blockers   142
    8.9   Muscarinic Receptor Blockers   145
    8.10 Cognitive Dysfunction: Alzheimer's Syndrome   147
        Bibliography   149
        Recommended Reading   149

9 The Central Nervous System. I: Psychotropic Agents   151

    9.1   Commonly Used Terms in Central Nervous
System Pharmacology   151
    9.2   Biochemistry and Physiology of Neurotransmitters in the
Central Nervous System   152
    9.3   Functions of Some Brain Regions   166
    9.4   Animal Screening of Psychotropic Drug Candidates   167
    9.5   Antidepressants   169
    9.6   Cocaine   171
    9.7   Cannabis   172
    9.8   Mood-Stabilizing Agent   174
    9.9   Antianxiety Agents   175
    9.10 Antipsychotic Drugs   177

9.11   Psychotomimetics   180
Bibliography   181
Recommended Reading   181

10 The Central Nervous System. II: Sedatives and Hypnotics   183

10.1   Definitions   183
10.2   Rapid Eye Movement Sleep   183
10.3   Ethanol   184
10.4   Non-Barbiturate, Non-Benzodiazepine Sedatives
and Hypnotics   185
10.5   Benzodiazepines   186
10.6   Barbiturates   187
10.7   Therapeutic Uses of Barbiturates   189
Bibliography   189
Recommended Reading   189

11 Analgesics. I: General Considerations and Non-Opioid Analgesics   191

11.1   Definition of Terms   191
11.2   Evaluation of Effectiveness of Analgesics in
Animals and Humans   192
11.3   Physiological and Biochemical Aspects of Pain
Production and Recognition   194
11.4   Non-Opioid Analgesics and Nonsteroidal
Antiinflammatory Analgesics   195
11.5   Coal Tar Analgesics   204
11.6   Capsaicin   205
11.7   Classification of Pain   206
Bibliography   206
Recommended Reading   206

12 Analgesics. II: Opioid Analgesics   208

12.1   Terminology   208
12.2   Morphine-Like Analgesics   208
12.3   Opioid Drug Antagonists   214
12.4   Analgesic Receptors   215
12.5   Endogenous Analgesic Receptor Agonists   217
12.6   Analgesic Mechanisms of Endorphins, Opioids,
and Some Other Drugs   219
12.7   Opioid and Analgesic Peptide Tolerance and Dependence   220
12.8   Physiological Implications of the Existence of Endogenous
Analgesic Substances   220
12.9   Mechanistically Novel Approaches to Analgesic
Drug Therapy   220

Bibliography   221
Recommended Reading   221

13 General and Local Anesthetics   222

13.1   Introduction   222
13.2   General Anesthetics   222
13.3   Local and Topical Anesthetics   227
       Recommended Reading   228

Part III: Pharmacology of Some Peripheral Organ Systems

14 The Cardiovascular System. I: Anatomy and Physiology   231

14.1   Aspects of Functional Anatomy of the Heart   231
14.2   Hypertension   233
14.3   Hyperlipidemia/Atherosclerosis   243
14.4   Myocardial Infarction   249
       Bibliography   253
       Recommended Reading   254

15 The Cardiovascular System. II: Arrythmias and Antiarrythmic Drugs   255

15.1   Arrhythmias   255
15.2   Myocardial Ischemia   260
       Bibliography   265
       Recommended Reading   265

16 The Cardiovascular System. III: Congestive Heart Failure and Diuretics   267

16.1   Congestive Heart Failure   267
16.2   Diuretics   271
       Bibliography   278
       Recommended Reading   278

17 Pharmacology of Some Histamine-Implicated Diseases   279

17.1   Allergy   279
17.2   Bronchial Asthma   283
17.3   Gastric Hypersecretion   287
       Recommended Reading   291

Glossary   292

Index   303

I

Chemical and Biological Bases
of Pharmacology

1
Some General Concepts of Pharmacology

Introduction

Pharmacology has been defined as the study of the manner in which the function of living systems is affected by chemical agents. It embraces all aspects of knowledge of drugs: chemistry, physiology, and biochemistry. Pharmacology is concerned with both drug effects and drug actions. *Drug effects* are drug-induced alterations in the normal function of an existing biological process, for example, a change in heart rate, a change in blood pressure, or alteration of the pain threshold. *Drug action* addresses where and how a drug ultimately produces its effect, that is, its biological mechanism. The combined study of drug actions and drug effects is called *pharmacodynamics*, as contrasted with *pharmacokinetics* which involves study of the rates of change of drug concentrations in the body as functions of rates of absorption, distribution, biotransformation (metabolism), and excretion of a drug.

The Recommended Reading section at the end of this chapter lists some useful literature sources for further reading in pharmacology. Some medicinal chemistry compendia present sizeable amounts of pharmacological discussions on specific topics. These works frequently have the advantage that the authors are chemists who are directing their writings to a chemistry constituency. Essential supplemental aids for studying and understanding pharmacology include textbooks of physiology and of anatomy and a medical dictionary.

Biological Membranes

1.2.1 Membrane Structure

Membranes are the thin layers of tissue that cover the surface or divide a space or an organ in a living organism. They are the "skins" of the cells of the body. Membranes hold the cell together, preventing loss of shape and loss of components of the cell to its environment. All food and nutrients must enter through the cell membrane, and waste products of the cell's metabolism must leave through the membrane. Additionally, some of the

important chemical functions of cells occur in or on the membrane. Some enzyme molecules are embedded in the membrane matrix, and they exert their catalytic effects there.
Many (but not all) of the receptor sites for drugs are located on or in the cell membrane.
Bacterial membranes are much more complex structurally than are mammalian
membranes; this permits bacteria to survive in various host organism environments.

Biological membranes can affect the extent and duration of the pharmacological
effect of a drug. Upon oral administration of a drug, it must pass through the membrane
lining of the gastrointestinal tract, cross the membranes that define the circulatory
system and, after transport in the blood, pass through the membranes of the capillaries
that nourish the cells that comprise an organ, and eventually reach those specific cells
where the drug exerts its desired pharmacological action. Membranes are intimately
involved in determining the absorption, distribution, and elimination patterns of drugs.
Membranes are composed chiefly of two classes of chemical compounds, proteins and
lipids. Among the membrane-associated proteins are some of the receptor sites for drugs.
A significant portion of membrane lipids is phospholipid material, glyceryl esters in
which the positions 1 and 2 hydroxyl groups of glycerin are esterified with long chain
fatty acids, and the carbon 3 hydroxyl is esterified with orthophosphoric acid which is
also esterified with an amino alcohol (such as ethanolamine) or with a quaternary amino
alcohol (such as choline), as illustrated in figure 1.1.

The third acidic function of the orthophosphoric acid moiety is unesterified, and
under physiological conditions it is dissociated to afford an anion. Thus, one portion
of the phospholipid (the di-ionic choline phosphate moiety) is extremely hydrophilic
and the other portion (bearing the fatty acid alkyl chains) is very lipophilic. The small
illustration at the right of the chemical structure illustrates the symbolic representation
of a phospholipid molecule. The circle represents the ionic (choline phosphate) portion
and the "strings" represent the long alkyl chains of the fatty acid residues.

The architecture of the biological membrane is shown diagramatically in figure 1.2.
The solid bodies are protein molecules that appear to be randomly distributed. Some of
these molecules are located on the outer or the inner surface of the membrane, and some
penetrate completely through the membrane structure. The body of the membrane
consists of a double layer of phospholipids ("lipid bilayer"), present in a highly organized
arrangement as illustrated. The alkyl chains participate in van der Waals interactions and
hydrophobic bonding phenomena, inter alia, and these interactive forces provide strength,
cohesiveness, and integrity to the membrane. Thus, the inner and outer surfaces of the
membrane are highly ionic and highly hydrophilic, and the inner portion of the membrane
is highly lipophilic.

1.2.2 Ion Channels

The structure of the membrane also includes pores or channels which extend through the
membrane and through which small polar molecules or ions may pass. These channels
are a part of the protein components and, by changes in the conformation (the shape)
of the protein molecules, the channel may open or close, as required by the physiology
of the body. For example, sodium ions are transported across nerve membranes via
sodium-specific channels (figure 1.3). The sodium channel contains a *gating component*
that opens or closes the channel and a *filter component*, 0.3–0.5 nm in diameter, lined
with carboxylate anions (parts of the amino acid components of the protein molecule)

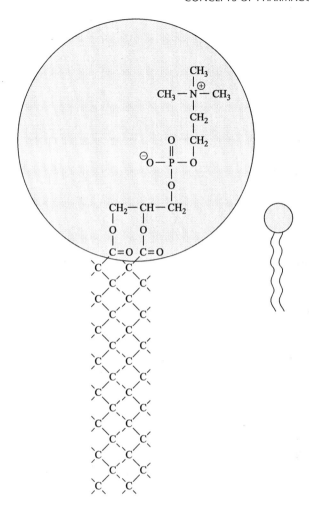

Figure 1.1 Phospholipid structure. (Reproduced with permission from reference 1. Copyright 2001 McGraw-Hill)

which attract the sodium cations but repel anions. These carboxylate anions also lower the free energy required to remove the shell of water molecules that surrounds the sodium ion. Only an unhydrated sodium cation fits the dimensions of the channel for passage across the membrane. Once inside the cell, the sodium cation regains its surrounding shell of water molecules. There are several other kinds of membrane channels selective, inter alia, for transport of potassium, chloride, calcium, or hydrogen ions or for other body constituents. The selectivity of an ion channel for a given ion is based, in part, on steric factors (the size of the ion involved and the diameter of the channel) and, in part, on a degree of selectivity for binding of an ion at a given channel. It is as yet imperfectly understood why most ion channels are so selective for the ion that they transport. Why, for example, does a potassium channel reject a sodium cation, even though Na^+ has a smaller diameter than K^+? It has been proposed that

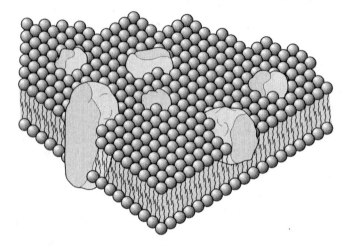

Figure 1.2 Lipid bilayer membrane model. (Reproduced with permission from reference 2. Copyright 2001 Morton Publishing)

the shape of the filter component of the potassium channel is splayed (funnel-shaped) to keep the carboxyl oxygens far enough apart to make the energy cost too high for coordinating a dehydrated Na⁺. One membrane channel has been identified which is not selective for ions of any species or charge, but admits ions simply by their size. The opening and closing of some ion channels are controlled by the effect of a nerve impulse. These are termed *voltage-gated* channels. Other ion channels are opened or closed by chemical interaction of the channel protein complex with some endogenous organic molecule. These are termed *ligand-gated* channels. There are many exogenous ligands that can control ion channels (even physiologically voltage-regulated ones). For this reason, many pharmacologists prefer the term "transmitter-gated", indicating the participation of an endogenous ligand) as a replacement for the term "ligand-gated".

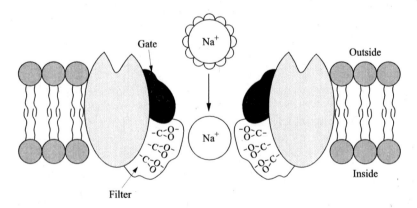

Figure 1.3 A sodium ion channel

1.2.3 Fluid Mosaic Membrane Model

This hypothesis derives from the proposal that biological membranes possess many of the properties of a fluid. The protein molecules embedded in the phospholipid matrix seem to float in it, much like corks in a pan of water. Thus, the arrangement and distribution of the protein molecules in the membrane are variable and are constantly changing.

1.2.4 Additional Chemical Details of Membrane Structure

The membrane representation in figure 1.2 is highly oversimplified. In addition to the phospholipid alkyl chains, the central portion of biologic membranes contains molecules of a lipophilic steroid, cholesterol (figure 1.4). The rigid cholesterol molecules interact with the alkyl chains of the phospholipids (via van der Waals and hydrophobic inter-actions) and keep them organized, thereby adding strength and cohesiveness to the membrane matrix structure. Cholesterol also restricts molecular motion in the hydrophobic portion of the membrane, thus diminishing the fluidity of the phospholipid matrix. A very small portion (2–5%) of mammalian membranes is carbohydrate material, covalently bound to the membrane lipid or protein components. The role(s) of these carbohydrates is/are imperfectly understood; however, it is established that antigen–antibody reactions involve carbohydrate material attached to membranes. Another important component of membranes is calcium cations which interact with the negative charges on the inner and outer membrane surfaces and add stability to the

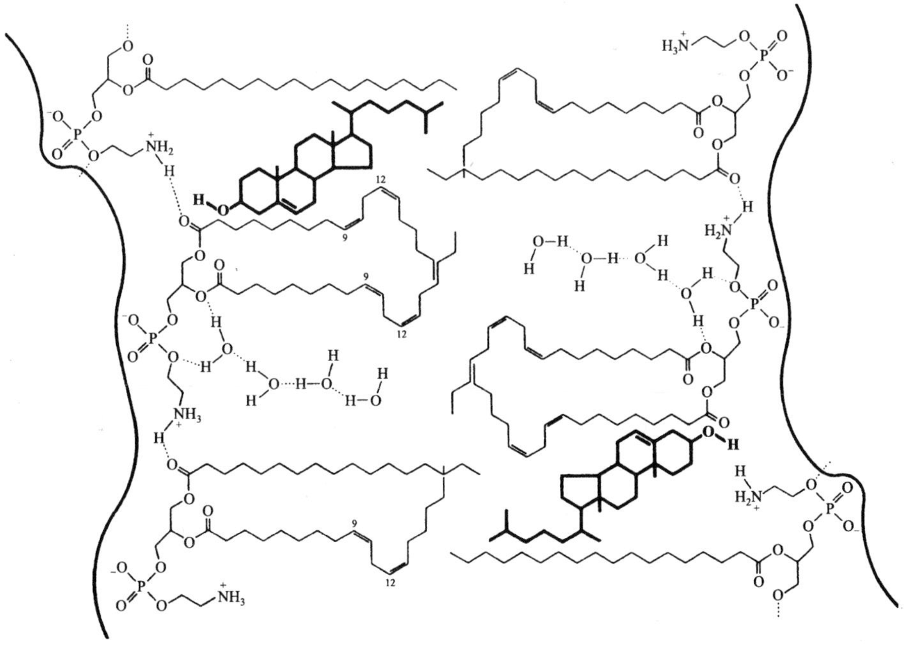

Figure 1.4 Phospholipid bilayer model of a membrane including cholesterol molecules (Reproduced with permission from reference 3. (Copyright 2000 W. B. Saunders)

membrane structure. Displacement of these calcium cations by a cationic drug molecule will lead to potentially significant changes in the organization and properties of the membrane.

Most biologic membranes are relatively permeable to water, either by simple diffusion or by the flow that results from hydrostatic or osmotic differences across the membrane. This kind of bulk flow of water can carry with it small, water-soluble substances such as urea. Indeed, most cell membranes permit passage only of water and urea by this mechanism. Note that this passage does not involve ion channels or pores.

1.3   Water: Structure and Pharmacological Significance

Water is the solvent and principal fluid in all living organisms. The pharmacological effects of drugs in the body occur in aqueous solution, and the physical nature of liquid water is significant in pharmacology. For many years it has been debated whether liquid water is a mixture of hydrogen-bonded species or is a continuum. Spectral studies lead to the conclusion that liquid water is a mixture of hydrogen-bonded molecular aggregates of varying size. A model for liquid water that seems useful in pharmacology is the *flickering cluster*, in which liquid water is depicted as a rapidly fluctuating mixture of variously hydrogen bonded aggregates of molecules (figure 1.5).

The arrangement of water molecules in the solid form (ice) is a three-dimensional crystal lattice in which each oxygen atom is surrounded tetrahedrally by four hydrogens. Hydrogen bonds hold the oxygens (and the crystal lattice) together. Because of the partially covalent character of the hydrogen bond, when a single hydrogen bond is formed with an oxygen atom, the remaining lone pair of electrons on the oxygen becomes more localized and more nearly sp^3 hybridized than would be the case if the oxygen were not hydrogen bonded. Accordingly, a single hydrogen-bonded water molecule is more susceptible to further hydrogen bonding, which in turn imparts additional stability to the existing system. The net result is that when one hydrogen bond forms, there is a tendency for several hydrogen bonds to form, producing a crystal lattice. When one hydrogen bond breaks, there is a tendency for an entire group of hydrogen bonds to break, causing the crystal lattice to collapse. Thus, short-lived flickering clusters of highly ordered, hydrogen-bonded water molecules are produced. According to this model, liquid water is not static, but it is highly dynamic. It is important to recognize that liquid water contains regions where the water molecules are in some "crystalline" arrangement.

Of special significance in pharmacology is the fact that biological membranes are closely associated with water. The inner and outer membrane surfaces, which contain large numbers of anions and cations, are hydrated and this water of hydration is highly associated and highly organized into a stable crystal lattice (termed an "iceberg"), of a thickness of one molecule of water or more. Before a drug can pass through a membrane or before it can attach to some receptor area on the membrane surface, it must disrupt and displace a portion of this shell of ordered iceberg water.

1.4   Drug Dissolution

Following oral administration of a solid drug, its *dissolution rate* is important in determining eventual levels of drug in the blood and elsewhere. A drug must be in solution for absorption phenomena to occur. If the drug dissolves too slowly in the

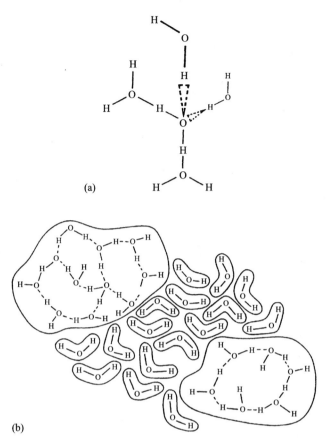

Figure 1.5 Schematic diagram of water structure. A. tetrahedral hydrogen bonding of a water molecule in ice; B. two-dimensional diagram of flickering clusters of water molecules in the liquid state. (Reproduced with permission from reference 4. Copyright 2003 Wiley interscience)

fluids of the gastrointestinal tract, it cannot efficiently diffuse to the wall of the gastrointestinal tract for absorption. The drug will be excreted in the feces. Variation in particle size and surface area of a solid drug can produce differences in rates of dissolution, as can chemical interactions between the drug and therapeutically inert binder or diluent in tablets and capsules. It is well established that polymorphic forms of the same chemical compound may have vastly different rates of dissolution.

1.5 Penetration of Membranes by Organic Molecules

1.5.1 Passive Diffusion

Organic drug molecules do not penetrate membranes by passing through ion or other channels. Most drug molecules are too large to fit the dimensions of the channels. The great majority of drugs crosses biological membranes by passive diffusion from a region of high concentration (e.g., the gastrointestinal tract for an orally administered drug) to a region of relatively lower concentration (the blood). Passive diffusion across a membrane requires that the drug molecule partition into the lipophilic membrane matrix, a process requiring that the drug molecule itself possesses some degree of lipophilic character. The rate of diffusion depends, in part, on the concentration gradient across the membrane; the greater the difference in concentration, the faster the rate of diffusion. The *percentage* of a drug dose absorbed at any time after administration, however, is the same, regardless of the size of the dose administered. More recently, it has been suggested that some drugs may be absorbed by passive diffusion through water-filled spaces at junctions between individual cells. This phenomenon is termed *paracellular absorption*, and it has been suggested to be important in absorption of very hydrophilic compounds.

1.5.2 Active Transport

Many dietary components cannot penetrate membranes by passive diffusion because of their highly hydrophilic character. An *active transport* mechanism is necessary for these substances. By definition, active transport involves a series of chemical interactions between the transported substance and component(s) of the membrane, which requires the expenditure of energy. Dietary sugars and amino acids are absorbed from the intestinal tract by highly specific transport mechanisms. Relatively few drugs can serve as substrates for existing physiological active transport mechanisms. However, as examples, levodopa, the antiparkinsonian drug, is actively transported across the intestinal wall by a mechanism used for absorbing dietary aromatic amino acids such as phenylalanine. The cytotoxic anticancer agent fluorouracil is actively transported across the intestinal wall by a mechanism used physiologically to absorb pyrimidine derivatives.

1.5.3 Kinetic Terminology

The kinetics of drug passage across a membrane can involve one or the other of two possibilities: the rate of transport may be constant and independent of the amount of drug to be absorbed (*zero-order kinetics*: characteristic of active transport); or the rate may be maximal at the instant when passage begins, and thereafter be constantly diminishing, always in proportion to the amount of drug yet to be absorbed (*exponential* or *first-order kinetics*: characteristic of passive diffusion).

1.5.4 Pinocytosis

A third membrane penetration process, *pinocytosis*, involves folding over of a part of the cell membrane and the formation of a small vesicle in which molecules of extracellular constituents are trapped ("vagination"). The vesicle wall can break and then the vesicle contents can be released within the cell, or they can be extruded from the other side of the cell. Pinocytosis is believed to be involved in the transport of insulin molecules from the blood stream into the brain. However, this process has not been demonstrated in the transport of small nonpeptide organic molecules across membrane barriers.

1.6   Lipophilicity, Hydrophilicity, and Partition Coefficient

As described previously, an organic molecule's penetration of membranes by passive diffusion is related to its lipid solubility and to its concentration gradient across the membrane. Passage of drugs across membranes of the oral cavity, of the gastrointestinal tract, through the skin, into the central nervous system (brain and spinal cord), into tissue cells, and through the kidney is related to the lipophilic character of the molecule. However, lipid solubility alone is not sufficient to implement absorption and distribution through the body. For example, oral administration of liquid petrolatum (mineral oil), a mixture of long chain (C_{18}–C_{24}) chiefly saturated hydrocarbons, does not result in absorption across the wall of the gastrointestinal tract. Hydrocarbons such as liquid petrolatum lack a significant degree of *water solubility* accompanying their lipid solubility. For optimum absorption and transport by passive diffusion processes, a molecule must possess a finite degree of *dual solubility*, in water and in lipophilic solvents. The *partition coefficient* of a drug is defined as the experimentally determined ratio of its concentration in a lipophilic solvent to its concentration in water or an aqueous buffer. In the simplest experimental procedure a weighed quantity of a drug is shaken with a system of accurately measured equal volumes of two immiscible liquids, water (or aqueous buffer) and an organic solvent. When the drug solute is equilibrated between the two solvents, the amount of drug in each solvent is determined.

Partition coefficient data are frequently expressed as the logarithm ("log P") of the fraction: solubility in nonpolar solvent/solubility in water. *n*-Octanol is frequently used as the water-immiscible solvent because of its superior solvent properties for a wide variety of organic molecules and because it has little tendency to form an emulsion when it is shaken vigorously with water. Drugs vary considerably in the numerical value of their partition coefficients. In general, the relationship between log P values of drugs and their pharmacological potency/activity is complex. There is no single optimal log P value for all drugs. However, in some specific chemical–pharmacological categories (e.g., barbiturate sedatives/hypnotics) the partition coefficients of all active compounds are numerically very similar. If the log P value of a barbiturate molecule falls between certain numerical limits, that molecule will be expected to depress the central nervous system. However, if the log P of a barbiturate molecule falls outside these numerical limits, almost always the compound will be found to be inactive as a sedative/hypnotic.

1.7   Drug Absorption and Transport

1.7.1 Pharmacological Significance of Acidity and Basicity of Drugs

Most drugs are either weak acids or weak bases. The development of a unit charge on a drug molecule depends on its pK_a and on the pH of its environment. The ability of a drug molecule to form an ion greatly affects its passage across membranes, because un-ionized molecules have greater lipid solubility and therefore the un-ionized form of a drug diffuses across the lipid barrier of the membrane much more easily than the ionized form. The fluids of the gastrointestinal tract display a wide pH range. The pH in the stomach (due to the secretion of hydrochloric acid as a part of the digestive process) is strongly acidic: between 1 and 3. Beyond the pyloric valve (leading from the stomach to the small intestine) in the duodenum (upper portion of the small intestine), the pH rises sharply and lies between 5 and 7. The pH of the intestinal fluids continues to rise gradually farther along the gastrointestinal tract, until a maximum pH of approximately 8 is attained in the colon. Therefore, the acidic stomach has the optimal pH for absorption (by passive diffusion) of weak organic acids, whereas absorption of organic bases from the stomach is minimal or non-existent. For example, the carboxyl group of a typical weakly acidic drug acetylsalicylic acid (aspirin) **1.1** is protonated to a lipophilic neutral species by the hydrochloric acid of the stomach. A modest portion of an oral dose of this drug is absorbed across the stomach wall.

COOH   COO$^-$

$-H^+$

$+H^+$

OCCH$_3$ OCCH$_3$

O   O

1.1   **1.2**

Amines are protonated by gastric hydrochloric acid, and the resulting hydrophilic cationic species lacks sufficient lipophilic character to permit gastric absorption by passive diffusion. At the higher pH in the duodenum, a certain percentage of molecules of organic bases is electrically neutral and these are lipophilic. Conditions favoring their absorption exist here. Ethanol, a neutral molecule, has an appropriate partition coefficient for a significant degree of absorption across the stomach wall as well as across the intestinal wall. Given that, at best, drugs are poorly absorbed from the stomach, gastric emptying time is an important factor. The faster the emptying time of the stomach, the faster the drug will arrive in the intestine for the most efficient absorption. Food is a factor in this process; passage of an orally administered drug into the intestine is much faster from an otherwise empty stomach. Some drugs slow the emptying time of the stomach (e.g., the muscarinic blocking drug propanthaline) and some drugs speed emptying time (e.g., the serotonin receptor stimulant metoclopramide). However, it must be borne in mind that, histologically and physiologically, the stomach wall is poorly designed for efficient absorption of its contents, including drugs.

1.7.2 Role of Villi

In the small intestine, that portion of the oral dose of a weakly acidic drug that was not absorbed from the stomach is at least in part in its ionized form (e.g., acetylsalicylate anion **1.2**), and it might be predicted that absorption of these anions would be highly disfavored in this higher pH environment. However, most weakly acidic drugs are absorbed rather efficiently across the duodenal wall, and there is an anatomic explanation for this phenomenon. The inner surface of the intestinal wall, although appearing smooth to the naked eye, is revealed by microscopic examination to be covered with small, hair-like protuberances, *villi* (figure 1.6). These structures are highly vascular, containing many tiny, thin-walled blood vessels. There are many villi per square inch of intestinal wall surface, and as a result the total surface area of the intestinal inner wall is increased by as much as 600-fold. The physiological function of villi is to facilitate absorption of dietary components that are not actively transported and that lack

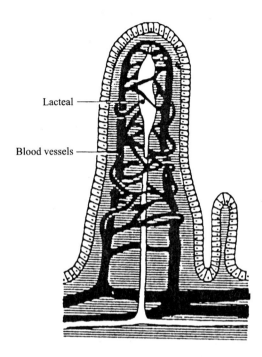

Lacteal —

Blood vessels —

Figure 1.6 An intestinal villus. (Reproduced with permission from reference 5. Copyright 2000 W. B. Saunders)

hydrophilic/lipophilic properties favoring passive diffusion processes. Villi serve this same purpose in the case of drug molecules. The large surface area created by the villi permits large *total* absorption of molecular species that have only an extremely poor diffusion tendency, and they also facilitate absorption of basic drug molecules which are not absorbed across the stomach wall.

1.7.3 Additional Factors Influencing Drug Absorption

Quaternary ammonium compounds generally have very low lipid solubility, and they have unfavorable partition coefficients for passive diffusion-based absorption anywhere along the gastrointestinal tract. Organic molecules bearing strongly acidic groups, such as a sulfonic acid moiety, are generally not well absorbed across gastrointestinal tract membranes. The sulfonic acid group is completely ionized over a wide pH range, and the resulting hydrophilic sulfonate anion inhibits diffusion across lipid barriers.

Another factor that affects absorption of some drugs from the intestine is the presence of materials in the lumen of the intestine with which the drug reacts chemically. For example, the tetracycline antibiotics chelate calcium ions to form a non-absorbable complex, and calcium-rich foods such a dairy products inhibit tetracycline absorption.

1.7.4 Other Possible Sites of Absorption

Absorption of drugs from the lungs, eyes, skin, and subcutaneous and intramuscular sites and from the oral and rectal cavities has been studied. All of these sites of drug absorption are governed by the same lipophilicity and partition coefficient requirements that were described for gastrointestinal tract absorption. The rectal route is often useful when the oral route is not practical (e.g., if the individual is vomiting or is unconscious). Approximately 50% of a dose of a drug that is absorbed from the rectum will bypass the liver and will escape first pass metabolic destruction (see chapter 3). Rectal administration is less common in the United States than elsewhere in the world, largely due to patient non-acceptance for aesthetic reasons.

The eye represents an especially intriguing and challenging site for drug administration. For some drugs, it has been determined that less than 1% of a dose instilled topically into the eye enters the target region(s) of the eye; the remainder enters the systemic circulation, having been absorbed in part across the membrane of the conjunctiva and in part through anatomic connections between the eye and the nasal cavity (nasolachrymal canal). The difficultly permeable surface of the cornea frequently inhibits significant absorption into the eye.

The empirical "rule of five" proposes that organic molecules are likely to have a poor absorption ability when their molecular weights are above 500; when their calculated log octanol/water partition coefficient is greater than 5; when the number of hydrogen bond *donors* is greater than 5; and when the number of hydrogen bond *acceptors* is greater than 10. This "rule" has been used by computational chemists and practitioners of (QSAR) in making decisions about design of new drug molecules. A limitation to the rule of five is that it is not applicable to those compounds that undergo active transport.

Currently, the entire concept of drug absorption is undergoing some profound changes in interpretation. Hydrophilic compounds are less soluble in hydrophobic environments such as the phospholipid bilayer of cellular membranes, and it seems to follow that such molecules cannot easily cross the membrane. In contrast, hydrophobic compounds easily diffuse into the membrane. These observations seem to reveal an inconsistency: if hydrophobic compounds prefer a lipid environment, then why do they diffuse through the cellular membrane into the highly hydrophilic cytoplasm of the cells? In the past few years it has been proposed (and in some instances, *established*) that so-called **transport proteins** play a major role in determining whether hydrophilic and hydrophobic compounds are allowed passage across the membrane barrier. Many drugs and/or their metabolites are now known to be substrates for membrane transporters. It has been estimated that humans possess 2000 or more different kinds of transporters. There are proposed to be three functional classes of these transporters: (1) *influx* transporters that facilitate movement of compounds into the cell; (2) *efflux* transporters that remove compounds from the cytoplasm of the cell; and (3) *exchangers* which permit simultaneous movement of solutes across membranes in opposite directions.

Further to confuse the issue, some transporters are "active" and they require energy to operate: the source of the energy is the ATP→ADP reaction. However, there appear to be transporters that are not coupled to an energy source, and they are passive. In some way, these facilitate the diffusion of molecules across the membrane, down their concentration gradients, to permit rapid equilibration across the membrane (ref. 12, Recommended Readings for Topics in Chapter 1).

If this entire transport protein concept is valid and is generally applicable, it will seem necessary to re-evaluate and re-interpret the entire concept of penetration of membranes by drugs and their metabolites, and this will also require re-assessment of at least some aspects of pharmacokinetics.

1.7.5 Drug Distribution

After a drug molecule reaches the blood stream, it must find its way to its appropriate site of biological action. The body is composed of a wide variety of tissues, and drug molecules have varying affinities for them. Distribution of drugs throughout the body is often rapid, and equilibrium between drug in the blood and drug in other body tissues is quickly established. However, there are instances in which distribution of a drug from the blood into other body compartments is not rapid, and a finite amount of time is required to establish equilibrium between blood and tissues. In this situation the concentration of drug in the blood cannot be correlated with its concentration at its site(s) of pharmacological action, and determination of blood levels of the drug will not be dependable in establishing a dosage regimen. If distribution to the site of action is very slow, extremely high blood concentrations may be needed to achieve a therapeutic concentration at the site of action. The incidence of side effects may be high under these circumstances, and the drug may be therapeutically unsafe.

1.7.6 Drug Binding by Blood Proteins

Many drugs, irrespective of the route of their administration, when transported in the blood stream, are not carried in simple solution in the aqueous phase. Rather, they are chemically and reversibly bound, to varying extents, to soluble proteins in the blood (primarily albumins). By this mechanism, the blood is capable of carrying relatively large amounts of water-insoluble molecules (e.g., steroids). Several specific domains have been identified within serum albumin molecules, which are involved in drug binding. The interaction between a drug molecule and the plasma protein molecule involves, inter alia, ion–ion and van der Waals interactions and hydrophobic bonding phenomena. However, only the free, unbound form of the drug produces a pharmacological response. Plasma-bound drug molecules do not enter cellular sites to be metabolized, nor are they filtered in the kidney to be excreted. Thus, a highly blood protein-bound drug generally has a longer duration of action, a lower apparent toxicity, and a lower perceptible biological effect than a similar drug with less affinity for plasma proteins.

Even the relative abundance of albumins in the blood does not alter the fact that the number of binding sites in the plasma is finite, and therefore the drug carrying capacity of the plasma proteins is limited. This, together with the fact that drugs differ in the tenacity with which they bind to the plasma proteins, is the basis for several "drug–drug interactions": the displacement of one drug from its blood protein binding sites by another drug that has a greater affinity for these sites. This phenomenon is illustrated by the anticoagulant drug *warfarin* **1.3**.

1.3

Overdosage with this drug can result in severe, even fatal, internal hemorrhage. Warfarin is highly blood protein-bound, yet it is easily displaced by some other drugs, such as aspirin or other salicylates. Assume that an individual ingests a 5-mg dose of warfarin, and that blood protein binding occurs to the extent of 98%. The therapeutic effect is produced by the 2% of the 5-mg dose (0.1 mg) that is in the free, unbound state in the body. Further assume that the individual subsequently ingests a normal therapeutic dose of aspirin to relieve a headache. The aspirin displaces a modest amount of the warfarin molecules from their blood protein-binding sites, liberating them into the blood in simple solution. When this has occurred, 96% of the dose of warfarin remains bound to the plasma proteins. Calculation of the amount of warfarin free and unbound in the circulatory system, that portion of the dose that is capable of being transported to sites of anticoagulant action (4% of the 5-mg dose) shows an increase from 0.1 mg (in the absence of aspirin) to 0.2 mg (in the presence of aspirin) — a 100% increase. This is the pharmacological equivalent of doubling the individual's dose of warfarin.

Many clinically significant protein displacement reactions involve acidic drugs. However, protein binding is not limited to acidic compounds. Basic drugs are bound by other soluble blood proteins, for example, glycoproteins (which also can bind neutral and acidic drug molecules and can act as a carrier for steroids) and β-globulin. Additional complicating factors are frequently involved in plasma protein binding; an individual may have a lower than normal albumin level in the plasma. This can be reflected in decreased binding capacity for a given drug, and the resulting increased amount of unbound drug in the circulatory system has the same effect as a greatly increased dosage. Such individuals may experience serious toxic symptoms unless the ingested dose is decreased. Table 1.1 lists the extent of blood protein binding by some common drugs.

1.8   The Blood–Brain Barrier

1.8.1 Physiology

The blood–brain barrier is a term applied to a complex, interlocking group of physiological–biochemical phenomena, the role of which is to deny passage of many chemical substances from the blood stream into the brain. The blood–brain barrier is a defense mechanism maintained by the body to protect the brain from noxious, toxic or otherwise undesirable substances that may have found their way into the blood stream. The blood–brain barrier should not be viewed simply as a membrane or any

Table 1. 1 Extent of blood protein binding of some common drugs

Drug	Percent of dose
Diazepam (anti-anxiety drug)	99
Thyroxine	99
Phenytoin (anti-epileptic)	90
Phenobarbital	50
Digoxin	20–40
Acetaminophen (Tylenol™)	20–30
Lithium cation	0

other single anatomic entity, even though the anatomic locale for the blood–brain barrier has been identified as the lining of the brain microcapillaries (tiny blood vessels) composed of a continuous layer of endothelial cells joined by tight junctions. In addition to the physical lipid barrier to penetration, there is also an *enzymatic* blood–brain barrier. A variety of metabolizing enzymes is located in the walls of the small blood vessels in the central nervous system. Some drug molecules are not merely repelled from the barrier because of their inappropriate solvent partition coefficient, but they are actually metabolically inactivated. Some molecules pass from the brain back into the systemic circulation by passive diffusion; for some chemical entities there are efflux pumps which transport hydrophilic molecules out of the brain.

Some portions of the brain are functionally outside the blood–brain barrier. For example, the chemoreceptor trigger zone/emetic center in the medulla oblongata is outside the barrier.

1.8.2 The Barrier and Small Organic Molecules

For a drug to exert an effect in the brain, it must penetrate and pass through the blood–brain barrier. For transport across the barrier by a passive diffusion process, a molecule must have a markedly lipophilic solvent partition coefficient. In the case of a drug whose pK_a is such that a sizeable proportion of the molecules in the peripheral blood is charged (either anionic or cationic) at physiological pH, penetration of the blood–brain barrier may be difficult. The molecular weight of a drug is also a factor; a large molecular weight diminishes ease of penetration. However, there seems to be no simple linear correlation between the magnitude of the molecular weight and ease or extent of penetration of the barrier.

For many of the normal physiological components of the body (e.g., sugars, amino acids) there exists a structurally specific active transport system, which moves a specific substrate from the blood into the brain. Active transport across the blood–brain barrier has been demonstrated for a few drugs, but for the great majority, passive diffusion from blood into the brain seems likely.

1.8.3 The Blood–Brain Barrier and Proteins or Peptides

Because proteins and peptides are hydrophilic molecules and moreover are extensively charged at physiological pH, it was believed for many years that penetration of the blood–brain barrier by this class of compounds is nonexistent or at least is highly nonphysiological. This is not necessarily true. Some systemically administered peptides and proteins (enkephalins, endorphins, thyrotropin-releasing hormone, and probably others) can (and do) elicit changes in central nervous system function. It is therefore assumed that they penetrate the blood–brain barrier. In some rare instances peptides appear to diffuse passively across the barrier, similar to the more lipophilic nonpeptide drug molecules. Pinocytosis,described previously, is the process by which molecules of insulin cross the blood–brain barrier. It is likely that other proteins and/or higher molecular weight peptides penetrate the barrier by pinocytosis. It has been speculated that there are active transport mechanisms by which peptides (even some exogenous, nonphysiological peptides) penetrate the blood–brain barrier. Experimental results present the possibility that the lipid bilayers of membranes are associated with some lipophilic low molecular weight peptides (12 amino acids). These peptides can bond covalently to the C-terminus of proteins that have molecular weights as great as 45 kDa, and the resulting complex is transported across the cell membrane.

1.8.4 External Factors Affecting the Blood–Brain Barrier

Animal experimental data and statistical studies in humans suggest that emotional and physical stress temporarily increase the permeability of the blood–brain barrier to certain lipophobic organic molecules. Some drugs enhance blood–brain barrier penetrability. Chronic use (or abuse) of amphetamine markedly enhances blood–brain barrier penetration by other solutes in the blood, regardless of their molecular weight and lipophilic character. These permeability increases, induced by amphetamine, are said to be reversible in approximately 4 h, and they are associated with behavioral changes, increased loco- motor activity, social withdrawal, and weight loss. Thus, some of the pharmacological effects that are described subsequently for amphetamines may not be produced only by amphetamine itself. The drug may be admitting a variety of other molecules into the brain to exert mischief.

1.9   Transport Across the Placental Membrane

The passage of drugs and their metabolites from mother to fetus occurs mainly by passive diffusion. In general, transport of drugs across the placental barrier, as for other membrane sites, depends upon the lipid solubility, dissociation constant, and partition coefficient of the drug. The shortest time possible for a drug to equilibrate between maternal blood and fetal tissue is estimated to be approximately 40 min. However, drugs with low partition coefficients may require hours, and they may not be detected in the fetus after a single dose to the mother. Fetal exposure to drugs that are rapidly eliminated by the mother is also likely to be small. Chronic medication presents the greatest concern. Because some degree of fetal exposure is likely with most drugs and because the consequences of such exposure are frequently unknown, many authorities advocate that drug administration during pregnancy be severely restricted. Ethanol readily penetrates the placenta and maternal ingestion of even moderate amounts increases the risk of fetal alcohol syndrome, a major cause of mental retardation and some physical abnormalities.

1.10   Storage Sites for Drugs and Their Metabolites

Drugs can be stored in the body in location other that those involved in blood protein binding. Some drugs may be stored in muscle tissue. Highly lipophilic molecules such as the insecticide DDT may accumulate in neutral body fats, remain there for long periods of time, and be very slowly mobilized from these sites to be metabolized and excreted. One authority has stated that, on the average, fat constitutes about 15% of the total body weight and its volume is approximately 25% of the volume of the total body water, although these values are highly variable. Thus, it has been stated, if a nonpolar drug molecule has a log octanol/water partition coefficient of 10, roughly 75% of the drug would, at equilibrium, be dissolved in the body fat, exerting no pharmacological effect, but forming a large reservoir of the drug in communication with the plasma compartment. It is important to recognize, however, that the body fat has a low blood supply (less than 2% of the cardiac output goes to body fat). Thus, drugs are delivered to the body fat rather slowly, so that the theoretical equilibrium distribution between fat and the body water is approached slowly. For practical purposes, therefore, partition into the body fat seems to be important only for a relatively few highly lipophic drugs such as general anesthetics and some psychoactive drugs (benzodiazepines) which are dosed chronically for long periods of time.

Lead is stored in teeth, hair, and bone and is slowly mobilized and excreted. Lead is actually incorporated into the crystal lattice of bone. Arsenic is stored in the liver, kidney, heart, and lung and for extremely long periods of time in hair, teeth, bone, and nails. Tetracycline antibiotics are divalent ion chelators and these compounds can adsorb onto the crystal surface and eventually be incorporated into the crystal lattice of bone. The propensity of the tetracycline antibiotics to produce permanent discoloration of the teeth, especially in children, is due to the formation of complex tetracycline–calcium–orthophosphate chelates in the teeth.

Bibliography

1. Hucho, F. *Neurochemistry.* VCH: Deerfield Beach, Fla, 1986; p. 54.
2. Singer, S. J. and Nicholson, G. L. *Science.* **1972**, *175*, 720.
3. Stenlake, J. *Foundations of Molecular Pharmacology.* Athlone Press: London, 1979; Vol. 2, p. 161.
4. Nogrady, T. *Medicinal Chemistry*, 2nd ed. Oxford University Press: New York, 1988; p. 4.
5. *Dorland's Illustrated Medical Dictionary*, 24th ed. Saunders: Philadelphia, Pa. 1965; p. 1690.

Recommended Reading

General References in Pharmacology and Physiology

1. Brunton, L., Lazo, J., Parker, K., Buxton, I. and Blumenthal, D., Eds. *Goodman and Gilman's The Pharmacological Basis of Therapeutics*, 11th ed. McGraw-Hill: New York, 2006.
2. Rang, H. P., Dale, M. M., Ritter, J. M. and Gardner, P. *Pharmacology*, 4th ed.; Morton Publishing: New York, 2001.
3. Guyton, A. C. and Hall, J. E. *Textbook of Medical Physiology*, 10th ed. W. B. Saunders: Philadelphia, Pa., 2000.
4. Abraham, D. J., Ed. *Burger's Medicinal Chemistry and Drug Discovery*, 6th ed. Wiley-Interscience: Hoboken, NJ, 2003; Vols 1–6.
5. Dorland, W. A., Ed. *Dorland's Illustrated Medical Dictionary (Standard Version)*, 29th ed. W. B. Saunders: Philadelphia, Pa., 2000.
6. *Annual Review of Pharmacology and Toxicology.* Serial published by Annual Reviews: Palo Alto, CA.

7. *Trends in Pharmacological Sciences* ("TIPS"). Serial published by Elsevier: Amsterdam (Netherlands).
8. Tortora, G. J. *Principles of Human Anatomy*, 8th ed. Addison Wesley Longman: Menlo Park, Calif., 1999.

Recommended Reading for Topics in Chapter 1

1. Audus, K. L., Chikhale, P. J., Miller, D. W., Thompson, S. E. and Borchardt, R. T. Brain Uptake of Drugs: The Influence of Chemical and Biological Factors. *Adv. Drug Res.* **1992**, *23*, 1–64.
2. Nogrady, T. Physicochemical Principles of Drug Action. In *Medicinal Chemistry. A Biochemical Approach*, 2nd ed. Oxford University Press: Oxford, UK, **1988**; pp. 3–57.
3. Tillement, J.-P., Houin, G., Zini, R., Urien, S., Albengres, E., Barré, J., *et al.* The Binding of Drugs to Blood Plasma Macromolecules: Recent Advances and Therapeutic Significance. *Adv. Drug Res.* **1984**, *13*, 59–94.
4. Swaan, P. W. Membrane Transport Proteins and Drug Transport. In *Burger's Medicinal Chemistry and Drug Discovery*, 6th ed. Abrahams, D. J., Ed. Wiley-Interscience: Hoboken, NJ, 2003; Vol. 2, pp. 249–293.
5. Olson, R. E. and Christ, D. D. Plasma Protein Binding of Drugs. *Annu. Rep. Med. Chem.* **1996**, *31*, 327–336.
6. Seydel, K. and Wiese, M. Function, Composition, and Organization of Membranes. In *Drug-Membrane Interactions. Analysis, Drug Distribution, Modeling*. Wiley-VCH: New York, 2002; pp. 1–33.
7. Begley, D. J., Ed. *The Blood–Brain Barrier and Drug Delivery to the CNS*. Marcel Dekker: New York, 2000.
8. Filmore, D. Breeching the Blood–Brain Barrier. *Modern Drug Discovery*, June, 2002, *June* 22–27.
9. Henry, C. M. Drug Delivery. *Chem. Eng. News* **2002**, *80*, 39–47.
10. de Boer, A. G., van der Sandt, I. C. J. and Gaillard, P. J. The Role of Drug Transporters at the Blood–Brain Barrier. *Annual Review of Pharmacology and Toxicology*. Annual Reviews: Palo Alto, CA. 2003; vol. 43, pp. 629–656.
11. Prokai, L. and Prokai-Tarrai, K., Eds. *Peptide Transport and Delivery into the Central Nervous System*. Progress in Drug Research, Vol. 61. Birkhäuser Verlag: Basel, Switzerland, 2003.
12. Dantzig, A. H., Hillgren, K. M. and de Alwis, D. P. Drug Transporters and Their Role in Tissue Distribution. *Annu. Rep. Med. Chem.* **2005**, *39*, 279–291.
13. Coleman, M. D. *Human Drug Metabolism. An Introduction*. John Wiley & Sons: Chichester, UK, 2005.

2

Pharmacokinetics

2.1 ### Introduction

For rational use of drugs, it is necessary to consider not only their pharmacological effects and actions, but also their rates of absorption, their distribution through the body, and their elimination. Pharmacokinetics, the branch of pharmacology that addresses these rate-related aspects, attempts to describe the fate of a drug in the body as a mathematical function of time and concentration of the drug. The pharmacokineticist describes this field of study by the acronym ADME: absorption, distribution, metabolism, excretion. A prominent worker in the field has remarked that, whereas classical pharmacology attempts to explain what a drug does to the body, pharmacokinetics attempts to explain what the body does with a drug. A fundamental premise in pharmacokinetics is that a relationship exists between a useful pharmacologic or a toxic effect of a drug and the concentration of the drug in some readily accessible locale in the body (e.g., the blood). However, while this assumption is valid for many drugs, it is not universally applicable. The pharmacokineticist utilizes experimentally derived data on the concentration of drug in blood or plasma, and attempts to utilize these values to estimate the concentration of drug in other relevant tissues of the body. It has been mentioned previously that the assumption that drug concentration in the blood or plasma parallels drug concentration in other body tissues is valid only for those drugs which equilibrate rapidly from the blood to those other tissues where appropriate receptor sites for the drug are located.

2.2 ### Bioavailability

The *bioavailability* of a drug is defined as the fraction of unchanged (unmetabolized) drug that is absorbed and then reaches the systemic circulation or its appropriate site of action, following administration by any route. For intravenous administration, the bioavailability of a drug is defined as unity. However, for drugs administered by other routes, bioavailability is often less than unity. Incomplete bioavailability may result from a wide variety of factors, among which are incomplete/faulty absorption, decomposition in the lumen of the stomach or intestine, first-pass metabolic inactivation, failure of the administered dosage form to release the drug efficiently and rapidly, and poor transport/distribution through the body.

Absolute bioavailability refers to the situation where the total quantity of drug reaching the systemic circulation is measured; it is determined by comparing the bioavailability of a test formulation (e.g., a tablet, capsule, or liquid product) of a drug with that of an intravenous dose where 100% bioavailability can be assumed. Note that complete bioavailability cannot be assumed for any orally administered solution. Absolute bioavailability can be calculated by the following equation:

$$\text{Absolute bioavailability} = \frac{AUC_{po}}{AUC_{iv}} \times \frac{Dose_{iv}}{Dose_{po}} \times 100\%$$

where AUC_{po} = area under the curve
following oral administration   (2.1)

AUC_{iv} = area under the curve
following intravenous administration.

The term *area under the curve* (AUC) is derived from the plot of plasma concentration (Y axis) versus time (X axis). AUC can be calculated by several mathematical methods, but more commonly direct integration and the trapezoidal rule are employed. The AUC is directly proportional to the total amount of the drug introduced into the plasma compartment (vide infra), irrespective of the rate at which it enters. Comparison of AUC following oral and intravenous administration can therefore be used to determine the fraction of the oral dose that enters the bloodstream, and thus be used to measure bioavailability. In other contexts, AUC describes the concentration of drug in the systemic circulation as a function of time (from zero to infinity).

Relative bioavailability refers to the situation where the total drug reaching the systemic circulation is not measured, but the bioavailability of the test formulation is compared to that of another formulation which is not administered directly into the systemic circulation, and where 100% bioavailability cannot be assumed (e.g., a comparison of two oral formulations). Relative bioavailability can be calculated by the following equation:

$$\text{Relative bioavailability} = \frac{AUC_{po}\ (test)}{AUC_{po}\ (standard)} \times \frac{Dose_{po}\ (standard)}{Dose_{po}\ (test)} \times 100\% \qquad (2.2)$$

2.3   Compartments

In pharmacokinetic terminology the body is divided into *compartments*, which are spaces into which a drug is assumed to be uniformly distributed. Compartments are considered to be bounded by membranes, and changes in drug concentration involve exchange between different compartments. "Transport" from one compartment to another may represent true transport from one location to another, or it may represent transformation from one chemical state to another within the same location (i.e., metabolic change of the drug molecule).

The total amount of water in the human body, as a percentage of body weight, varies from 50 to 70% and is somewhat less in women than in men. Body water is distributed into four main compartments:

1. *Extracellular fluid*: the blood plasma (approximately 4.5% of body weight)
2. *Interstitial fluid*: that water in the spaces between tissues or cells (16% of body weight) and the water in lymph (approximately 1.2% of body weight)
3. *Intracellular fluid*: the water inside the cells of the body (30-40% of body weight)
4. *Transcellular fluid*: the cerebrospinal, intraocular, pleural, and synovial fluids, and the fluids of the digestive tract (2.5% of body weight)

A simple, frequently applicable assumption is that the rates of transport from one compartment to another are proportional to their concentrations, in which case they can be described by using first-order rate constants. However, this simple assumption is not always applicable. The transport between compartments may involve a saturable carrier (active transport), and in this case the transport rates generally cease to be proportional to the drug concentration.

2.4   Clearance

Clearance (CL) is a measure of the body's ability to eliminate a drug. Figure 2.1 is a simplified graphic representation of the physiology of drug elimination. The rate of presentation of a drug to a drug elimination organ is a function of organ blood flow Q and the concentration (C_A) of drug in the arterial blood entering the organ. The rate of exit of a drug from the drug eliminating organ is a function of the organ blood flow Q and the concentration (C_V) of the drug in the venous blood leaving the organ. Thus, the rate of elimination of drug by the organ is the difference between the rate of presentation and the rate of exit. For example, renal clearance, CL_R, may be calculated by the following expression:

$$CL_R = \frac{C_u V_u}{C_p}$$

where CL_R = renal clearance

C_u = urinary concentration

C_p = plasma concentration   (2.3)

V_u = rate of flow of urine

The units calculated here for renal clearance are volume (milliliters) of blood cleared of drug per unit of time (minutes).

Clearance is the most important concept to be considered when a dosage regimen is to be established for long-term administration to a patient. The clinician will wish to maintain "steady state concentrations" of the drug within a known therapeutic range of dose sizes. If one assumes complete bioavailability, the steady state will be achieved when the rate of drug elimination equals the rate of drug administration. Thus, if the desired steady state concentration of the drug in the blood is known, the rate of clearance of drug by the patient's body will dictate the rate (i.e., the interval between doses)

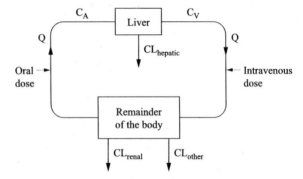

Figure 2.1 Schematic representation of concentration–clearance relationship. (Reproduced with permission from reference 1. Copyright 1995 Wiley)

at which the drug should be administered. Clearance is meaningful only if the *volume of distribution* (vide infra) is known.

In the case of a drug that is efficiently removed from the body, for example by hepatic metabolic conversion to a new molecule or by excretion of unchanged drug in the bile, the concentration of unchanged drug in the blood leaving the liver will be low and the clearance of drug from the body's pool of blood will be determined by the rate of hepatic blood flow. By rather arbitrary definition, drugs which are classed as being rapidly removed by the liver have hepatic clearances greater than 6 mL of blood per minute per kilogram of body weight. Examples of such drugs are chlorpromazine, morphine, and propranolol. For drugs that have a very high liver extraction ratio, clearance will be affected only to a moderate degree by such factors as enzyme induction or hepatic disease.

It is obvious that utilization of clearance in pharmacokinetic studies requires laboratory measurement of blood concentrations of drug. These analytical determinations are often carried out on plasma, due to the relative ease of sample handling and processing. However, the drug may also be bound to some extent to the formed elements of the blood, so that determination of drug concentration only in the plasma may produce incorrect values. However, measurement of "blood-to-plasma ratio" can permit the pharmacokineticist to utilize appropriate mathematics to convert clearance values determined in plasma to their corresponding "blood" values. This strategy is based on the premise that drug bound to the blood cells is in equilibrium with drug in the plasma.

2.5   Volume of Distribution

The volume of distribution, V_d, relates the amount of a given drug in the body to the concentration, C, of the drug in the blood or plasma:

$$V_d = \text{amount of drug in the body}/C \tag{2.4}$$

This value does not necessarily reflect a "real" physiological volume, but rather it indicates the fluid volume that would be required to contain all of the drug in the body at the same concentration in which it is found in the blood or plasma. For a man weighing 68 Kg (150 lb) the plasma volume is approximately 3 L, the blood volume is approximately 5.5 L, the volume of the extracellular fluid exclusive of the plasma is approximately 12 L, and the total body water is approximately 42 L. However, many drugs exhibit volumes of distribution that are much larger than these fluid volumes. For a 68–kg man who is given a 500-μg dose of the cardiac glycoside digoxin, the plasma concentration of the drug will be approximately 0.7 ng L^{-1}. Thus, the calculated volume of distribution for digoxin in this patient is 700 L! The lipophilic digoxin molecule exhibits a very large V_d because it distributes extensively into the body tissues (chiefly the skeletal muscle) leaving only a small amount in the blood, where the concentration of the drug can be conveniently measured. Drugs that are extensively bound to plasma proteins (such as the anticoagulant drug warfarin) remain largely in the plasma compartment, and hence they demonstrate a numerically small volume of distribution.

The volume of distribution for drugs is dependent upon a large number of parameters, including the pK_a of the drug, the degree of binding to plasma proteins, the partition coefficient of the drug into the body fat, and the degree of binding to other tissues of the body. As might be anticipated, the volume of distribution for a given drug can change, as a function of the individual's age, gender, disease state, and body composition.

In a two compartment model, there is more than one volume of distribution: (1) the volume in the central compartment; and (2) the volume of a peripheral compartment. Comparisons of volumes of distribution are generally made using the *volume of distribution at steady state* (V_{ss}). This parameter represents the sum of the volumes of distribution of all the compartments into which the drug may be distributed, and there are mathematical strategies for calculating it.

2.6 First-Order Elimination Kinetics

It is customary to postulate a minimum number of compartments, consistent with a reasonable description of events. Approximations are frequently used to prevent the mathematics involved from becoming impossibly difficult. Thus, if a drug is injected into the bloodstream, subsequently diffuses into the extracellular spaces, and finally is excreted, the system strictly involves at least two body compartments, the blood and the extracellular fluid. However, these are often treated as a single compartment if drug distribution between the two compartments is rapid relative to the rate of drug elimination from the body. Another frequently adopted approximation is to neglect "deep" compartments, such as fat or bone, which communicate with the extracellular fluid but do not rapidly equilibrate with it.

In the simplest ("single-compartment") model, a drug which has been injected intravenously is assumed to be instantaneously and uniformly distributed and to be removed at a rate proportional to its concentration. The rate of elimination of the drug under these conditions is described by the equation:

$$C = C_0 e^{-kt}$$

where C = concentration of the drug at time t

C_0 = initial concentration of the drug   (2.5)

t = time

e = 2.718 (natural log, base e)

The logarithmic form of this equation is:

$$LnC = LnC_0 - kt \tag{2.6}$$

Thus, if a drug is exponentially eliminated (first order kinetics: rate is proportional to the amount of drug still to be eliminated), a plot of log concentration versus time should give a straight line (figure 2.2). This graph shows the relationship to be expected for an exponentially cleared drug, between dose (or initial concentration) of the drug and its duration of effect, as measured by the time required to reach the threshold drug concentration in the body (the dotted horizontal line). The duration of effect of the drug varies with the *logarithm* of the dose. Thus, if a 10-unit dose of the drug produces an effect lasting for 1 day, a 100 (10^2) unit dose produces an effect lasting for 2 days, and a 1000 (10^3) unit dose produces an effect lasting for 3 days. This graph implies that it is impossible to obtain a prolonged effect from a drug merely by administering a massive dose, if that drug is rapidly excreted. Here it is seen that multiplying the original dose by 100 extends the duration of effect only three-fold. Nearly all drugs are limited as to the amount of drug that can be introduced into the body at any one time and hence massive doses, besides being ineffective at prolonging the duration of effect, are dangerous. Methods which prolong the duration of effect of a drug that is normally rapidly cleared from the body involve strategies to slow absorption or to diminish the rate of metabolic inactivation or the rate of excretion.

The *two-compartment model* is a more complicated pharmacokinetic situation, in which the body tissues are considered to be combined into one peripheral compartment which drug molecules can enter and leave via a central compartment, frequently considered to be the plasma.

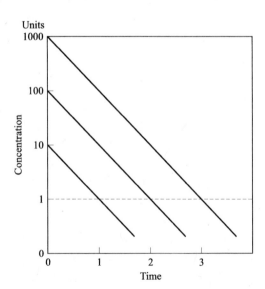

Figure 2.2 Exponential elimination of a drug. Relationship between dosage and duration of effect. (Reproduced with permission from reference 2. Copyright 1974 W. B. Saunders)

2.7   Biological Half-Life

Many drugs are eliminated at an apparently exponential rate, so that their first-order rate constants and corresponding half-lives can be computed. The *biological half-life* of a drug is the time required for its concentration in the body to fall to one half the initial concentration. Table 2.1 lists half-lives of a number of drugs. This tabulation shows that half-lives vary greatly, and this must be taken into account in administering drugs.

It must be emphasized that a graph of plasma concentration versus time will be linear only in the simplest case of a one-compartment model. In such a situation the half-life can be read directly from the straight line of the graph. However, more frequently, a more complex multi-compartment model is operative and determination of half-life is more complex and difficult. Equations for calculating half-lives are described in the literature as providing an "estimated half-life value", using estimates of clearance and volume of distribution.

Multiple dosing half-life refers to the half-life for a drug which is equivalent to the chosen dosing interval, so that plasma concentrations or amounts of drug in the body show a 50% decrease during the dosing interval at *steady state* (rate of presentation of the drug is equal to the rate of its excretion). These parameters are defined in terms of mean residence time in the central compartment (MRTC) and mean residence time in the body (MRT):

$$t_{1/2\mathrm{MD}}^{\mathrm{plasma}} = 0.693 \times \mathrm{MRTC} \tag{2.7}$$

$$t_{1/2\mathrm{MD}}^{\mathrm{plasma}} = 0.693 \times \mathrm{MRT} \tag{2.8}$$

MRTC in a one-compartment body model is the inverse of the rate constant for elimination. In a multiple compartment model where the multiple dosing plasma

Table 2.1   Half-lives of selected drugs in humans

Drug	*Half-life (h)*
Azathioprine	0.16
Acetylsalicylic acid	0.25
Lansoprazole	0.9
Amoxicillin	1.7
Morphine	1.9
Tubocurarine	2.0
Acetaminophen	2.0
Dextromethorphan	2.2
Sulfisoxazole	6.6
Alprazolam	12
Valproic acid	14
Amantadine	16
Amitryptyline	21
Mefloquine	20 days

half-life is useful, MRTC is given by the volume of the central compartment where drug concentrations are measured, divided by the clearance.

2.8   Model for Combined Absorption and Elimination

When absorption and elimination of a drug are taken into account simultaneously, a more complex graphic model results (figure 2.3). The solid curve is the theoretical curve showing blood level after an oral dose of the drug during simultaneous absorption and elimination (dashed curves). The absorption curve shows the theoretical blood level to be expected following intravenous administration of the same size dose as the oral one. The variable t_{max} is the time to maximum blood level and Y_{max} is the maximum blood level. This treatment of pharmacokinetic data has the following mathematical consequences which are significant in therapy Firstly, t_{max}, after which the blood concentration of drug becomes maximal, is independent of concentration. Note that this is valid for those drugs whose effect is not dose-dependent. For drugs that have dose-dependent kinetics (e.g., saturation of a metabolic pathway at higher doses) this is not valid. Given a particular drug and method of administration, the maximal blood level occurs at the same time, whether the dose is large or small. Secondly, the concentration maximum Y_{max} is proportional to the size of the dose. Thus, if the dose is doubled, the maximum blood concentration is also doubled.

2.9   Nonlinear Pharmacokinetics

Nonlinearity in pharmacokinetics (changes in such parameters as clearance, volume of distribution, and half-life as a function of dose or concentration of the drug) usually

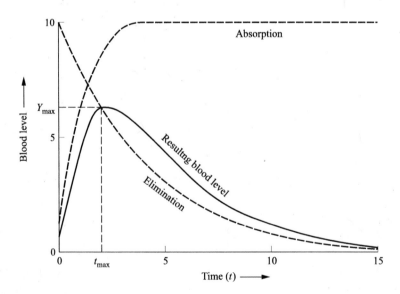

Figure 2.3 Absorption and elimination curves. Combination shows resulting blood level in human. (Reproduced with permission from reference 2. Copyright 1974 W. B. Saunders)

results from saturation of protein binding, metabolic pathways, and/or active renal transport of the drug. In protein binding, as the concentration of drug increases, the fraction unbound must also increase as all binding sites become occupied. For a drug that is metabolized by the liver with a low extraction ratio, the saturation of plasma proteins increases both the volume of distribution and clearance as drug concentration increases. Thus, the half-life may remain constant. Concentration at steady state does not increase linearly as the rate of drug administration increases. For drugs that are cleared into the liver with high extraction ratios, the steady-state concentration can remain linearly proportional to the rate of administration.

All metabolic processes are probably saturable, but they will appear to be linear if the drug concentrations are much less than the plasma concentration at which half of the maximal rate of elimination (K_m) is reached. When plasma concentration exceeds the K_m nonlinear kinetics is observed. Saturable metabolism makes first-pass metabolism less than expected.

2.10   Loading Dose

In clinical practice, "loading dose" is sometimes significant. This is defined as the size dose that may be given at the onset of therapy, with the aim of achieving the target concentration in the blood rapidly. One equation presented as calculation of "the appropriate magnitude" of the loading dose is as follows:

$$\text{loading dose} = \frac{C_{p.target} \times V_{ss}}{F}$$

where $C_{p.\,target}$ = target concentration of drug in plasma   (2.9)
V_{ss} = volume of distribution at steady state
F = fractional bioavailability

Most pharmacokineticists concede that this calculation will, at best, provide a numerical value that is only an approximation. The use of a loading dose calculation may present dangers. A particularly sensitive individual may be exposed abruptly to a toxic concentration of the drug. Also, if the drug has a long half-life, a prolonged period of time will be required for the concentration in the plasma to fall, if the initial level achieved by the loading dose was excessive. Loading doses tend to be large and they are often administered rapidly intravenously. This practice can be dangerous if toxic effects occur as a result of actions of the drug at sites that are in rapid equilibrium with the blood.

Bibliography

1. Benet, L. A. and Perotti, B. Y. Drug Absorption, Distribution, and Elimination. In *Burger's Medicinal Chemistry and Drug Discovery*, 5th ed. Wolff, M. E., Ed. Wiley: New York, 1995; Vol. 1, p. 116.
2. Modell, W., Schild, H. O. and Wilson, A. *Applied Pharmacology*. W. B. Saunders: Philadelphia, Pa, 1974; pp. 73, 75.

Recommended Reading

1. Wilkinson, G. R. Pharmacokinetics. In *Goodman and Gilman's Pharmacological Basis of Therapeutics*, 10th ed. Hardman, J. G., Limbird, L. E. and Gilman, A. G., Eds. McGrawHill: New York, 2001; pp. 3–29.
2. Benet, L. Z., Perotti, B. Y. and Hardy, L. Drug Absorption, Distribution, and Elimination. In *Burger's Medicinal Chemistry and Drug Discovery*, 6th ed. Abraham, D. J., Ed. Wiley Interscience: Hoboken, N. J. 2003; Vol. 2, pp. 633–647.
3. Rang, H. P., Dale, M. M., Ritter, J. M. and Gardner, P. *Pharmacology*, 4th ed. Churchill Livingstone: Philadelphia, Pa., 2001; pp. 87–92.
4. Pharmacokinetic Data: Tabulations of clearance rate, volume of distribution, half-life, urinary excretion, percentage of plasma binding, peak time, and peak concentration for a large number of drugs. Recommended Reading, 1, pp. 1924–2023.
5. Fichtl, B., Nieciecki, A. V. and Walter, K. Tissue Binding versus Plasma Binding of Drugs: General Principles and Pharmacokinetic Consequences. In *Advances in Drug Research*, Testa, B., Ed. Academic Press: London, 1991; Vol. 20, pp. 117–166.
6. Schoenwald, R. D., Ed. *Pharmacokinetics in Drug Discovery and Development*. CRC Press: Boca Raton, Fla., 2002.
7. Ings, R. M. I. Pharmacokinetics. In *Medicinal Chemistry Principles and Practice*. King, F. D., Ed. Royal Society of Chemistry: Cambridge, UK, 1994; pp. 67–85.

3

Drug Metabolism

3.1   Metabolism of Xenobiotics

The body has a remarkable variety of physiological systems available as defense mechanisms against foreign substances ("xenobiotics") which may have been introduced into it. Most (but not all) drugs are metabolized in the body, being converted into other chemical entities. The strategy of the body in these chemical conversions is to attempt to change the drug or other foreign molecule into a species that is more water-soluble and less lipid-soluble, that is, to alter the partition coefficient of the molecule. This promotes excretion of the resulting product from the body. Commonly, a given drug will not be metabolized to a single chemical entity, but rather to a number of different metabolites. The metabolizing enzymes of the body are generally substrate nonspecific and the chemical nature of the in vivo alteration(s) of molecules of a given drug depends upon which metabolizing enzyme(s) the drug molecule encounters first. Rather empirically, it seems that the body's metabolism apparatus is designed to prepare drug molecules for excretion via the kidneys in the urine.

3.2   Aspects of Functional Microanatomy of the Kidney

Figure 3.1 illustrates, somewhat diagrammatically, the functional unit of the kidney, the *nephron*. There are approximately 1 million microscopic-sized nephrons in each kidney. The small capillaries in the glomerulus network are *fenestrated* (contain "windows"). The size of these "windows" or pores permits the passage of water and small solute molecules. Glomerular filtration operates on principles of Newtonian physics and is believed to be analogous to a chemical laboratory filtration process. As blood from the systemic circulation enters through the afferent renal arteriole and flows through the network of thin walled capillaries comprising the glomerulus, approximately 20% of the liquid portion of the blood is forced out into the hollow portion of the nephron, *Bowman's capsule* (the funnel). The filtering force in this process is the hydrostatic pressure of the blood which derives from the work of the heart. Additionally, the diameter of the afferent arterioles leading into the glomerular network is larger than that of the efferent arterioles leading out of the glomerulus and as a result a *glomerular filtration pressure*, estimated to be 50 mm Hg, is established. The remaining 80% of

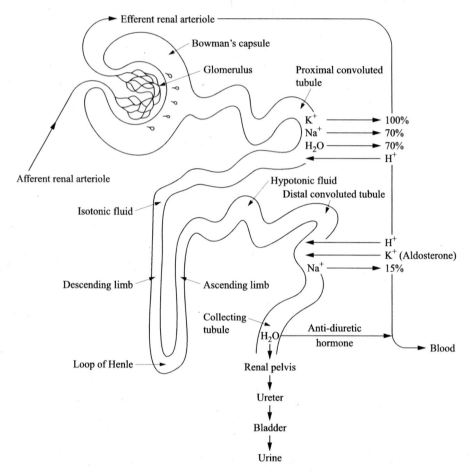

Figure 3.1 The nephron, the functional unit of the kidney

the blood passing through the glomerular network flows back into the general systemic circulation.

The filtrate that enters Bowman's capsule is water which contains the inorganic solutes and some of the organic solutes of the blood (including drug molecules that are not bound to plasma proteins). The molecules of the soluble plasma proteins are too large (above molecular weight 20 000) to pass through the fenestrations in the capillary walls. The formed elements of the blood (red cells, white cells, and blood platelets) do not pass through the filtration system because they also are too large. The human kidneys filter approximately 170 L of fluid per 24-h period. Because the daily output of urine for the average adult is on the order of 1.5 L, this leads to the conclusion that 99% of the volume of the glomerular filtrate is reabsorbed. The glomerular filtrate moves from Bowman's capsule into the tubule system, as illustrated in figure 3.1. The renal tubules are surrounded by a network of capillaries which act to reabsorb most of the water and certain of the solutes back into the general circulation. Some biochemically useful filtrate components, such as sodium and potassium cations, glucose, and amino acids are reabsorbed into the blood from the tubular filtrate by active transport mechanisms. Thus, the glomerular filtrate is concentrated and eventually, urine is formed which moves down to the urinary bladder for excretion.

Role of the Kidney in Drug Excretion

3.3.1 Glomerular Filtration and Tubular Reabsorption

Drug molecules that have been filtered from the blood in the glomerulus move, in solution, into the tubules which lead ultimately to the urinary bladder. However, the cells of the walls of the tubules are physiologically very active and they can extract certain solute molecules from the tubular filtrate as it moves through the tubules. These solutes are returned to the general systemic circulation. Although, as mentioned previously, some specific solutes in the filtrate are reabsorbed by active transport mechanisms, almost all drug molecules (which are foreign substances in the body) are reabsorbed across the walls of the tubules into the general circulation by passive diffusion. The more lipophilic a drug molecule, the more efficiently it will be expected to be reabsorbed. For markedly lipophilic drugs, as much as 99% of the filtered drug is passively reabsorbed. Therefore, the body attempts to convert lipophilic drug molecules into very hydrophilic, lipophobic ones which have a diminished tendency to be reabsorbed across the tubule walls, but rather they remain in the tubular solution to be conveyed to the urinary bladder for excretion. A drug molecule that is bound to a plasma protein is not filtered in the glomerulus but it remains in that portion of the blood that returns to the general circulation. However, because blood protein-bound drug is in equilibrium with free drug dissolved in the aqueous phase of the blood (which can be pass through the fenestrations), even a drug that is extensively plasma protein-bound eventually will be filtered at the glomerulus for excretion.

3.3.2 Tubular Secretion

An added physiological complication is that the cells of the tubular walls can extract certain solute molecules from the systemically circulating blood and can then secrete these molecules directly into the tubular filtrate solution, whence they are conveyed to the urinary bladder. Some drugs are in part (in some instances, virtually totally) excreted from the body by this tubular wall secretion mechanism. This process is not simple passive diffusion, but rather it is an active transport phenomenon. The drug molecule, having been secreted into the lumen of the tubule, has little or no tendency to migrate in the opposite direction.

An example of tubular secretion of a drug is benzyl penicillin ("penicillin-G") **3.1** whose short duration of pharmacological effect was originally ascribed to facile metabolic cleavage of the β-lactam ring, but is now known to be caused by its rapid and efficient secretion across the tubular walls into the tubular filtrate. *Probenecid* **3.2** competes with benzyl penicillin for the tubular active transport mechanism, and thus in the presence of probenecid, the effect of a dose of benzyl penicillin is prolonged.

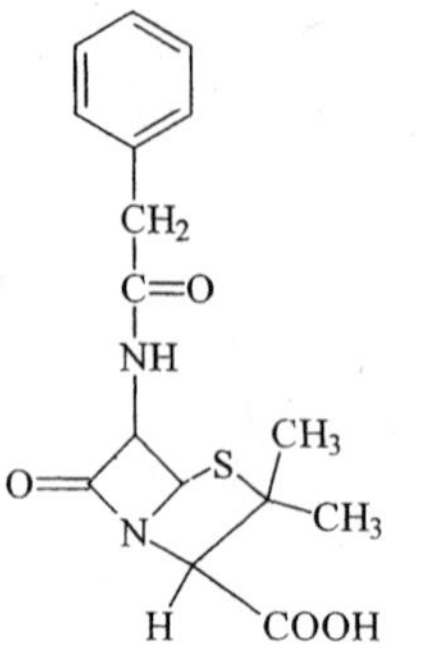

3.1

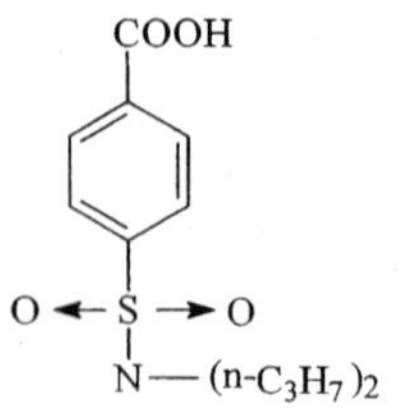

3.2

In sum, hydrophilic molecules are rapidly and efficiently excreted via the kidneys in the urine, and more lipophilic molecules, even though they are removed from the circulating blood by glomerular filtration, may be efficiently reabsorbed back into the general circulation and overall they may be rather slowly excreted from the body. Thus, it might be predicted that the duration of effect of lipophilic drugs is prolonged, and this is often the case.

3.3.3 Effect of pH on Tubular Reabsorption

Renal tubular reabsorption by passive diffusion is pH- dependent. When the tubular urine is alkaline, weakly acidic molecules are excreted more rapidly because they are ionized and thus are less lipophilic. The opposite is true for weakly basic solutes. Urinary drug excretion can be influenced by manipulation of the pH of the tubular solution. Large intravenous doses of sodium bicarbonate render the tubular contents slightly alkaline, and similar large doses of ammonium chloride or ascorbic acid render the tubular contents slightly acidic. In the case of ammonium chloride, the ammonia is converted into urea in the liver, leaving chloride anions and hydrogen cations. The chloride displaces bicarbonate which assumes a hydrogen ion to form carbonic acid, which is subsequently dissipated as carbon dioxide and water. The remaining chloride anion, with accompanying sodium cation, appears in the glomerular filtrate. Base-conserving mechanisms in the tubules are activated, hydrogen ions are exchanged for sodium, and an acidic urine is generated. As will be described in subsequent chapters, alteration of tubular pH can be employed clinically.

3.4   Extrarenal Routes of Drug and/or Drug Metabolite Excretion

Although the most important excretory route is via the urine, some drugs and their metabolites may be excreted from the body by routes not involving the kidney.

3.4.1 In the Feces

This route is frequently linked with secretion of the drug from the gall bladder into the duodenum (upper region of the small intestine) as a component of bile (enterohepatic circulation). The liver cells have the ability to extract various solutes (including some drug molecules) from the plasma and to deposit them in the bile. The active transport systems involved are said to be similar to those described for the walls of the renal tubules. In some instances (e.g., the cardiac glycoside digoxin and the now obsolete cathartic agent phenolphthalein) the drug constituent of the secreted bile is absorbed, in part, back across the intestinal wall into the systemic circulation, thus delaying removal of the drug from the body and prolonging its pharmacological effect. Some other drugs are not reabsorbed and they appear as fecal constituents. For an individual who has compromised liver function, the extraction of drugs from the blood by the liver may be seriously inhibited, and this can result in prolonged, elevated levels of the drug in the blood.

3.4.2 In Sweat

Lithium cation, lead, arsenic metabolites such as methyl arsenite are excreted by this route.

3.4.3 In Expired Air

Ethanol, inhalation anesthetics, volatile metabolites of arsenic are excreted in expired air. Approximately 3% of ingested ethanol is excreted unchanged in the expired breath, and the partition coefficient for ethanol between blood and expired air is 2000:1. Thus, the amount of ethanol in expired air can be measured (by analyzing the gas mixture collected by breathing into a balloon or into a "breathalyzer") and the corresponding blood level can be calculated.

3.4.4 Milk of Lactating Women

Since breast milk is more acidic than plasma, basic drugs may be slightly concentrated in this fluid, and the concentration of acidic drugs, lower. Ethanol will be in the same concentration in milk as in plasma.

3.5 Undesirable Metabolic Consequences

3.5.1 Lethal Synthesis

The metabolic detoxication strategies of the body are not infallible. Some drug metabolites are more pharmacologically potent or more toxic than the originally administered drug. An example of *lethal synthesis* is fluoroacetic acid **3.3**, which is extremely toxic and is employed commercially as a rodenticide. In the body, fluoroacetate reacts with coenzyme A to form fluoroacetylcoenzyme A. This unnatural metabolite is converted into fluorocitric acid **3.4**, in which a fluoroacetate residue has replaced the physiologically normal acetate. Fluorocitric acid inactivates aconitase, a key enzyme in the Krebs tricarboxylic acid cycle, an essential component of the body's cellular energy generating process. In the absence of functional enzyme, citric acid **3.5** cannot be converted into *cis*-aconitic acid **3.6**, and the Krebs cycle is disrupted, with a lethal result.

F-CH$_2$-COOH $\xrightarrow[\text{steps}]{\text{Multiple}}$

$$\begin{array}{c} \text{F---C---COOH} \\ | \\ \text{OH---C---COOH} \\ | \\ \text{H---C---COOH} \\ | \\ \text{H} \end{array}$$

$$\begin{array}{c} \text{H} \\ | \\ \text{H---C---COOH} \\ | \\ \text{HO---C---COOH} \\ | \\ \text{H---C---COOH} \\ | \\ \text{H} \end{array}$$

$\xrightarrow{\text{Actonitase}}$

3.3   **3.4**   **3.5**   **3.6**

3.5.2 Poorly Soluble Metabolites

Metabolism does not always result in products that are more water soluble. Sulfanilamide **3.7** undergoes metabolic acetylation of the aromatic amino group. This metabolite **3.8** is much less water-soluble than is sulfanilamide itself. The metabolite in solution in plasma undergoes glomerular filtration in the kidney, and as its dilute solution in the glomerular filtrate passes along the tubules, reabsorption of water results in concentrating the N^4-acetylsulfanilamide solute to the point where its solubility is exceeded and it crystallizes in the lumen of the tubules. These crystals can completely obstruct the tubule, causing severe physical damage and malfunction of the kidney. Early clinical attempts to minimize this hazard involved forcing a high intake of water to increase urine production and to maintain the tubular concentration of N^4-acetylsulfanilamide as low as possible. Large intravenous doses of sodium bicarbonate were also administered to maintain the pH of the tubular urine on the alkaline side of neutrality. In an alkaline environment, a larger proportion of the metabolite molecules will be in the anionic form **3.9**, which is more water-soluble than the protonated neutral molecules. The introduction of newer antibacterial sulfonamides having different solubility properties and/or different metabolic pathways has largely corrected the clinical problems associated with sulfanilamide, which is rarely administered internally.

3.7   **3.8**   **3.9**

3.6   Differences in the Metabolic Fate of Enantiomers

Not only can the pharmacological actions and effects of enantiomers differ, but also the way in which the body disposes of the mirror image isomers may differ significantly. The β-adrenocepter blocker propranolol **3.10** is marketed as the racemate. The less pharmacologically active R-(+) enantiomer is more rapidly metabolized than is the more active S-(−) enantiomer.

The sedative/hypnotic agent *thalidomide* **3.11**, marketed as the racemate, was prescribed in Europe, about 40 years ago, for pregnant women. The sedative effect in thalidomide resides in the *R*-(+) antipode, which is metabolized to nontoxic products. The *S*-(−) antipode is metabolized in part to *N*-phthaloylglutamine **3.12**, and some data indicate that this metabolite is the teratogen that causes the rare birth defect *phocomelia* in humans.

3.10   **3.11**

3.12

The clinical value of these findings is complicated by the demonstration that both pure enantiomers of thalidomide are rapidly metabolically racemized in humans. Thus, administration of the pharmacologically active, presumably non-teratogenic enantiomer can give rise to the teratogenic metabolite. Due to these metabolic phenomena, there appears to be no way to render thalidomide safe for pregnant women. Remarkably, thalidomide is utilized therapeutically in treatment of certain types of leprosy and is widely used in the treatment of certain immunological conditions. However, it can be used for these purposes only in men and in postmenopausal women.

3.7   In Vivo Drug Metabolism

3.7.1 Liver Microsomal Metabolism

The chemical changes that the body induces on a drug molecule may occur in several organs or tissues: blood, skin, lung, lumen and walls of the intestine, kidney, or brain, but the liver plays a dominant role in drug metabolism and inactivation. The liver contains an extensive array of enzyme systems which utilize many kinds of functional groups on many kinds of molecules as substrates. Much (but not all) of the hepatic metabolism of drugs is catalyzed by *microsomal enzymes*, which are located on the *endoplasmic reticulum*, an anatomic component of the cytoplasm of liver cells. The endoplasmic reticulum is a lipid bilayer–protein complex, similar to the chemical composition of the cell membrane. Under magnification, one form of the endoplasmic reticulum resembles a long, folded ribbon (figure 3.2).

Metabolizing enzymes are located on the surface of the endoplasmic reticulum. Microsomes are not true anatomic structures, but rather they are small pieces of tissue formed by fragmentation of the endoplasmic reticulum. To prepare these microsomes, liver tissue is homogenized with aqueous buffer and a certain portion of the homogenate is separated by ultracentrifugation. The principle of ultracentrifugation is that the weight of suspended particles is a function of the speed at which the centrifuge spins.

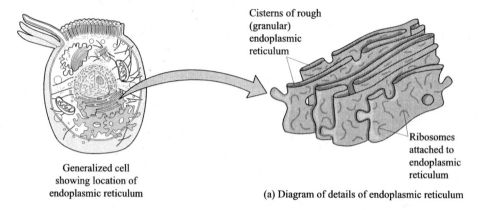

Figure 3.2 Endoplasmic reticulum. (Reproduced with permission from reference 2. Copyright 1999 Wiley)

Larger, heavier tissue fragments settle from the suspension, and smaller fragments remain suspended in the liquid phase. Thus, tissue particles of different weights can be separated by controlling the spinning rate. One such suspended fraction contains the microsomes of the endoplasmic reticulum which bear the metabolizing enzymes that are unaltered by the preparative process and retain their catalytic ability. In general, these liver microsomal enzymes lack a high degree of substrate specificity or selectivity and they can metabolize a great variety of compounds and functional groups. However, they do not accept endogenous hydrophilic body components such as sugars or amino acids as substrates, probably because they best catalyze reactions on molecules that are decidedly lipophilic.

3.7.2 First-Pass Metabolism

Oral administration of some drugs results in less bioavailability (and less of a pharmacologic effect) than occurs following intravenous administration of the same size dose. In certain instances, the decreased drug availability is explained, not by poor absorption from the gastrointestinal tract, but rather by the fact that the blood flow (portal venous system) from the stomach and the intestine goes directly to the liver. Materials absorbed from the gastrointestinal tract into the blood pass through the liver before entering into the general circulation. Thus, in the case of drugs that are rapidly and efficiently metabolized by liver enzymes, a greater proportion of an oral dose is metabolized during the journey through the liver than would occur following intravenous administration, which does not provide for the direct passage of the entire dose through the liver before distribution throughout the body. This phenomenon is termed the *first-pass effect*, and for some drugs it is quite important. For example, the bioavailability of oral preparations of morphine may be as small as 25% of the total dose because of first-pass metabolic inactivation. When morphine is to be given orally it is necessary to adjust the dose to compensate for first pass metabolic inactivation. A drug which is administered sublingually (under the tongue) or in the buccal (cheek) cavity and is absorbed across the mucosa of the mouth passes directly into the general circulation without entering the portal venous system, and it escapes first-pass metabolism. Approximately 50% of the dose of a drug that is absorbed from the rectum will bypass the liver. Thus, the potential for first-pass metabolism is less for administration by the rectal route than by the oral route.

3.7.3 Enzyme Induction

Some drugs increase the catalytic activity of certain nonspecific metabolizing enzymes in the liver by as much as several-fold by a process termed *enzyme induction*. The increased enzyme activity is caused in part by increased de novo synthesis of enzyme molecules. However, not all enzyme systems in the body are inducible. Because many drug metabolites are biologically less active or even inert, enzyme induction frequently decreases the intensity and the duration of a drug's effects. The sedative/hypnotic and seizure control drug phenobarbital is an especially potent liver enzyme inducer. Regular dosage for several days produces a maximal enzyme-inducing effect. Such a dosage regimen of phenobarbital in an individual who is also receiving certain anticoagulant drugs greatly speeds the metabolic inactivation of the anticoagulant, and its efficacy is diminished. A larger dose of the anticoagulant must be administered to achieve the desired level of clotting time extension. However, if the individual terminates the phenobarbital dosage, the catalytic activity of the anticoagulant-metabolizing enzymes slows to the normal level and if the individual's dosage of the anticoagulant drug is not also lowered, a possibly fatal overdose effect from the drug may result. The relative ineffectiveness of certain oral contraceptives in women who take phenobarbital concurrently has been attributed to enhanced metabolic inactivation of estrogens due to enzyme induction.

The mechanism by which phenobarbital induces some enzymes is as yet incompletely understood. There is a tremendous inter-individual variability in induction and inhibition of metabolizing enzymes, such as the cytochrome oxidase P-450 family. A more contemporary example of a hepatic enzyme inducer is the antiviral agent rifampin.

St. John's wort is an herbal product which has been widely publicized for relief, inter alia, of the symptoms of anxiety. One of the chemical constituents of St. John's wort, *hyperforin*, induces the expression of a P-450 cytochrome oxidase enzyme (CYP3A4). This enzyme is responsible for metabolic inactivation of certain oral contraceptives and of the immunosuppresant drug cyclosporin. Thus, chronic ingestion of St. John's wort may result in diminished duration (and magnitude) of effect of these drugs.

3.7.4 Drug Disposition Tolerance

Drugs may induce the liver enzymes involved in their own metabolism. The observation that ethanol is metabolized twice as rapidly by alcoholics who have been drinking as by nondrinkers is ascribed to enzyme-induction phenomena. Thus, the folklorish contention that experienced drinkers can "hold their liquor" has a physiological basis. Phenobarbital induces the enzymes which inactivate it. These are examples of *drug disposition tolerance*. Ethanol or phenobarbital, even when ingested alone, requires continuously larger doses to attain the desired level of pharmacological effect. Enzyme induction may be involved, among other factors, in the habituation (addiction?) that exists to ethanol and the barbiturates. A consistently high chronic ingestion of ethanol enhances ethanol oxidation and also increases metabolic inactivation of barbiturates. However, the metabolic activity of the liver is markedly reduced in the temporary presence of high levels of ethanol. This perhaps explains both the increased tolerance of alcoholics to barbiturates when they are sober, and their *enhanced* response to the same hypnotics when they are drinking heavily.

It must be noted that a significant portion of the tolerance that develops to ethanol and to the barbiturates is due to some sort of cellular adaptation (pharmacologic or tissue tolerance). The tissue simply becomes refractory to the effect of the drug. This effect has been stated as being due to "multiple mechanisms".

3.7.5 Inhibition of Metabolizing Enzymes by Drugs

Certain chemical substances inhibit enzyme systems which metabolically inactivate drugs. In this situation, the effect of the drug is magnified and prolonged. The phenomenon is exemplified by the so-called *cheese syndrome*. Among the early agents proposed for relief of the symptoms of clinical depression was an inhibitor of the enzyme monoamine oxidase: *pargyline* **3.13**. The basis for this therapeutic approach was the theory that clinical depression is caused by hypofunction of norepinephrine-mediated pathways in the brain. Because monoamine oxidase is one of the enzymes involved in metabolic inactivation of norepinephrine, it was speculated that a drug which inhibits this enzyme would protect norepinephrine from inactivation, resulting in increased levels of the neurotransmitter in the brain, and thus relief from the depression syndrome would result. A small number of patients on a dosage regimen of the monoamine oxidase inhibitor drug experienced acute hypertension, severe headaches, and in a few instances, patients suffered fatal intracranial hemorrhage. It was eventually determined that a common factor in these tragic occurrences was that the affected individuals had ingested a sizeable amount of ripened cheese before the onset of hypertensive symptoms. This type of cheese contains sizeable amounts of a potent hypertensive agent, tyramine **3.14**, a normal product of the fermentation process involved in the cheese production. Normally, dietary tyramine is harmless because it is efficiently metabolically inactivated in the intestinal wall by monoamine oxidase, prior to absorption. However, if this enzyme is inactivated by the antidepressant drug, tyramine will be protected from metabolic destruction. It will be absorbed across the intestinal wall; it will accumulate in various parts of the body, and produce the blood pressure elevation described.

3.13   3.14

3.8   Chemical Aspects of Drug Metabolism

Metabolic processes involved in the inactivation of drug molecules are subdivided into two categories: *phase I* (functionalization reactions: oxidation, reduction, or hydrolysis) and *phase II* (conjugation reactions). A survey of the following metabolic reactions reinforces the earlier comment that, regardless of the specific chemistry involved, the body's overall metabolic strategy is aimed at altering the partition coefficient of the foreign organic molecule toward greater hydrophilicity (water solubility).

3.8.1 Phase I Drug Metabolism (Functionalization Reactions)

Oxidation

This is probably the most common of the functionalization reactions. Included in this category are introduction of a hydroxyl group onto an aromatic ring or into an aliphatic position; epoxidation of carbon–carbon double or triple bonds; N-oxidation of an amine; oxidation of an alcohol to carbonyl; oxidation of an aldehyde to a carboxylic acid; oxidative dealkylation of oxygen or nitrogen; and oxidative deamination. A complex series of nonspecific enzymes in the liver catalyzes oxidation of a great variety of organic substrates, and key enzymes in this oxidase system are microsomal enzymes, the *cytochrome oxidase* P-450 family (CYP 450) which, like other cytochromes, consist of a combination of heme (the iron-containing, oxygen-binding component of hemoglobin) with a protein different from that in hemoglobin (figure 3.3).

Cytochrome oxidase P-450 enzymes bind oxygen (at the iron atom) and transfer it to a wide variety of substrates. As illustrated in figure 3.4, the ferric iron of the P-450 enzyme combines with a molecule of the drug substrate (DH); it receives an electron from NADPH–P450 reductase, reducing the iron to the divalent state. This complex combines with molecular oxygen, then with a proton, and with a second electron (either from NADPH reductase or from cytochrome b_5) to form the $Fe^{3+}OOH$–DH complex which combines with another proton to yield an equivalent of water and $(FeO)^{3+}$–DH complex. The ferrene oxide $(FeO)^{3+}$ extracts a proton from DH with the formation of short-lived free radicals D· and $Fe^{3+}OH·$. The drug free radical acquires the bound ·OH free radical to form the hydroxylated drug metabolite, DOH, and the P-450 enzyme is regenerated in its original state.

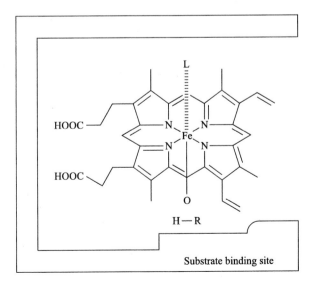

Figure 3.3 Cytochrome P-450 active site

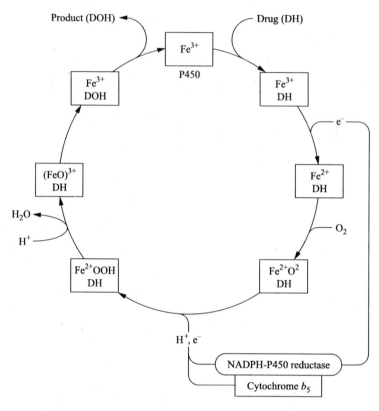

Figure 3.4 Monooxygenase P-450 cycle. (Reproduced with permission from reference 3. Copyright 2001 Churchill Livingston)

The cytochrome oxidase P-450 family (of which there are as many as 500 isoenzymes, not all of which are found in the human body) is probably the major drug metabolizing system in the body and it plays a role in both detoxication and toxication processes. A number of liver microsomal P-450 enzymes are involved in fatty acid and steroid biosynthesis and catabolism. Of the approximately 30 human liver P-450 enzymes, only a modest number (six to eight) is involved in drug metabolism. The liver is not the only site of action of the P-450 cytochromes; members of the family are also found, inter alia, in the respiratory tract, the kidney, and the gastrointestinal tract.

With respect to first-pass metabolic inactivation of orally administered drugs, it has been estimated that oxidation by the cytochrome P-450 (CYP3A) enzyme system is involved in 60% of these drugs. Compounds that inhibit this enzyme system can diminish the extent of first-pass metabolic inactivation. The ability of grapefruit juice to reduce first-pass effects has been recognized for several years. *Spiro*-ortho esters have been isolated from grapefruit and these have been identified as the substances that inhibit the P-450 enzymes. Contemporary, ongoing research suggests an important therapeutic role for these grapefruit constituents, to improve bioavailability of a variety of orally administered drugs.

As examples of metabolic hydroxylation of drugs: acetanilide **3.15** is converted into the *p*-hydroxy derivative **3.16**. Toluene **3.17** is oxidized at the ring methyl to benzyl alcohol **3.18**. Metabolic activity does not stop at this point, however. The alcohol group of **3.18** is oxidized to the aldehyde, thence to the carboxylic acid **3.19** which is excreted in the urine. The reputed lower toxic hazard of toluene compared to that of benzene as laboratory solvents is based on the premise that metabolism of toluene follows the path illustrated, giving rise to relatively nontoxic metabolites, whereas benzene undergoes a much more complex sequence of metabolic alterations (figure 3.5). The first-formed metabolite is believed to be the arene oxide **3.20** which undergoes further metabolic conversions to chemically reactive semiquinones such as **3.24** (a radical anion) and quinones such as **3.25** which form yet other free radical products that disrupt a variety of physiological processes.

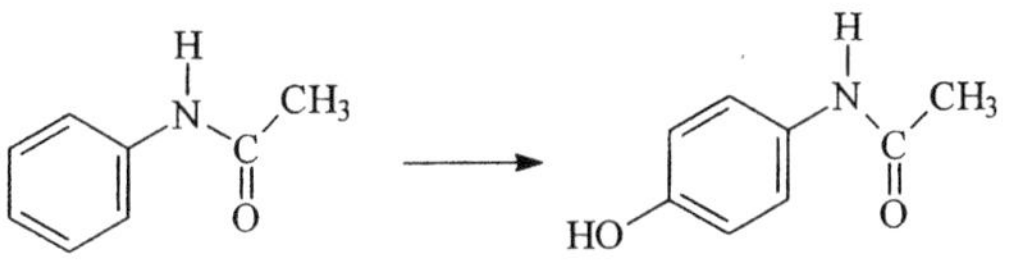

3.15   3.16

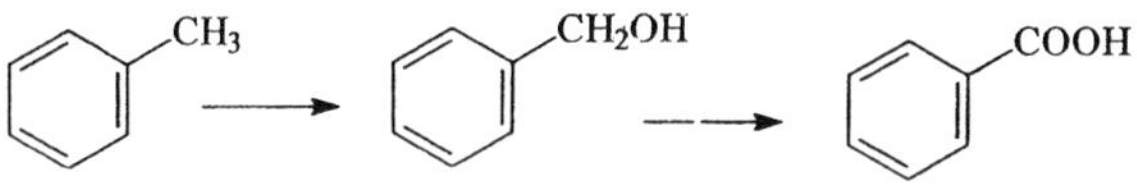

3.17   3.18   3.19

Hydroxylation of a secondary aliphatic carbon in meprobamate (**3.26** → **3.27**) is an example of a rather common metabolic functionalization. Oxidase enzymes prefer to attack secondary aliphatic carbon atoms.

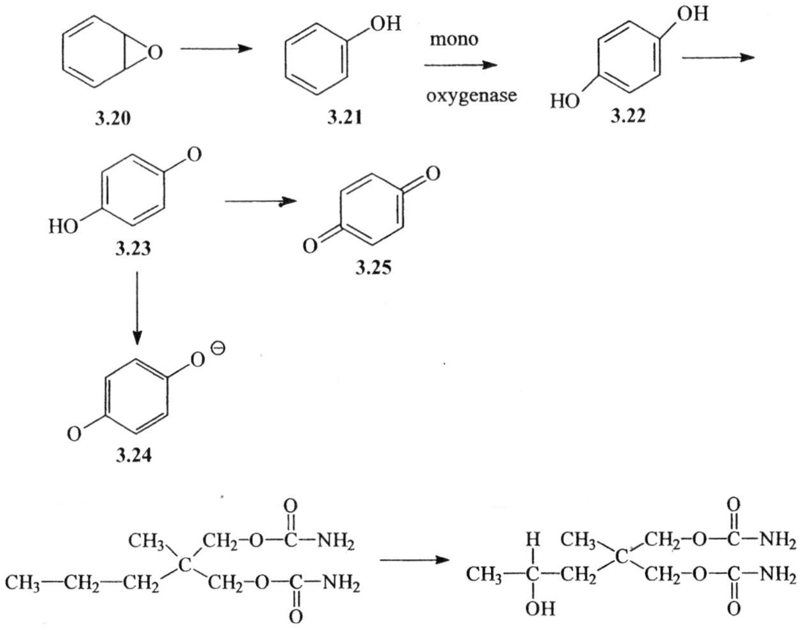

3.20   3.21   mono oxygenase   3.22

3.23   3.25

3.24

3.26   3.27

Figure 3.5 Metabolic fate of benzene

Both aliphatic and aromatic amino groups can undergo a variety of oxidative changes. Many tertiary amines, exemplified by trimethylamine **3.28**, are converted by cytochrome oxidases into *N*-oxides **3.29** which, by virtue of their dual charge character, are highly hydrophilic. One of the metabolic pathways of aniline **3.30**, a typical aromatic amine, is stepwise oxidation to phenylhydroxylamine **3.31**, thence to nitrosobenzene **3.32**.

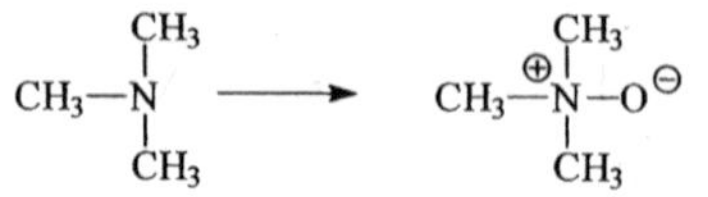

3.28   **3.29**

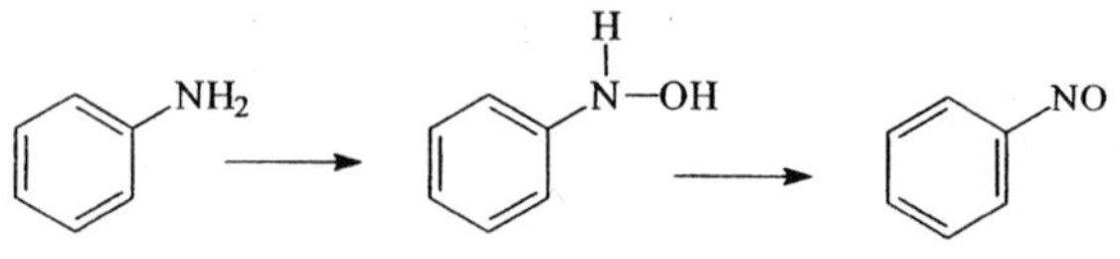

3.30   **3.31**   **3.32**

Oxidative O-dealkylation (ether cleavage) is shown in conversion of acetophenetidine **3.33** to *p*-acetylaminophenol **3.35**. The oxidative step involves enzyme-mediated introduction of a hydroxyl group onto the α position of the ethyl moiety of the ether. The resulting hemiacetal product **3.34** spontaneously collapses to liberate the free phenolic group and the elements of acetaldehyde.

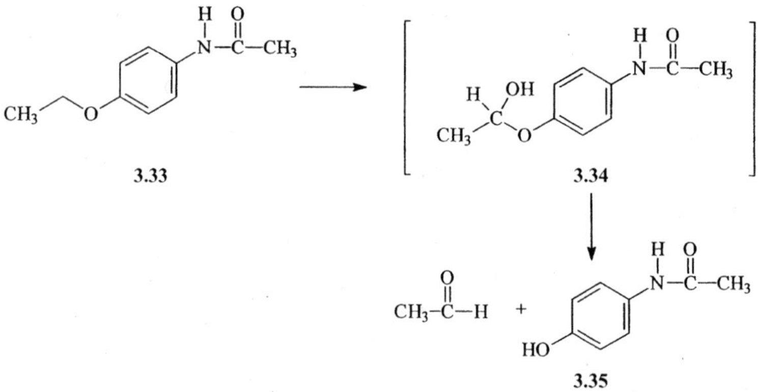

3.33   **3.34**

3.35

Analogous reactions occur in metabolic N-dealkylation, typified by desoxyephedrine **3.36**. Oxidase enzymes hydroxylate the *N*-methyl group, and the resulting carbinolamine (hemiaminal) **3.37** expels the elements of formaldehyde to form the primary amine **3.38**. Further oxidative metabolism can occur on this first-formed metabolic product. Enzyme-mediated hydroxylation of the amine-bearing carbon gives rise to a second carbinolamine **3.39** which collapses to eliminate ammonia and to form a nitrogen-free product, phenylacetone **3.40**.

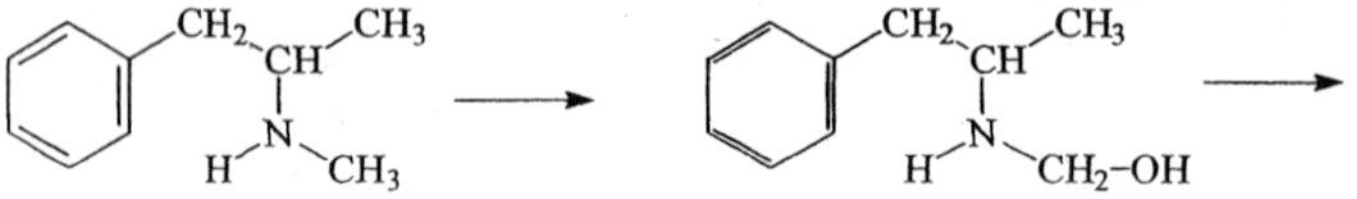

3.36   **3.37**

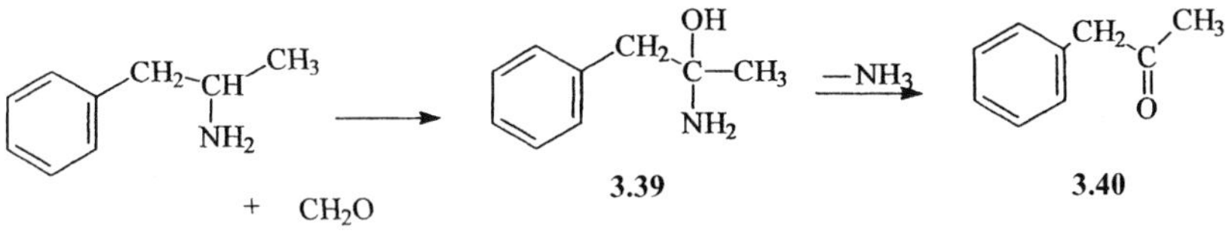

3.38

The metabolic fate of codeine (figure 3.6) illustrates a phenomenon cited earlier, namely that the body frequently converts a drug molecule into several different products. Indeed, each of the metabolic products of codeine shown in figure 3.6 is subject to further metabolic transformations producing additional numbers of products. Some additional major metabolites of codeine involve glucuronidation of the 6-OH group.

Ethanol is initially attacked in the liver by a soluble cytoplasmic enzyme, alcohol dehydrogenase (equation 3.1):

$$CH_3CH_2OH \quad \xrightarrow[\text{dehydrogenase}]{\text{alcohol}} \quad CH_3-\overset{\overset{\text{O}}{\|}}{C}-H \quad \xrightarrow[\text{dehydrogenase}]{\text{aldehyde}} \quad CH_3-COOH \qquad (3.1)$$

Ethanol / Acetaldehyde

The product of this reaction, acetaldehyde, is acted upon in the liver by aldehyde dehydrogenase to form acetic acid which can then enter the Krebs tricarboxylic acid cycle and be converted into carbon dioxide and water. Metabolites of *disulfiram* **3.41** (one of which is the *N,N*-diethylthiomethyl carbamate **3.42**) inhibit several forms of aldehyde dehydrogenase. As a result, acetaldehyde from the metabolism of ethanol is not rapidly metabolized to acetic acid, and the blood acetaldehyde level rises from five to ten times higher than in an untreated individual. The resulting *acetaldehyde syndrome* is manifested within 5 to 10 min and is characterized by flushing of the face and neck, an intense pulsing headache, blurred vision, respiratory difficulties, nausea and vomiting,

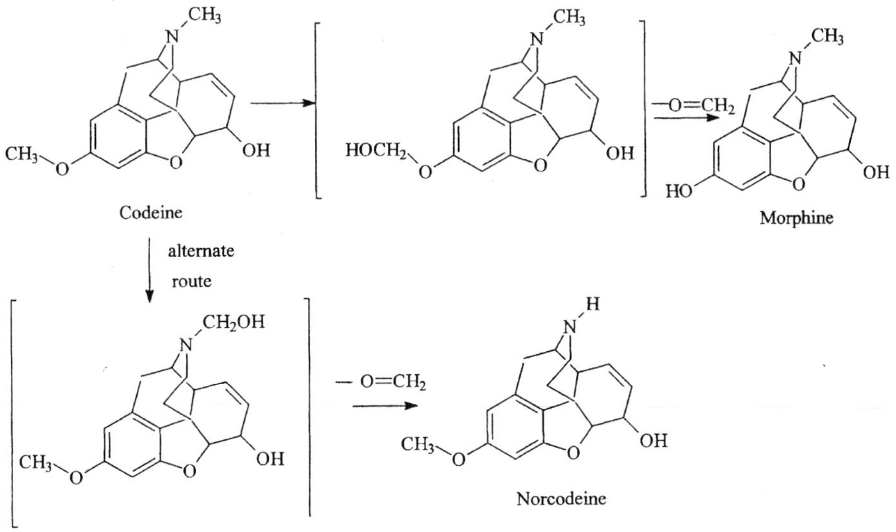

Figure 3.6 Metabolic fate of codeine

sweating, thirst, chest pain, weakness, dizziness, and confusion. The flushing of the face may be replaced by pallor, and the blood pressure may fall to shock levels. Sensitization may last as long as 14 days. Disulfiram ("antabuse") has been employed in treatment of chronic alcoholism, to render the ingestion of ethanol intolerable to the user. However, the disulfiram strategy has not been widely successful due to poor acceptance and compliance by the alcoholic individual.

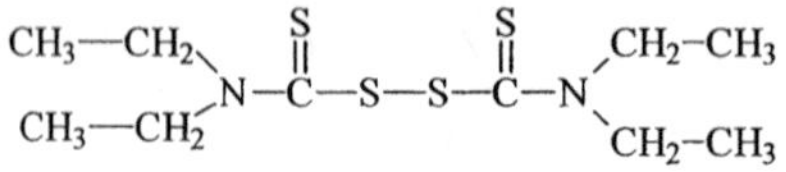

3.41 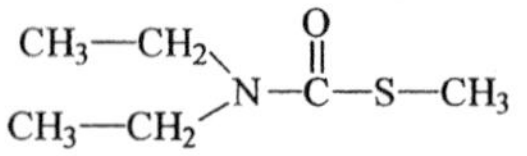   **3.42**

Methanol toxicity results from ingestion of large amounts of methanol as a beverage. Methanol metabolism follows a pathway similar to that of ethanol. However, the methanol metabolites, formaldehyde and formic acid, are much more toxic than acetaldehyde and acetic acid. Unlike acetic acid, which undergoes rapid metabolic destruction leading to carbon dioxide and water, there is no facile, efficient in vivo method for removal of formic acid and it accumulates in the body to produce systemic acidosis (decrease in pH of body fluids) which is believed to cause the severe symptoms of methanol poisoning, including permanent blindness. One treatment for acute methanol intoxication is the administration of ethanol, which competes with methanol for the catalytic site(s) of the alcohol metabolizing enzymes.

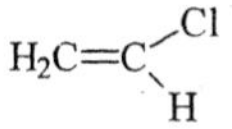

3.43 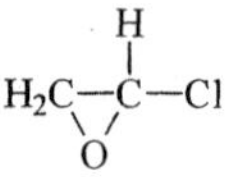   **3.44**

A common metabolic path for many olefins and acetylenic compounds is epoxidation, illustrated for vinyl chloride **3.43**. The antiseizure agent carbamazepine **3.45** is converted into its 10,11-epoxide **3.46**. This product is pharmacologically active; it is further metabolized to the dihydrodiol which is inactive. Epoxide metabolites are frequently very reactive chemically, and some of them have been implicated in the toxicity and carcinogenicity of some chemicals.

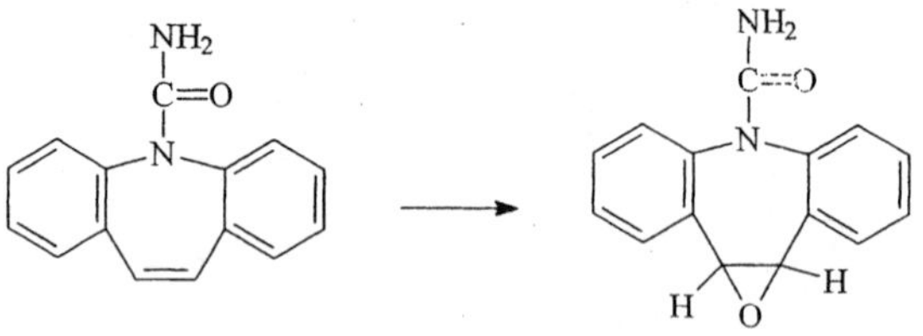

3.45   **3.46**

Dimethyl sulfoxide **3.47** is oxidized to dimethyl sulfone **3.48** which is excreted via the kidney. A minor metabolite is dimethyl sulfide **3.49**, a reduction product. This low-boiling metabolite is excreted in part in the expired air which imparts a characteristic unpleasant odor to the breath.

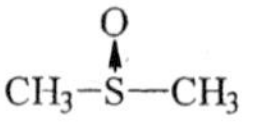

3.47 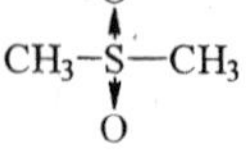   **3.48** 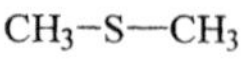   **3.49**

Reduction

There are several reductase systems in the liver, which is the most important site for this type of metabolic reaction. The metabolic conversion of chloral hydrate **3.50**, the hydrated derivative of trichloroacetaldehyde, into trichloroethanol **3.51** produces the product which is responsible for the observed sedative/hypnotic effects of chloral hydrate.

$$Cl_3C-\underset{\underset{OH}{|}}{\overset{\overset{H}{|}}{C}}-OH \longrightarrow Cl_3C-CH_2OH$$

3.50   **3.51**

Aromatic nitro compounds, typified by nitrobenzene **3.52**, are sequentially reduced to nitrosobenzene **3.53**, thence to phenylhydroxylamine **3.54**, and finally to aniline **3.55**.

From early times toxicologists recognized that the gross signs and symptoms of nitrobenzene intoxication are almost identical with those of aniline poisoning. This is explained on the basis that the two molecules are metabolized, in part, to common products. Phenylhydroxylamine **3.54** has been implicated as a causative factor in aniline and nitrobenzene-derived methemoglobinemia.

Reductive dehalogenation is an important pathway in metabolism of halogenated hydrocarbons such as carbon tetrachloride (which is carcinogenic) and the inhalation anesthetic halothane **3.56** (figure 3.7).

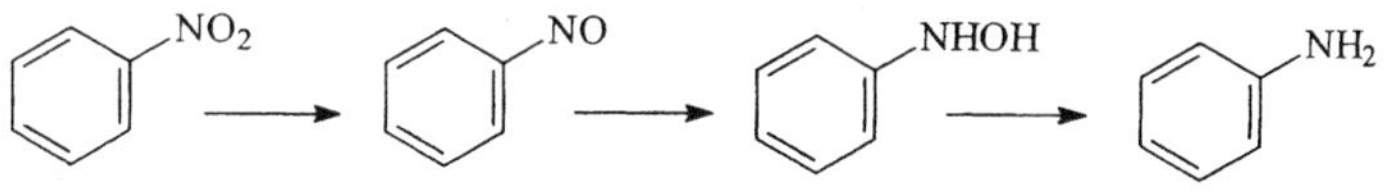

3.52   **3.53**   **3.54**   **3.55**

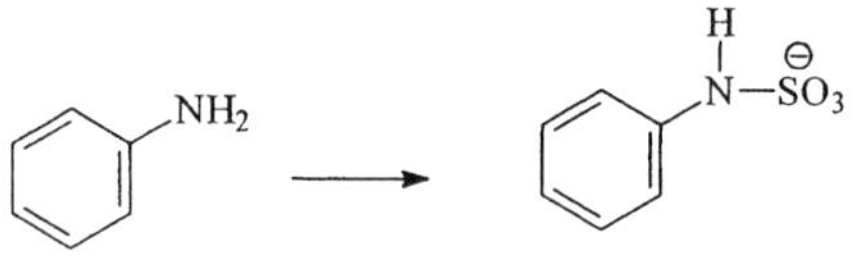

3.55 Aniline   Aniline–sulfuric
                                   acid conjugate

Chemically reactive free radicals that result from homolytic cleavage of the halogen–carbon bond, can initiate undesirable chemical events in the body.

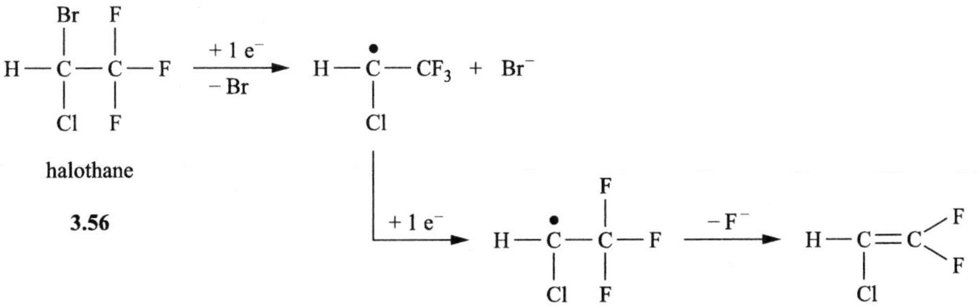

halothane

3.56

Figure 3.7 Metabolic fate of halothane

Hydrolytic cleavage of esters and amides

Carboxylic esters are hydrolytically cleaved by nonspecific enzymes in the blood, kidneys, liver, and several other organs and tissues. Because of the prime role of the blood in transporting drug molecules throughout the body, ester hydrolysis in this tissue is especially significant. The local anesthetic procaine **3.57** and the myoneural blocking agent succinyl dicholine **3.58** undergo extensive hydrolytic cleavage in the blood and thereby are inactivated pharmacologically.

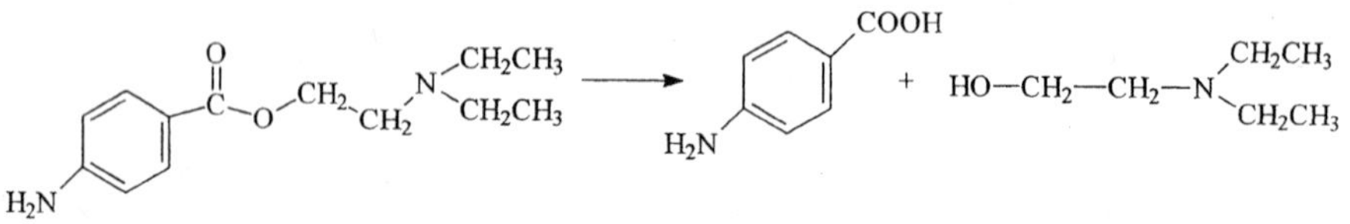

3.57

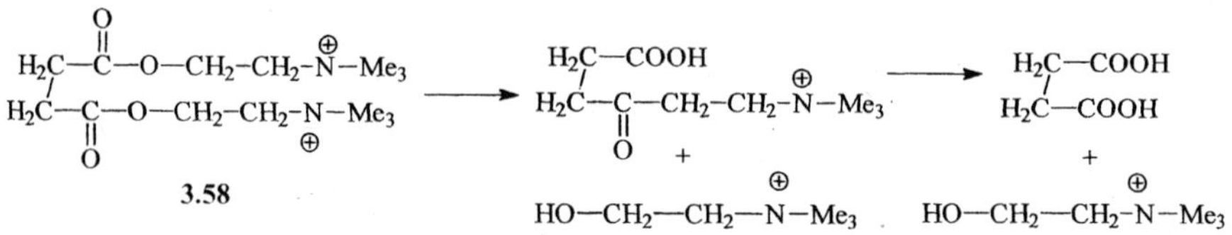

3.58

Sterically hindered esters may not be as readily cleaved in the body. Approximately 50% of a dose of atropine **3.59** is excreted in the urine with the ester moiety unaltered. Some drugs (e.g., meperidine **3.60**) resist cleavage by blood esterases, but they are substrates for liver esterases. Many esters of inorganic acids, e.g., glyceryl trinitrate **3.61** are also enzymatically cleaved. A significant percentage of some labile ester moieties (also typified by glyceryl trinitrate) is cleaved nonenzymatically in vivo, a function of body temperature and of the pH character of the various tissue fluids.

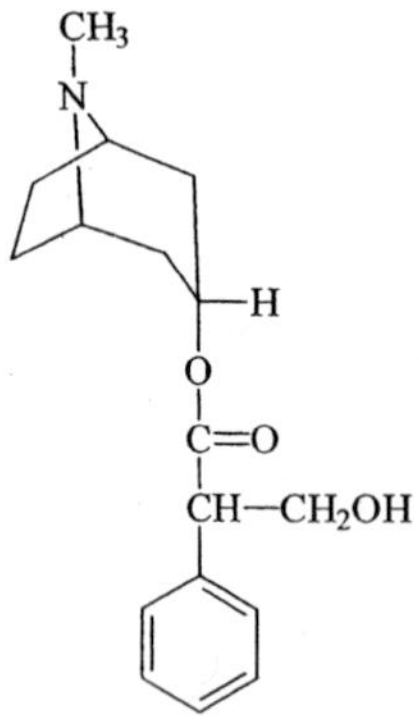

3.59 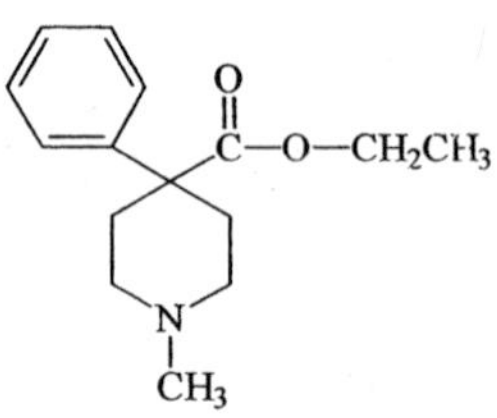   3.60 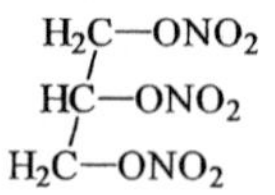   3.61

Amides are more resistant to hydrolysis than esters, and some amide-derived drugs are excreted with their amide moiety unchanged. However, liver amidases cleave the carboxamide linkages of succinylsulfathiazole **3.62** and of the local anesthetic/ antiarrhythmic agent lidocaine **3.63**. The relative ease of in vivo enzyme-mediated hydrolysis of lidocaine contrasts with its resistance to laboratory hydrolytic conditions (heating in alcoholic potassium hydroxide solution).

3.62   **3.63**

3.8.2 Phase II Drug Metabolism (Conjugation Reactions)

Acetylation

Enzyme-mediated acetylation occurs mainly in the liver and is aimed primarily at the nitrogen of aromatic or aliphatic primary amino groups, hydrazines, sulfonamides, and amino acids. The source of the acetyl moiety is acetyl coenzyme A; a number of N-acetyltransferases catalyze the reaction. The therapeutic problem of diminished water solubility of certain acetylated sulfonamide metabolites has been described previously.

Glucuronidation and sulfate conjugation

Conjugation with these hydrophilic moieties is frequently the major metabolic pathway for drugs that bear aliphatic or aromatic hydroxyl, amino, or sulfhydryl groups. Examples are the glucuronidation of the phenolic group of morphine **3.64** and sulfation of the amino group of aniline **3.55** and of the hydroxy group of phenol **3.65**. It is noteworthy that the O-3 (phenolic) glucuronidation of morphine **3.64** abolishes analgesic effect, whereas O-6 glucuronidation permits retention of analgesic effect. Glucuronidation and sulfation often occur side-by-side in the body, competing for the same functional group on a drug molecule.

3.64   **3.65**

Amino acid conjugation

This is an important metabolic route for carboxylic acids. Endogenous glycine forms conjugates with aliphatic, aromatic, and heterocyclic carboxylic acids. The resulting conjugates are readily excreted in urine and bile. Glycine conjugation is illustrated by a formerly popular introductory organic chemistry laboratory experiment, in which students ingested benzoic acid orally, and subsequently collected their urine. Upon acidifying the urine, attractive crystals of the benzoic acid–glycine conjugate (hippuric acid) **3.66** separated in virtually quantitative yield. In addition to glycine, several other amino acids (e.g., glutamine, taurine, ornithine) conjugate with carboxyl-containing xenobiotics.

Benzoic
acid

3.66
Glycine conjugate of benzoic acid
(hippuric acid)

Glutathione conjugation

Glutathione **3.67**, a thiol-containing tripeptide which is widely distributed among the tissues of the body, is of considerable significance in metabolism of endogenous compounds, as well as drugs and other xenobiotics. Glutathione exists in two oxidation states in equilibrium with each other, a reduced form G-SH and an oxidized (disulfide) form G-S-S-G. The reduced form (as its strongly nucleophilic sulfide anion) is involved in conjugation reactions. It reacts with such electrophilic functional groups as benzyl halides, aliphatic nitrate esters, epoxides, and quinones (figure 3.8).

3.67

Glutathione conjugation reactions may be nonenzymatic or enzymatic (glutathione *S*-transferase). The epoxide and quinone reactants with glutathione may themselves be newly formed metabolic products of drugs or other xenobiotics. The first-formed glutathione conjugates are rarely excreted as such, but they undergo further and frequently chemically complex biotransformations prior to fecal excretion via the bile. One such subsequent pathway is illustrated in figure 3.9. Usually the reaction of an electrophile with glutathione is a detoxication process, but some carcinogens have been activated by conjugation with glutathione. The G-SH: G-S-S-G equilibrium redox system has been implicated in drug metabolism.

3.9   Animal Species Differences in Drug Metabolism

Significant differences exist in the metabolism of drugs in different animal species, and this is an important factor in attempting to extrapolate pharmacological results from animal studies to humans. In some instances, the species differences are quantitative differences in *rates* of metabolism. Generally, the smaller the animal, the higher the metabolic rate. Thus, a certain barbiturate maintains unconsciousness in a human for several hours at a dose level of 50 mg kg^{-1} of body weight, but this same dose level in a mouse produces sleeping time of only minutes.

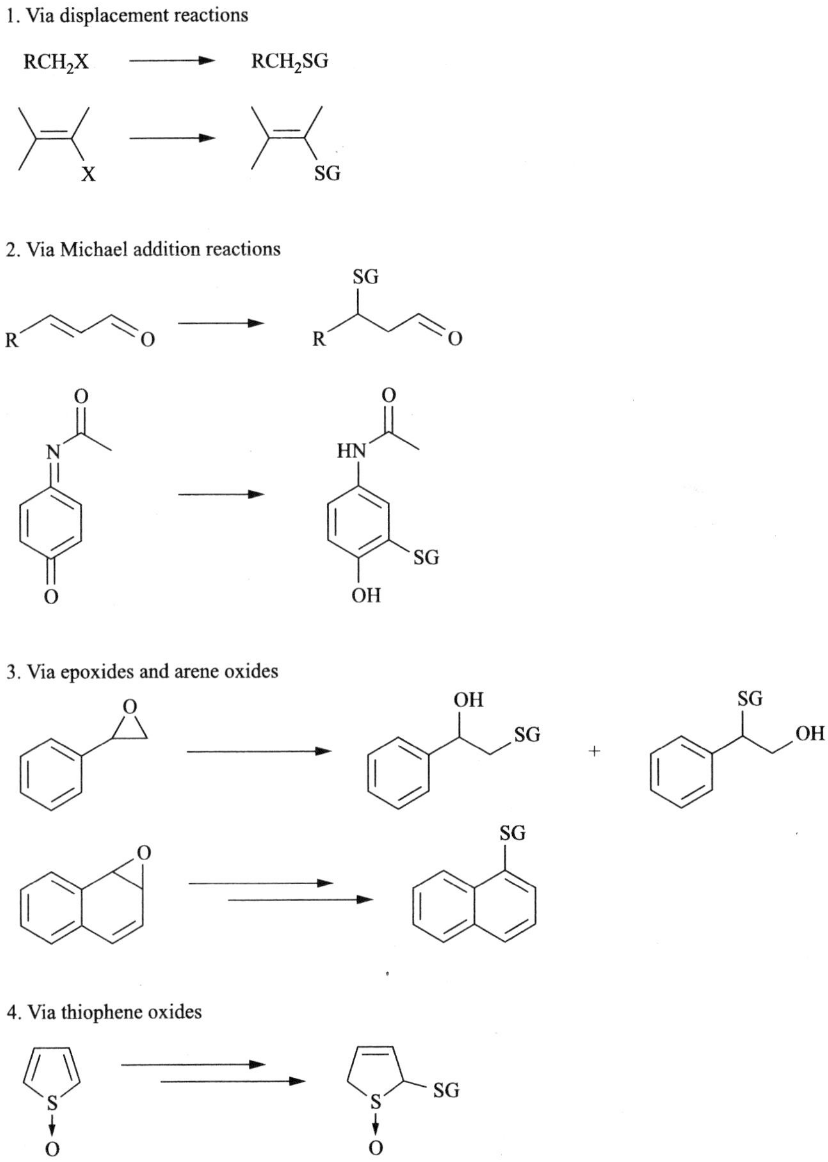

1. Via displacement reactions

$$RCH_2X \longrightarrow RCH_2SG$$

2. Via Michael addition reactions

3. Via epoxides and arene oxides

4. Via thiophene oxides

Figure 3.8 Reactions that lead to glutathione adducts

More significant species differences in the pattern of drug metabolism occur when there are qualitative differences in the metabolic pathway(s). The insecticidal metabolite (maloxon) of *malathion* **3.68** (figure 3.10) inactivates a vital enzyme system in the insect body by forming a covalent bond with a portion of the catalytic surface, which is a lethal effect. However, in mammals the malathion molecule is efficiently attacked by blood esterases which hydrolytically cleave the ester linkages. The resulting metabolite,

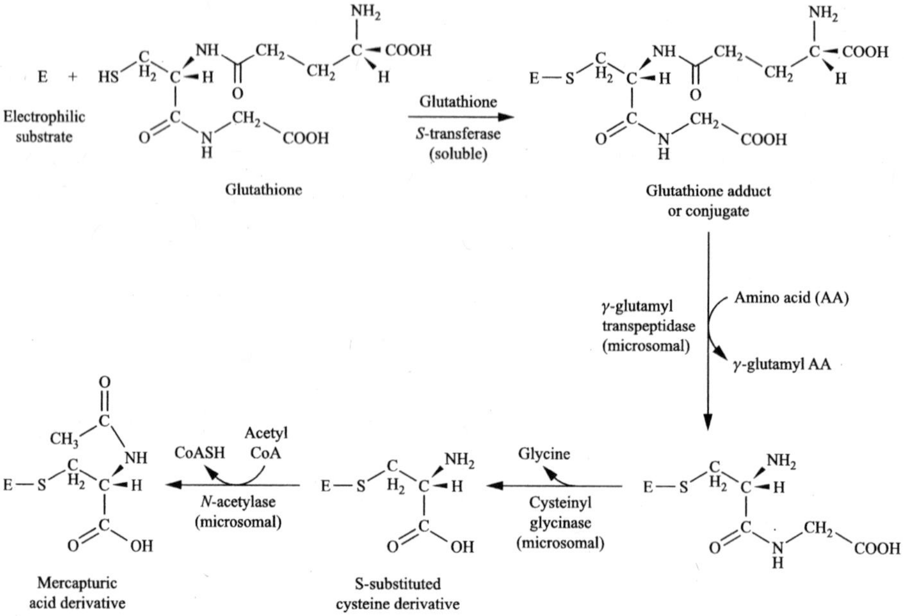

Figure 3.9 Metabolic fate of glutathione conjugate

the free dicarboxylic acid **3.69**, is much more hydrophilic than the intact malathion diester, and the diacid is rapidly excreted in the urine. An analogous esterase system in most insect bodies is much less effective, and the insect has no powerful defense against malathione. In addition, it appears that the free dicarboxylic acid has a greatly diminished ability to react with the target enzyme molecule as shown in figure 3.3e for malathione. It is intrinsically less toxic. A malathion lotion has been employed therapeutically by direct application to the scalp for combating head lice infestation.

3.10   Human Genetic Variation in Drug Metabolism

The rate of drug metabolism is frequently affected by genetic variation. For example, succinyl dicholine **3.58**, a short-acting neuromuscular blocking agent, owes its short duration of action to rapid hydrolytic cleavage of its ester links by blood esterases. In a small proportion of the population, this esterase system is deficient or is hypoactive, a genetic trait, and succinyl dicholine molecules are not as rapidly inactivated. In these individuals, the effect of a "normal" dose of succinyl dicholine is prolonged and is more potent, even to the point of producing toxic symptoms.

The antituberculosis drug *isoniazid* **3.70** is metabolically inactivated by acetylation of the hydrazide nitrogen **3.71**. Humans can be classed as *slow* or *fast acetylators*, depending upon the amount and/or efficiency of N-acetyltransferase in the liver. For a slow acetylator, the circulating levels of isoniazid produced by a "normal" dose are significantly higher, and these individuals may exhibit toxic side effects from the drug.

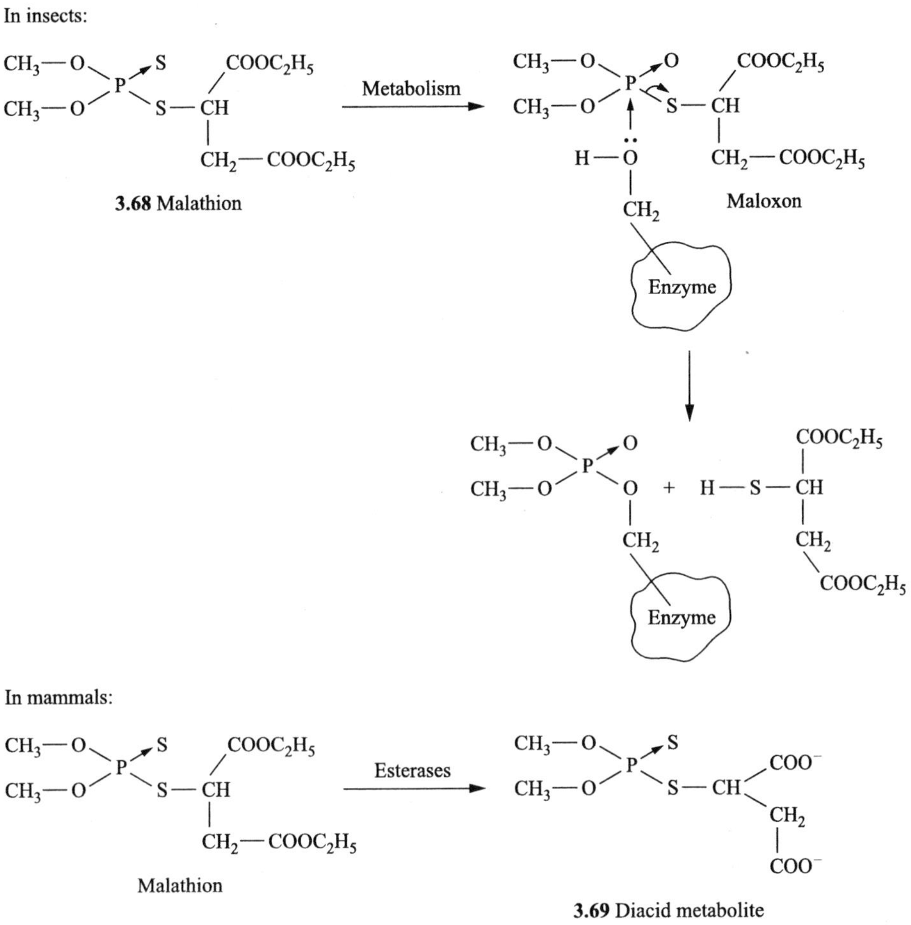

In insects:

In mammals:

Figure 3.10 Species differences in drug metabolism in insects and in mammals

The distribution of slow and fast acetylators in the world population shows interesting variations in ethnic groups. Approximately 50% of Caucasians and African—Americans of are slow acetylators, but approximately 90% of ethnic Japanese and Eskimos are fast acetylators. Racial differences in drug metabolism are manifested also with respect to ethanol. Many ethnic Chinese (and other east Asians, said to be as much as 50% of this population) lack an efficient aldehyde dehydrogenase system. In these individuals, ethanol ingestion results in a higher level of acetaldehyde in the tissues, which can cause some of the symptoms the acetaldehyde syndrome described previously.

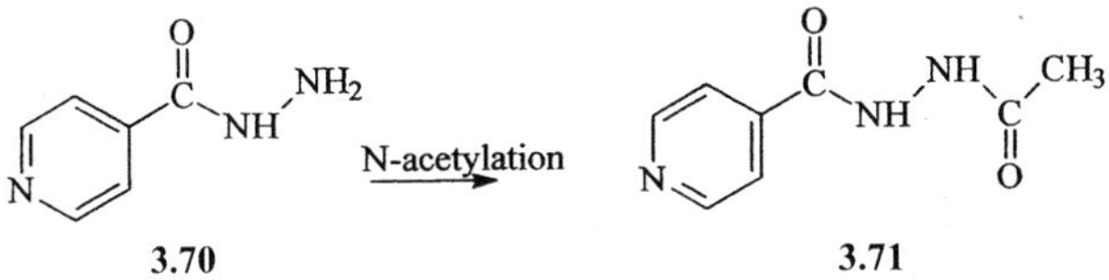

In recognition of the role of genetic variation with respect to drugs, a new term has been coined: *pharmacogenomics*. This is the study of how genetic variations affect the way that individuals respond to drugs.

3.11   Age and Gender Differences in Drug Metabolism

In the newborn, several metabolizing enzymes such as liver microsomal oxidases, glucuronyl transferase, acetyl transferase, and plasma esterases have low activity. However, in most infants, after approximately 8 weeks these enzymes attain adult-level catalytic activity. The markedly reduced ability of the newborn to effect conjugation reactions on drugs is one reason that morphine is considered to be a poor choice for analgesia in labor. The drug can pass from the mother's circulatory system into that of the soon-to-be-born infant. Morphine is metabolically inactivated by conjugation with glucuronic acid, and in the newborn the effect of a "normal" dose of morphine is magnified and prolonged. This metabolic deficit may be coupled with the newborn's poorly developed blood–brain barrier, further magnifying the undesirable effects of morphine in the infant.

The activity of liver microsomal enzymes sometimes decreases with age, but this effect is highly variable, and it is not possible to state general trends. Some elderly individuals exhibit drug metabolizing activities comparable to those of young adults and some elderly individuals demonstrate grossly reduced rates of drug metabolism.

Significant gender differences in the rate and manner of drug metabolism have been noted in animals, but in the past these have been considered to be relatively unimportant in humans. This conclusion seems to have resulted, at least in part, from the fact that no one sought to determine or to evaluate differences in metabolism between men and women. However, there is now a new interest in sex differences in drug metabolism (and in drug actions and effects). It seems likely that gender differences in human drug metabolism are due, at least in part, to the influence that sex hormones have on the activity of metabolizing enzymes. It must be anticipated that vigorous pursuit of this aspect of human pharmacology will lead to improved therapeutic strategies. (see Recommended Reading, ref. 16).

3.12   Prodrugs and Latentiated Drugs

A different aspect of metabolism of xenobiotics involves *prodrugs*. Some molecules may require in vivo enzymatic catalysis to convert them from a pharmacologically inactive form into a pharmacologically active one. The discovery of prodrugs through metabolism studies has frequently been purely serendipitous. Nevertheless, advantages and objectives of the use of prodrug molecules in therapy can be cited:

1. Improved chemical stability
2. Improved physical properties for formulation of dosage forms
3. Improved patient acceptance and compliance (e.g., overcoming an unpleasant taste)
4. Improved bioavailability
5. Prolonged duration of action
6. Decreased side effects

The metabolic reduction of chloral hydrate **3.50** to trichloroethanol **3.51** (illustrated earlier in this chapter), the conversion of minoxidil **3.72** into its sulfate ester **3.73**, and the production of nitric oxide (NO) from glyceryl trinitrate **3.61** are illustrative.

3.72   Sulfotransferase   **3.73**

The anticancer drug cyclophosphamide **3.74** illustrates a prodrug whose activation in the body is partly enzyme-mediated and is partly non-enzymatic (figure 3.11). Cyclophosphamide is acted upon in the liver by a cytochrome P-450 system which triggers a cascade of chemical changes. As illustrated, the aziridinium derivative **3.75** opens to generate a carbocation **3.76** which is believed to be the pharmacologically active species.

The prodrug strategy should be distinguished from *drug latentiation* in which the active form of a drug is derivatized to afford a compound that will be enzymatically reconverted in the body into the original molecule. The advantages and objectives of latentiation are much the same as those that were cited above for prodrugs. A classic example of latentiation is the esterification of morphine with acetic acid to form heroin (see chapter 12).

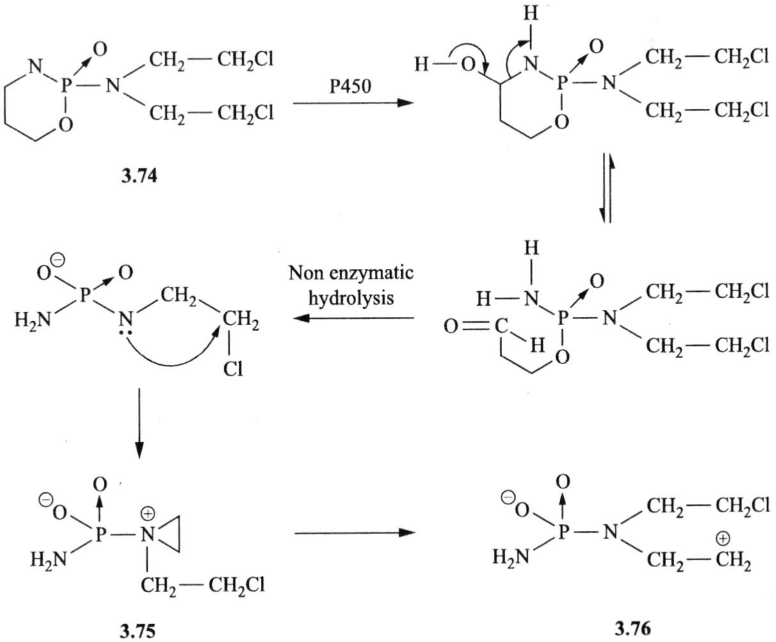

Figure 3.11 Metabolic activation of cyclophosphamide

Bibliography

1. Daniels, T. C. and Jorgensen, E. C. In *Wilson and Gisvold's Textbook of Organic Medicinal and Pharmaceutical Chemistry*, 8th ed. Doerge, R. F., Ed. Lippincott: Philadelphia, Pa., 1982; p. 500.
2. Tortora, G. J. *Principles of Human Anatomy*, 8th ed. Wiley: Hoboken, N.J. 1999; p. 39.
3. Rang, H. P., Dale, M. M., Ritter, J. M. and Gardner, P. *Pharmacology*, 4th ed., Churchill Livingstone: Philadelphia, Pa., 2001; p. 81.

Recommended Reading

1. Testa, B. and Soine, W. Principles of Drug Metabolism. In *Burger's Medicinal Chemistry and Drug Discovery*, 6th ed. Abraham, D. J., Ed. Wiley-Interscience: Hoboken, N.J., 2003; Vol. 2, pp. 431–498.
2. Mesnil, M. and Testa. B. Xenobiotic Metabolism by Brain Monooxygenases and Other Cerebral Enzymes. *Adv. Drug Res.* **1984**, *13*, 95–207.
3. Lennard, M. S., Tucker, G. T. and Woods, H. F. Stereoselectivity in Pharmacokinetics and Drug Metabolism. In *Comprehensive Medicinal Chemistry*. Hansch, C., Ed. Pergamon Press: Oxford, UK, 1991; Vol. 5, pp. 187–204.
4. Breckenridge, A. M. and Park, B. K. Enzyme Induction and Inhibition. In Recommended Reading ref. 3, pp. 205–217.
5. Juchau, M. R. Species Differences in Drug Metabolism. In Recommended Reading ref. 3, pp. 219–235.
6. Juchau, M. R. Developmental Drug Metabolism. In Recommended Reading ref. 3, pp. 237–249.
7. Commandeur, J. N. M., Stijntjes, G. J. and Vermeulen, N. P. Enzymes and Transport Systems Involved in the Formation and Disposition of Glutathione S-Conjugates. *Pharmacol. Rev.* **1995**, *47*, 271–330.
8. Moore, L. B., Goodwin, B., Jones, S. A., Wisely, G. B, Serabjit-Singh, C. J., Willson, T. M., *et al*. St. John's Wort Induces Hepatic Drug Metabolism Through Activation of the Pregnane X Receptor. *Proc. Soc. Exptl. Biol. Med.* **2000**, *97*, 7500.
9. Ioannides, C., Ed. *Enzyme Systems That Metabolise Drugs and Other Xenobiotics*. Wiley: New York, 2002.
10. Kalf, G. F. Recent Advances in the Metabolism and Toxicity of Benzene. *Crit. Rev. Toxicol.* **1987**, *18*, 141–159.
11. Ding, X. and Kaminsky, L. S. Human Extrahepatic Cytochromes P450. In *Annual Review of Pharmacology and Toxicology*; Cho, A. K., Blaschke, T. F.; Insel, P. A. and Loh, H. H., Eds., Annual Reviews: Palo Alto, CAL., 2003; Vol. 43, pp. 149–173.
12. Rouhi, A. M. Citrus Chemistry Boosts Drugs. *Chem. Eng. News.* **2003**, *81*, 38–39.
13. Testa, B. and Mayer, J. M. *Hydrolysis in Drug and Prodrug Metabolism. Chemistry, Biochemistry, and Enzymology*. Wiley-VCH: Zürich Switzerland, 2003.
14. Evans, D. C., Hartley, D. P. and Evers, R. Enzyme Induction. Mechanisms, Assays, and Relevance to Drug Discovery and Development. *Annu. Rep. Med. Chem.* **2003**, *38*, 315–331.
15. Gandhi, M., Aweeka, F., Greenblatt, R. M. and Blaschke, T. F. Sex Differences in Pharmacokinetics and Pharmacodynamics. *Annu. Rev. Pharmacol. Toxicol.* **2004**, *44*, 499–523

4

Drug Receptors

4.1 Receptor Sites and Drug Binding Sites

4.1.1 Definition of Terms

Most drugs exhibit biological effects in minute amounts and concentrations. The effects produced by these drugs are attributed to their chemical interactions with a specific substance, a cellular component, which is termed a receptor. The term *receptor* has been formally defined as a cellular macromolecule that is concerned directly and specifically with chemical signaling between and within cells. The combination of an appropriate ligand with its receptor(s) initiates a change in cell function. The term *recognition site* refers to the region(s) on the receptor macromolecule to which endogenous ligands bind, whether or not this binding produces a pharmacological response. It is assumed that structurally specific drugs present a high degree of molecular complementarity toward the receptor site with which they interact. An interaction of this type has been compared to the interaction of a substrate molecule with the catalytic region of an enzyme. However, in the case of a drug–receptor interaction, the drug molecule does not necessarily undergo a permanent structural change. For a drug to be useful therapeutically, it must act selectively on particular cells or portions of cells. It should exhibit binding site selectivity. There may be other loci in the body where a drug molecule interacts and binds, but these interactions do not produce the desired effect, or perhaps they produce no discernible effect of any kind. These nonproductive interaction sites are termed *sites of loss*. Thus, there is a physiological difference between the terms receptor site and drug-binding site. Interaction of a drug molecule with a receptor site results in a specific pharmacological response. It must be emphasized that no drug interacts only with those specific populations of receptor sites involved in producing the desired pharmacological response. Every drug must be expected to interact with other receptor systems in the body, and thus produce *side effects*, unwanted pharmacological effects accompanying the desired response. There is probably no drug that has only one pharmacological effect. The art and science of drug design involves selecting some desired pharmacological response and making it the most prominent effect of that drug molecule within a specific range of dose levels.

4.1.2 Isolation of Receptors

Understanding pharmacological phenomena would be greatly facilitated if it were possible to isolate pure receptor site molecules for in vitro studies with drugs. However, many drug receptors are an integral part of the organized structure of the cell (for example, the cellular membrane) and therefore these receptor molecules are not easily isolable chemically by presently available techniques. It is a challenging endeavor to separate a receptor macromolecule, which is a part of the membrane structure, from the membrane matrix without disrupting or damaging the structure/shape of the receptor molecule. Although some success has been reported in this area (e.g., with nicotinic acetylcholine receptors and some types of analgesic receptors), membrane-bound receptor isolation is not a trivial or routine undertaking. However, advances in genetic engineering methodology have made possible the cloning of many types and subtypes of receptor protein molecules, and this technique has greatly facilitated pharmacological studies.

4.1.3 Chemical Nature of Receptors

Many drug receptors are proteins. The chemical heterogeneity of proteins and the presence of many kinds of functional groups in their amino acid components provide a multiplicity of different chemical sites for drug binding. However, this restricted definition of drug receptors must be expanded to include other types of biopolymers essential to the physiology of the body. Nucleic acids, especially DNA **4.1** (figure 4.1), contain a variety of functional groups, and these molecules also are well adapted to bind a great variety of exogenous molecules, including antibacterial agents, antimalarials, and carcinostatic drugs. Some agents used, inter alia, to control malignancies bind reversibly to double-stranded DNA by *intercalation*. This phenomenon, illustrated in figure 4.2 for adriamycin, is essentially a squeezing of the drug molecule between adjacent base pairs of the DNA helix system, resulting in a local distortion of the helical structure. A structural requirement for such drugs is that the molecule be planar, and these molecules most frequently contain aromatic rings. The planar moieties are held between the planar purine and pyrimidine rings of the DNA by van der Waals and charge transfer complexation interactions.

Polysaccharides are involved, among other functions, in stimulation of the immune response, and it seems likely that drug receptor sites are also to be found in these and other complex carbohydrate molecules of the body.

4.1.4 Drug–Receptor Interactions

As stated previously, the interaction between a drug and its in vivo receptor alters the physiological/biochemical processes in which the receptor molecule normally participates. The net observable change is the pharmacological effect of the drug. The interaction between a drug and its receptor produces subtle changes; for example, in the conformation of an enzyme molecule, which may be reflected in alteration of the catalytic activity of the enzyme. If the receptor protein is a part of a reversible ion channel through a membrane, a drug-induced conformational change in the protein may cause the channel to open or to close or it may prevent the opening or closing ("locking" the channel), changing the duration of a particular open/closed state, which will affect the ability of ions to migrate into or out of the cell.

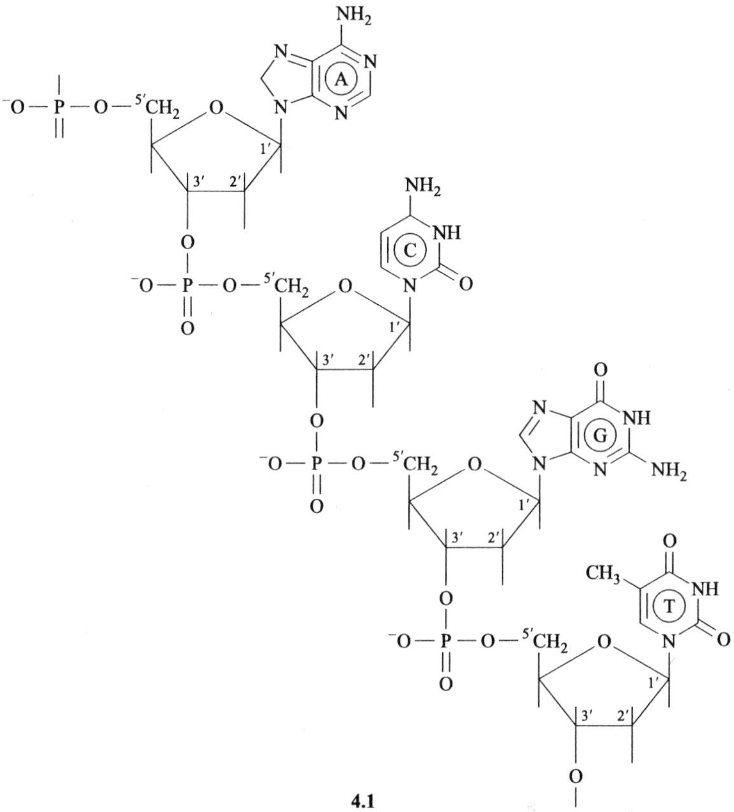

4.1

Figure 4.1 Section of a DNA chain. A = adenine; G = guanine; C = cytosine; U = uracil; and T = thymine

When a drug (the ligand) is dissolved in aqueous solution, it is solvated by the water molecules. It is likely that polar residues on the receptor are also solvated. When drug and receptor interact, the binding of the of the ligand drug molecule to the receptor displaces some or all of these solvating water molecules from both species. Thus, the ligand-binding process may be viewed as an exchange reaction whereby the solvated ligand (L) and receptor (R) give up all or a part of their solvation in exchange for interaction with each other.

$$R-H_2O + L-H_2O \rightleftharpoons R-L + 2H_2O \tag{4.1}$$

Chemical types of drug–receptor interactions

The interactions between a drug or other xenobiotic substance and its receptors are the same kinds that are encountered in the chemistry laboratory (in descending order of interactive strength):

1. Covalent bonding
2. Ionic bonding
3. Hydrogen bonding
4. Charge transfer complexation

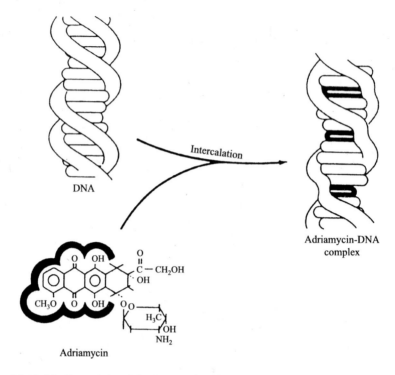

Figure 4.2 Model of intercalation of the planar portion of the adriamycin molecule into DNA, illustrating local unwinding of the helical structure. (Reproduced with permission from reference 4. Copyright 1964)

5. Ion–dipole interaction
6. Dipole–dipole interaction
7. van der Waals interaction
8. Hydrophobic bonding

Covalent bonding

This is the strongest of these interactions, but relatively few drugs bind covalently to their receptors. Absence of covalent attachment is a desirable property for many drugs. Formation of a covalent bond is a difficultly reversible (or essentially irreversible) process. Thus, a drug that binds covalently to its receptor is expected to exhibit a prolonged duration of action, which may be undesirable because control over the magnitude and duration of effect is lost. Pharmacologically and clinically, it is usually preferable that a drug–receptor interaction be a readily reversible process.

Hydrophobic bonding

This is one of the weakest forces of intermolecular association, but it is one of the most important drug–receptor interactions. In aqueous solution, a nonpolar region of an organic molecule (e.g., an alkyl chain) cannot be solvated by the water molecules, and as a consequence a remarkable phenomenon occurs. The water molecules in the region of the nonpolar alkyl group become highly associated with each other, and they become highly ordered into a quasi-crystalline lattice (figure 4.3). Thus, the lipophilic alkyl chain is surrounded by a layer of highly ordered crystal lattice water molecules. The nonpolar segment of the molecule induces a higher degree of order in the surrounding water molecules than is found in the body of the water phase, which is composed of highly dynamic flickering clusters. The entropy of this stable, highly ordered crystal lattice water is low, in accord with the third law of thermodynamics: "The entropy of a perfect crystal at $0°$ K is zero". When two lipophilic alkyl chains (one from a drug molecule and one from a receptor molecule), each surrounded by a lattice of crystalline water molecules, come close together, these alkyl chains are shielded to a greater extent from interaction with water molecules. The two alkyl chains squeeze out the water molecules between them and a new crystal lattice forms around the two alkyl chains. Some of the crystal lattice water structure collapses when the two chains come together, and overall a smaller total number of crystal lattice water molecules is needed to surround these two closely adjacent alkyl chains than when they were separate. Thus, the entropy of the system increases because there is less total ordering of water molecules. The entropy gain stabilizes the interaction between the two alkyl chains, and this is termed the hydrophobic bond.

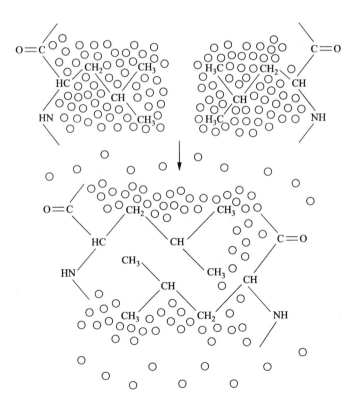

Figure 4.3 Hydrophobic interactions between two alkyl chains. (Reproduced with permission from reference 1. Copyright 1988 Oxford University Press)

Empirical data suggest that the energy contribution due to hydrophobic interaction is very small (e.g., 0.7 kcal mol^{-1} for one CH_2 group and 2 kcal mol^{-1} for a benzene ring). However, it is important to recognize that a drug is expected to interact with its receptor, not just at one functional group via one type of chemical interaction, but rather by a series of interactions of different kinds involving multiple portions of the drug molecule and of the receptor molecule. Thus, even though many (most) of the interactive forces cited above are individually quite weak, in the aggregate these effects are significant in binding the drug to its receptor. For many drugs, these weak interactive forces are more important than stronger types of interaction (such as covalent bonding or ionic bonding). Ionic bonding and the weaker forces enumerated previously provide a reversible drug–receptor interaction, governed by the mass action law.

4.1.5 Asymmetric Character of Receptors: Three-Point Attachment Hypothesis

Receptors tend to be highly stereospecific with respect to enantiomers of chiral drug molecules. This is not surprising; receptor proteins are composed of optically active amino acids; carbohydrates are optically active; and the helices of nucleic acids possess no plane nor point of symmetry. Nature is asymmetric.

Figure 4.4 illustrates the three-point attachment hypothesis (Easson–Stedman hypothesis) which proposes that, for a drug molecule to produce a pharmacological response, three positions (groups) on the drug molecule must interact simultaneously with three opposing complementary positions on the receptor molecule. Because three points are required to define an absolute configuration of an asymmetric carbon, the biological inequality of enantiomers can be rationalized. Only one of the enantiomers possesses the correct absolute configuration to permit its simultaneous interaction with all three of the opposing complementary receptor subunits. Recent experimental data suggest that in some instances, the three-point attachment hypothesis is inadequate. According to this evidence, if the agonist binding sites on a protein molecule are in a cleft or are on protruding residues, four subsites may be required for the macromolecule to discriminate between enantiomers. The three-point model is applicable only if the chiral molecule can approach the protein binding domain from just one direction,

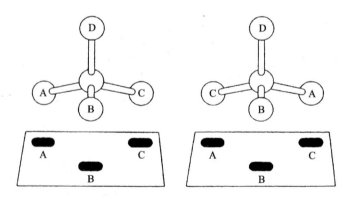

Figure 4.4 Three-point attachment hypothesis for enantiomers

that is, if the receptor area is planar. The work invoked in defining a four-point model for stereospecific binding to a protein also provides a rationalization for the observation that chiral analogs of some optically inactive neurotransmitters (e.g., acetylcholine, dopamine) display a high degree of stereospecificity in their interactions with their protein receptors. The prochiral center in acetylcholine and dopamine can indeed be shown to interact with its receptor subsites in a highly stereospecific manner (see Recommended Reading, ref. 8).

It does not necessarily follow that one enantiomer of a chiral drug molecule possesses pharmacological activity and the other enantiomer is pharmacologically inert. The literature is replete with examples of chiral molecules, both of whose enantiomers display marked pharmacological effects. However, the effects of the enantiomers differ, frequently involving different receptor systems or, less frequently, involving interactions at different domains on the same receptor molecule.

Structurally Nonspecific and Structurally Specific Drugs

The corpus of pharmacological actions of drugs can be subdivided grossly into two categories, as follows.

4.2.1 Drug Actions That Do Not Depend on Chemical Structure

In this group, the actions of a drug reflect a summation of the physical chemical properties of the molecule and are not directly related to the details of its chemical structure. This describes a *structurally nonspecific drug*, and modest alterations of the chemical structure of such drugs frequently have little or no effect on their pharmacological actions. There seems to be no specific, discrete "receptor site" for drugs in this category. It is characteristic of these drugs that they are administered in large doses (grams). Examples of such drugs are chloral hydrate, ethanol, and some inhalation anesthetics. One theory of the mode of action of general inhalation anesthetics suggests that the structurally nonspecific volatile anesthetics (which are lipophilic) dissolve in the lipophilic portions of nerve cell membranes in the central nervous system (and perhaps elsewhere also). The resulting increased fluidity of the membrane matrix is reflected in perturbation of the membrane's protein ion channels which in turn causes abnormalities in ion flow across the membrane, leading ultimately to a diminution of nerve impulse transmission and anesthesia. It has also been speculated that one component of the depressant action of ethanol on the central nervous system may involve a similar solution of ethanol in the lipid phase of central nervous system membranes. Experimental evidence suggests that the interactions of ethanol and related compounds are not uniform throughout the lipid bilayer, which reflects a non-uniform distribution of various phospholipids and cholesterol within the membrane. This effect of altering the fluidity of membranes may also be related to ethanol's reported ability to disrupt the physiological activities of membrane ion channels.

Insofar as its pharmacology is concerned, there seems to be nothing unique about the structure of ethanol. Most low molecular weight aliphatic alcohols (e.g., methanol, propanol) have similar abilities to act as "pure" CNS depressants, very similar to ethanol. The obvious advantage of ethanol as a beverage is its taste.

4.2.2 Drug Actions that Depend on Chemical Structure

In this group the actions of a drug depend primarily on its chemical structure and even modest changes in the molecular structure (including stereochemical changes) may result in dramatic changes in pharmacological actions (increase, decrease, completely change qualitatively). A specific, discrete in vivo receptor site interaction with the drug is suggested, rather than involvement of a broad expanse of the cell surface, as was hypothesized for structurally nonspecific drugs. Because a large area of the cell surface is not involved, fewer drug molecules are required to elicit an effect, and a smaller dose of such a drug is needed. It has been estimated that the area covered by a dose of acetylcholine sufficiently large to reduce the heart rate of the toad by 50% would be only 0.016% of the total surface area of the ventricular cells. This reflects a profound structural specificity for the acetylcholine molecule and strongly supports the validity of the existence of receptor sites.

4.3   Agonists and Antagonists: Occupancy Theory of Drug Action

4.3.1 Definition of Occupancy

There is no single, unified theory that explains or is applicable to all in vivo drug actions. However, certain theories provide useful rationalizations for aspects of the actions of some drugs. *Occupancy theory* proposes that the intensity of a pharmacological effect is directly proportional to the total numbers of receptors occupied by the drug. The drug effect becomes more intense as the number of occupied receptors increases. Hence, the maximal pharmacological response corresponds to occupation of all receptors. A graph of the magnitude of drug effect (Y axis) versus the size of dose (X axis) (figure 4.5) demonstrates that a level of dosage can be reached beyond which no further increase in drug effect is noted. According to occupancy theory, at maximal response all receptors are occupied, so that additional amounts of drug have no place to attach to produce a further effect.

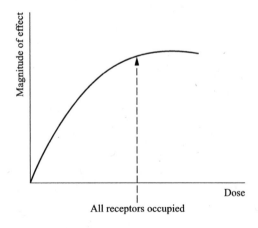

Figure 4.5 Plot of dose versus magnitude of effect

4.3.2 Affinity and Efficacy

Extension of occupancy theory led to a further proposal that a drug–receptor interaction comprises two stages, illustrated by the following equation:

$$R + D \underset{k_2}{\overset{k_1}{\rightleftharpoons}} RD \longrightarrow E \tag{4.2}$$

where R = receptor
D = drug
RD = drug – receptor complex
E = pharmacologic effect
k_1, k_2 = rate constants for adsorption and desorption

The two stages may be described as (1) complexation of the drug with its specific receptor, described as a reversible process frequently following the law of mass action; and (2) production of the drug effect, usually considered to be an irreversible step. For a molecule to manifest biological action, it must have *affinity* for the receptor (owing to complementary structural characteristics) and it must also have *intrinsic activity* or *efficacy*, which is a measure of the ability of the drug–receptor complex to produce the biological effect. Accordingly, *agonists* and *antagonists* have strong affinity for the receptor, and both form a drug–receptor complex. However, only the agonist has the ability to trigger a pharmacological response: only the agonist possesses intrinsic activity. The term *agonist* (derived from the Greek: a performer or "doer") is defined as a molecule that binds to a receptor and produces a pharmacological effect.

A plot of the intensity of pharmacological response or effect (*Y* axis) versus the logarithm of the dose (*X* axis), figure 4.6a,b, can be used to compare the *potencies* and *efficacies* of drugs. In figure 4.6a, drugs A, B, and C differ in the size dose required to produce their maximum effect. Drug A requires a smaller dose than B which in turn requires a somewhat smaller dose than drug C. However, A, B, and C produce the same magnitude of maximum response. The curves become level at the same high effect. Therefore, all three drugs possess similar intrinsic activity, but they differ in *potency*, the amount of drug required to achieve the maximal possible response. A drug is said to be potent when it demonstrates high biological response per unit of weight. In contrast, in figure 4.6b, drugs A, B, and C differ in the level of their maximal response. A produces a greater response than B, and C produces the lowest maximal response.

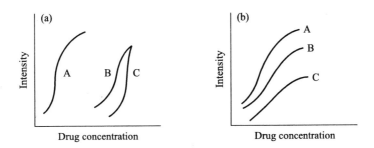

Figure 4.6 Dose–response curves with (a) equal and (b) different intrinsic activities

A has the greatest intrinsic activity. All three drugs demonstrate approximately the same same potency; they produce their maximal response at approximately the same dose. Pharmacological assays almost always compare the potencies of two drugs on the basis of doses that produce the same level of response, rather than by comparing magnitudes of response produced by the same dose of two drugs.

By definition, drugs that have a marked affinity for the receptor and little or no intrinsic activity are *antagonists*. Antagonists can be used to counteract the effects of drugs that have a greater intrinsic activity by denying them access to the receptor. The terms potency and efficacy (or activity) are not synonymous. Potency is related to the size of dose necessary to produce a desired pharmacological effect. Activity (or efficacy) refers to the magnitude of this effect, regardless of the size dose required to achieve it.

4.3.3 Limitations of Occupancy Theory

Occupancy theory fails to explain several important aspects of drug action, such as why agonists produce a pharmacological response and antagonists do not, even though they presumably occupy the same receptor. Efficacy remains a mysterious and elusive property of a drug molecule. Also, it is widely accepted that many antagonists do not interact at the same macromolecular receptor domain to which the agonist binds. Many antagonists have been shown to interact at allosteric sites.

Classical occupancy theory does not elucidate a mechanism of drug action at the molecular level in terms of chemical structure. Also, it has been established for many drugs that maximal pharmacological response does not require occupancy of all available receptors. For these drugs there are "spare" receptors, and occupancy theory is not applicable. Indeed, contemporary thought presents the conclusion that some full agonists can produce their maximum effect by occupying as few as 2% of the total number of receptors available.

4.4 Competitive and Noncompetitive Antagonists

In competitive antagonism, simultaneous bindings of the agonist and the antagonist are mutually exclusive. There are three possible explanations for this: (1) the agonist and the antagonist compete for the same binding site; (2) the agonist and the antagonist bind to adjacent sites that overlap; or (3) different sites are involved, but they influence the receptor macromolecule so that the agonist and the antagonist cannot be bound at the same time (an allosteric effect). In noncompetitive antagonism, the agonist and the antagonist can be bound simultaneously, but the antagonist binding reduces or prevents the action of the agonist (also an allosteric effect). Antagonists are frequently categorized as surmountable or insurmountable.

4.5 Induced Fit

It appears likely for protein receptor sites (and possibly for other chemical categories of drug receptors) that the "normal" (resting, ground state) geometry of the receptor domain need not be complementary to that of the agonist drug molecule for the drug to interact appropriately and productively with the receptor. The protein molecule and its receptor site domain should be visualized as being flexible rather than rigid, and the protein molecule can, to a considerable extent, alter its conformation (its molecular shape) to achieve an optimal complementary fit for binding with the agonist molecule. Drug–receptor complexes can best be considered as interactions between two floppy components, a structurally flexible drug molecule and a structurally flexible receptor molecule; this is *induced fit*. A drug molecule may interact with the receptor in a conformation other than the drug's lowest energy conformation. The energy expended on the drug molecule by its assuming a higher energy conformation is compensated for by the energy advantage of the formation of the drug–receptor complex. As will be illustrated numerous times subsequently, the ability of a drug molecule to induce a conformational change, a change in the shape of its macromolecular receptor, may be the basis for the mechanism of action of the drug.

4.6   Partial Agonists and Inverse Agonists

A *full agonist* is a drug that elicits the largest pharmacological response that the receptor system/tissue can give. A *partial agonist* at any dose level cannot produce the same maximal biological response as a full agonist, even though the partial agonist is presumed to bind as tightly and as well to the receptor as the full agonist. If the maximal response to a full agonist is expressed as unity, the maximal response to a partial agonist is expressed as being greater than zero, but less than unity. In sum, a partial agonist may have high affinity for its receptor, but it possesses low intrinsic activity. Drugs B and C in figure 4.6b may be viewed as partial agonists. A rationalization of partial agonism is illustrated in figure 4.7. Receptors are proposed to be capable of existing in two states, a resting state and an activated state, which are in equilibrium with each other. In the absence of ligand molecules, the equilibrium favors the resting state. A full agonist binds preferentially to the activated state and shifts the equilibrium toward activation. A partial agonist shows a weaker preference and it shifts the equilibrium to a smaller extent, even when the receptors are fully occupied. An antagonist shows no preference and it does not shift the equilibrium, although it reduces the effect of an agonist by preventing the agonist from binding to the receptors.

An *inverse* agonist is a molecule that is considered to be a full agonist, but it produces a pharmacological effect at the receptor opposite to that of a "normal" agonist. For example, a "normal" agonist may lower the blood pressure, and the inverse agonist will raise the blood pressure. One rationalization of the concept of inverse agonism proposes that the inverse agonist molecule binds preferentially to the resting state of the receptor, thus producing the so-called *negative efficacy*. However, this phenomenon is significant only in some comparatively rare situations.

4.7   Enzymes as Drug Receptors

4.7.1 Types of Enzyme Inhibition

Some drugs utilize the catalytic surface(s) of enzymes as their receptor site(s), and frequently the net effect of the drug–enzyme interaction is inactivation of the enzyme.

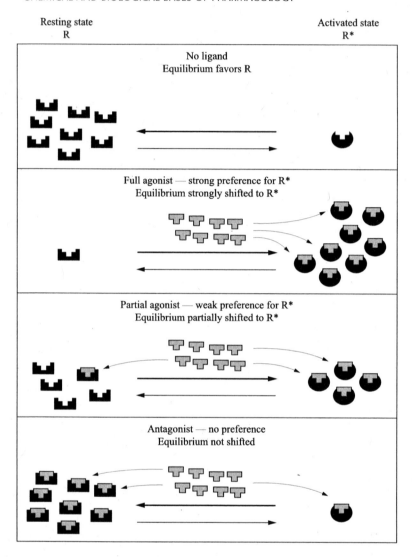

Figure 4.7 Representation of effects of ligands on receptor activation. Agonists, partial agonists, and antagonists. (Reproduced with permission from reference 3. Copyright 2001 Churchill Livingstone)

Approximately half of the 20 most widely sold drugs worldwide are enzyme inhibitors. Inhibition is commonly designated as reversible or irreversible, but these terms are purely descriptive of the process, and irreversible inhibition does not imply destruction of the enzymatic protein. A commonly applied criterion for reversibility is regeneration of complete catalytic activity after dialysis of the inhibited enzyme. Reversible inhibition implies an equilibrium between the enzyme and the inhibitor, with a measurable inhibitor constant, K_i. This is a quantitative expression of the affinity of the enzyme for the inhibitor. The degree of inhibition depends on the concentration of the inhibitor, and after equilibrium is attained, the effect is independent of time. In contrast, irreversible inhibition increases with time, and the effectiveness of the inhibitor drug reflects the rate constant for the reaction.

Enzyme inhibition is broadly classified as competitive or noncompetitive. Some drugs do not fall exclusively into either category, and these are designated as *mixed inhibitors*. A competitive inhibitor resembles the natural substrate of the enzyme sufficiently that it binds at the catalytic surface to form an enzyme–substrate complex. This complex may be completely incapable of dissociating into products ("dead-end" inhibition), or the complex may provide an abnormal reaction product. In noncompetitive inhibition, enzyme–substrate binding is unaffected because the inhibitor is bound at an alternative (allosteric) site. Competition may arise because the enzyme–inhibitor–substrate complex cannot dissociate, or the complex may break down to give the normal product(s), but at a highly reduced rate.

4.7.2 Enzyme Kinetics in Pharmacology

Drug–enzyme interactions can be studied by invoking principles of enzyme kinetics. A fundamental premise in enzyme chemistry is that the catalytic process is a stepwise phenomenon:

$$E + S \underset{k_2}{\overset{k_1}{\rightleftharpoons}} ES \xrightarrow{k_3} E + P \tag{4.3}$$

where k_1, k_2, k_3 = rate constants
E = enzyme
S = substrate
P = product

The formation of the enzyme–substrate complex, a necessary part of the reaction, is reversible. The collapse of this complex to release the product is essentially irreversible.

The rate of an enzyme-catalyzed reaction is a function of the concentration of the substrate. This relationship can be expressed as follows:

$$(V_{max} - V) [S] = K \tag{4.4}$$

Where V_{max} = maximal velocity of the reaction
V = thereaction velocity
$[S]$ = substrate concentration
K = a constant

It follows that the reaction is saturable with respect to the concentration of the substrate.

The Michaelis–Menten equation (4.5) derives from equations (4.3) and (4.4):

$$V = \frac{V_{max} \times [S]}{[S] + K_m} \tag{4.5}$$

where K_m is the Michaelis–Menten constant. By definition, K_m is the concentration of the substrate that gives rise to a half-maximal reaction rate. V_{max} and K_m are the two most important constants used to characterize an enzyme reaction in kinetic terms and hence to assess the effect of a drug on the enzyme's catalytic ability. A Michaelis–Menten plot of V (Y axis) versus S (X axis) yields a rectangular hyperbola. However, by taking the reciprocals of both sides of the Michaelis-Menten equation, the plot of $1/V$ (Y axis) versus $1/S$ (X axis) (a so-called Lineweaver–Burk plot) produces a straight line.

As suggested previously, reversible enzyme inhibitors can be categorized as follows:

1. *Competitive inhibitors* are those that decrease substrate affinity (i.e., increase K_m). They bind reversibly to the active site of the enzyme. The effect of the inhibitor can be reversed by increasing the concentration of the normal substrate.
2. *Noncompetitive inhibitors* are those that decrease the V_{max}. This type of inhibitor cannot be reversed by increasing the concentration of substrate because the inhibition depends upon binding of the inhibitor to a site on the enzyme other than the active site (allosteric site).
3. Some *reversible inhibitors* affect both the K_m and the V_{max}. These are the mixed or *uncompetitive* inhibitors.

4.7.3 Transition State Analogs

A special case of reversible inhibition is represented by *transition state analogs*. A chemical reaction passes through a higher state of energy than that of the mixture of reactants before forming the reaction products. The peak of the energy barrier is termed the transition state. In terms of energy, an enzyme E decreases the activation energy by strongly binding to the substrate S; the transition state of the substrate is represented by S|:ES|. Determining the structure of the substrate transition state is important for understanding the mechanism of the catalytic process, and for designing enzyme inhibitors. The concentration of ES| in any reaction is extremely small, so that direct structural analysis of ES| is usually impossible, but an indirect approach is provided by the transition analog theory. This predicts that the enzyme binds to the substrate transition state much more tightly than to the substrate, and a transition state analog inhibitor also binds more tightly to the active site of the enzyme than a simple substrate analog inhibitor. Transition state analog-type inhibitors are not transformed into products. When a hypothetical transition state structure can be proposed, it may be possible to design an inhibitor which is a transition state analog. The transition state analog enzyme inhibitor strategy has been applied to adenosine deaminase which catalyzes the conversion of adenosine **4.2** into inosine **4.4**. It is believed that the catalysis proceeds via a tetrahedral intermediate **4.3**. The product of the reaction, inosine, is a good inhibitor of the enzyme, but the antibiotic coformycin **4.5**, which more closely resembles the transition state intermediate **4.3**, **is** many orders of magnitude superior as an inhibitor.

4.7.4 Active Site-Directed Irreversible Inhibitors

These drugs (also called affinity labels) are chemically active analogs of the target enzyme's substrate. The drug forms a covalent bond with some element of the enzyme's catalytic surface and destroys its catalytic ability. Generally, the affinity label has some electrophilic substituent that can form a stable covalent bond with a nucleophilic group on the enzyme. The strategy of active site-directed inhibition presents two

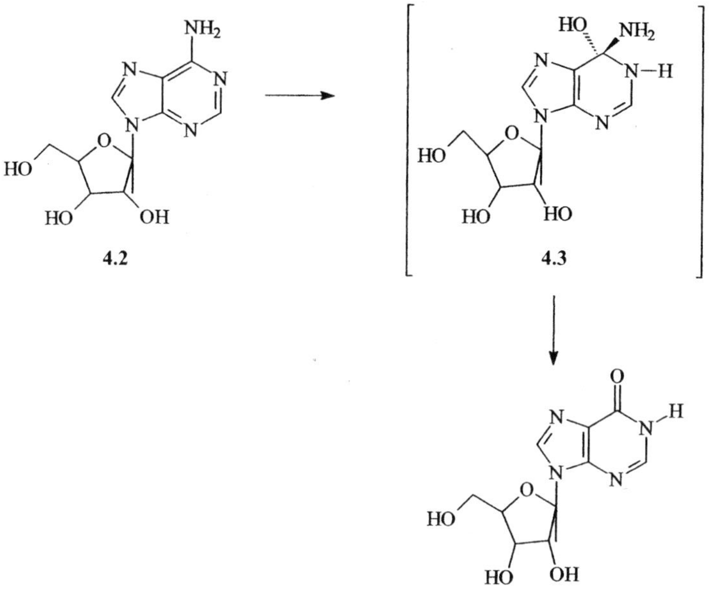

4.2

4.3

4.4

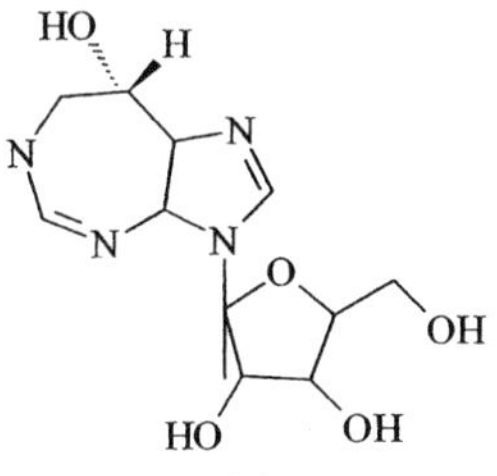

4.5

possible disadvantages. First, such drugs are intrinsically reactive molecules, and a large portion of an intended dose may be simply hydrolyzed by the aqueous solution in which it is administered. Second, the molecules may act nonspecifically and nonproductively with other moieties on protein molecules. Specific examples of covalent inhibition of enzymes are found in chapter 8 (phosphorus-derived acetylcholinesterase inhibitors).

4.7.5 Suicide Substrate Inhibitors

These drugs (also termed k_{cat} substrates) are designed to provide greater specificity of enzyme inhibition. Drugs of this type are themselves inactive when they are administered, but they are activated when they serve as a substrate for the enzyme that they were designed to inhibit. This enzymatic reaction produces a reactive product molecule which further reacts to form a covalent bond with the enzyme and inhibits it. Thus, the

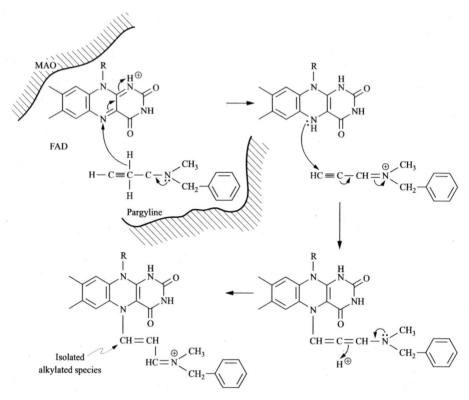

Figure 4.8 Suicide enzyme inhibition: momoamine oxidase and pargyline. (Reproduced with permission from Recommended Reading reference 5, p. 547. Copyright 1996 Springer-Verlag)

enzyme itself produces its own inhibitor from the originally harmless substrate molecule and it is perceived to have "committed suicide". To design a suicide substrate molecule, the catalytic mechanism of the enzyme and the nature of the functional groups at the enzyme's catalytic site must be known and well understood. Figure 4.8 illustrates the reaction of flavin-linked monoamine oxidase with a potent, irreversible inhibitor of the enzyme, the drug *pargyline*. The inhibitor ultimately forms a covalent bond with the enzyme system via the flavin cofactor.

Bibliography

1. Nogrady, T. *Medicinal Chemistry*, 2nd ed. Oxford University Press: New York, 1988; p. 42.
2. Goth, A. *Medical Pharmacology*, 10th ed. Mosby: St. Louis, Mo., 1981, p. 8.
3. Rang, H. P., Dale, M. M., Ritter, J. M. and Gardner, P. *Pharmacology*, 4th ed. Churchill Livingstone: Philadelphia, Pa., 2001; p. 11.
4. Lehrman, L. S. *J. Cell. Comp. Physiol.* **1964**, 64(Suppl. 1), 27.
5. Jambhekar, S. S. Biopharmaceutical Properties of Drug Substances. In *Principles of Medicinal Chemistry*, 4th ed. Foye, W. O., Lemke, T. L. and Williams, D. A., Eds. Williams & Wilkins: Baltimore, Md., 1995; p. 27.

Recommended Reading

1. Williams, M., Deecher, D. C. and Sullivan, J. P. Drug Receptors. In *Burger's Medicinal Chemistry and Drug Discovery*, 5th ed. Wolff, M. E., Ed. Wiley–Interscience: New York, 1995; Vol. 1, pp. 349–397.
2. Nogrady, T. Receptor–Effector Theories. In *Medicinal Chemistry: A Biochemical Approach*, 2nd ed. Oxford University Press: Oxford, UK, 1988; pp. 58–91.
3. Milligan, G., Bond, R. A. and Lee, M. Inverse Agonism: Pharmacological Curiosity or Potential Therapeutic Strategy. *Trends Pharmacol. Sci.* **1995**, *16*, 1–13.
4. Larsen, I. K. Intercalating Agents. In *A Textbook of Drug Design and Development*. Krogsgaard-Larsen, P. and Bundgaard, H., Eds. Harwood: Chur, Switzerland, 1991; pp. 214–223.
5. Dugas, H. Suicide Enzyme Inactivators and Affinity Labels. In *Bioorganic Chemistry*, 3rd ed. Springer Verlag: New York, 1996; pp. 542–560.
6. Muscate, A. and Kenyon, G. L. Approaches to the Rational Design of Enzyme Inhibitors. In *Burger's Medicinal Chemistry and Drug Discovery*, 5th ed. Wolff, M. E., Ed. Wiley: New York, 1995; Vol. 1, pp. 733–782.
7. Anfelt-Rønne, I. Enzymes. Recommended Reading ref. 4, pp. 274–334.
8. Mesecar, A. D. and Koshland, D. E., Jr. A New Model for Protein Stereospecificity. *Nature* **2000**, *403*, 614.
9. Transition State Analogs. Recommended Reading ref. 5, pp. 79–85.
10. Foreman, J. C. and Johansen, T., Eds. *Textbook of Receptor Pharmacology*. CRC Press: Boca Raton, Fla., 2002.

5

Principles of Pharmacological Assays

Affinity (Binding) Assays

5.1.1 Principles

Affinity or binding assays are frequently among the first that are performed on a newly synthesized or isolated chemical compound. Preliminary information concerning the ability of the compound to bind to a specific receptor (i.e., the affinity of the compound for the receptor) is a valuable prelude to more extensive pharmacological assays described later. The challenge in receptor binding studies is illustrated by the estimate that any one receptor type in the central nervous system contributes as little as one-millionth of the total weight of the brain. The principal requirements for a successful binding assay are that the ligand to be evaluated pharmacologically (either an agonist or an antagonist) for a specific receptor must bind with high affinity and specificity. It must not be metabolized under the conditions of the binding assay, and the ligand molecule must be capable of being radiolabeled (commonly with ^3H, ^{14}C, ^{35}S, or ^{125}I) with high specific activity so that extremely small amounts of receptor-bound radiolabeled ligand are detectable. Samples of biological tissue containing the receptors are incubated in an aqueous environment with varying concentrations of the radiolabeled ligand. After equilibrium is established, the biological tissue is separated by centrifugation or filtration, and the amount of radioactivity in the tissue is measured using appropriate scintillation counting instrumentation.

A major problem in binding assays is that the crude biological materials used (tissue, membranes) contain chemical domains other than the specific receptors, which also bind ligand molecules: *sites of loss*. Quantitative estimation of nonspecific binding in the assay involves performing a parallel binding experiment using an excess concentration (>100×) of nonradioactive ligand that specifically competes with the radiolabeled ligand for the receptor site. In the presence of excess nonradiolabeled ligand, the nonspecific binding can be determined, and in this situation, the only radioligand that remains attached to the tissue represents that which is bound to components other than the receptor (nonspecific binding). The amount of nonspecific binding subtracted from the total binding gives an estimate of the amount of specific binding to the receptor.

74

Quantitative receptor binding data are expressed as the IC_{50} value, the concentration of the unlabeled ligand being studied which is required to inhibit 50% of the specific binding of the radioligand.

To establish that the binding of a radioligand to a tissue reflects a true receptor–ligand interaction rather than nonspecific one(s) of the radioligand with the tissue, it has been proposed (Recommended Reading, ref. 12) that the following criteria must be met:

1. Binding of the radioligand to the target tissue should be saturable, which indicates a finite number of specific binding sites.
2. Binding affinity must be high (dissociation constant of ligand–receptor complex, $K_d \sim 10^{-10}$ to 10^{-8}). The generally useful concentration range in high affinity binding assays is picomolar through nanomolar.
3. Radioligand binding should be readily reversible, consistent with physiological conditions.
4. The distribution of binding sites within the tissue or cells should be consistent with the physiological role of the ligand recognition site.
5. The pharmacology of the binding site should have an agonist/antagonist rank order potency similar to that observed for the natural ligand in functional test procedures.
6. A simultaneous correlation of binding data with biological dose/concentration curves in identical tissue preparations should be generated.

Binding assays provide useful information about specific receptors or receptor subtypes; they are simple and rapid to perform; they require relatively simple equipment and apparatus; and they are eminently adaptable to automated robotic procedures. However, binding studies give information only about affinity of a ligand for the receptor. They provide no information about the intrinsic activity (efficacy) of the ligand nor whether it is an agonist, a partial agonist, an inverse agonist, or an antagonist. Also, binding assay conditions are usually nonphysiological.

5.1.2 Radioimmunoassays

Radioimmuno assays represent an extremely sensitive isotopic method. They can be utilized for almost any drug that can form a *hapten* (a synthetic antigen). The drug is first coupled to a protein such as a serum albumin, thus forming an antigen. Antibodies to this antigen are produced in an appropriate laboratory animal. A solution of this antibody material is then equilibrated with an isotopically labeled analog of the drug under investigation, after which the free radioactive ligand is removed. Upon subsequent addition of "cold" (nonradiolabeled) drug ligand, re-equilbration occurs and some of the bound radiolabeled ligand is displaced and freed. After removal of both "hot" and "cold" antigen–antibody complexes, the increase in free radioactive ligand is measured. This gives a very sensitive measure of the concentration of "cold" ligand, which is detectable in amounts that would be impossible to quantify using more traditional assay procedures.

5.1.3 High Throughput Automated Assays

Robotic microscale synthetic chemistry techniques ("combinatorial chemistry"), originally developed for peptide synthesis, have been adapted to small, nonpeptide organic molecules. As a result it is possible for a single laboratory worker to synthesize literally hundreds (perhaps thousands) of chemical entities, and it is not possible to predict what pharmacological actions might be demonstrated by individual members of this population of compounds. Efficient accomplishment of initial pharmacological screening of this great number of compounds has necessitated revolutionary new techniques of screening: rapid and cost effective methods for assessing biological activity, with the goal of identifying leads (trivially called "hits"), using radioligand binding assays involving as large a number of different receptors as is possible and practicable. Some automated, coordinated combinatorial chemistry-targeted screening operations in the pharmaceutical industry are capable of screening 100 000 compounds through 10 to 30 assays, feats unattainable by conventional synthetic chemistry protocols and pharmacological screening methods.

There may be serious problems attendant to this strategy, in addition to the binding study limitations cited above: establishing that the receptor–ligand interaction is selective and the extended time required for data generation in binding studies. Selectivity of the receptor–ligand interaction can be assessed if the compound is assayed in a sizeable number (10 to 15) of different assays. If the compound is active in the same concentration range in three or more of the assays, it may be concluded that the effects are nonselective for the receptors involved. Scintillation counting as the final step in binding assays requires as long as 2 min per sample. It is obvious that screening several thousand compounds in duplicate in 10 to 15 binding assays, even at only one dose level, represents a massive, unacceptable amount of time. Recommended Reading ref. 14, addresses possible solutions to these problems.

5.1.4 Cell-Based Assays

These assays, utilizing isolated intact cells, have the potential to identify the same compounds as in a receptor binding assay, but they may also uncover the effects of compounds that act at other known and unknown sites in the cell. A perceived added advantage to cell-based assays is that the culture medium, supplemented with some blood serum, presents a more physiologically relevant environment which allows the test compound to interact with the medium components and also with the cell membrane. A disadvantage to cell-based assays is that, once a promising compound has been identified, considerable additional study is often necessary to determine the molecular site of action in the cell. Also, for reproducibility of results, the experimental parameters of a cell-based assay must be rigorously controlled. Frequently, cell-based assays are significantly more expensive than binding assays.

5.2   Quantification of Biological Responses

5.2.1 Dose–Response Curves

Binding assays on a newly available chemical entity are frequently followed by a variety of *biological screens*, which are rapid, routine biological tests aimed first at exploiting leads obtained from the binding assay data and involving a qualitative assessment of the biological effects of the compound, often in an intact animal. The nature of the animal's response may suggest use(s) in therapy. However, the qualitative fact of an attractive drug response is not enough: *quantification* of the biological response is essential. A relationship must be established between the magnitude of the biological response and the size of the dose given, and this information can be plotted as a graph known as a *dose–response curve*.

Regardless of how potent a drug may be, there is a dose below which it will not induce a response: the *threshold dose*. Similarly, there will be a dose above which more intense effects cannot be induced, the *ceiling dose*. Between these two extremes, the threshold and the ceiling, progressively larger doses generally produce progressively greater biological responses.

5.2.2 Biological Variation

In a pharmacological assay based upon a biological response as the indicator, it is recognized that these responses are inherently variable. The thrust of all of the ensuing discussion of pharmacological assays might be summarized in one statement: "No two animals are exactly alike." *Biological variation* or *variance* is defined as the appearance of differences in magnitude of response among individuals in the same population who are given the same drug dose. Factors such as rate of absorption of the drug, rate and/or ease of transport in the body, rate of metabolism, and rate of excretion may differ significantly from individual to individual, and the sum total of these factors contributes to biological variation. Biological variation is as significant in therapy of human patients as it is in pharmacological assays in animals. Pharmacological assays can be devised to take biological variation into account, and assays can be designed to give almost any degree of accuracy, by repeating the biological tests and by using statistical methods to interpret the results. A well-planned assay should furnish evidence of the pharmacological activity of the test preparation and also of the limits of error of the biological test. It should answer the question: How dependable/believable are the data?

5.2.3 Types and Uses of Pharmacological Assays

Bioassays, which usually depend upon a comparison of the sample to be tested with a standard preparation of a known drug, are of two kinds: analytical dilution assays and comparative assays.

Analytical dilution assays

Here there is only a quantitative difference between the two drug preparations; the test preparation contains an unknown quantity of the standard. Analytical dilution assays are used only when no adequate chemical methods are available, as for example, insulin preparations for which chemical assays (combustion analysis, acid–base titration, chromatographic analysis, spectral analysis) provide no information about the biological potency of the product. Provided that the standard and the unknown have the same chemical composition (i.e., that they are both insulin) the analytical dilution assay result does not depend upon the exact assay method used nor upon the animal species used.

Comparative assays

In these assays the test drug and the standard are chemically not the same. These assays are used to assess the activity and potency of new drug candidates. For example, a new analgesic candidate may be compared with morphine or with aspirin, or a new topical anesthetic candidate may be compared with cocaine. In a comparative assay, the result is highly dependent upon the assay method and the animal species used. The relative activities of the standard drug and the unknown, determined in animal experiments, give only a preliminary and tentative indication of their relative activity and potency in humans.

Three bioassay designs are commonly used in comparative assays: (1) direct assays, (2) indirect quantitative assays, and (3) indirect quantal assays. The basis for direct assays is adjustment of the drug dose until a desired effect is produced. For example, in the biological standardization of digitalis preparations, a solution of the drug is slowly infused intravenously into a test animal (guinea pig) until the heart stops. Direct assays have certain disadvantages, the most important being that they are not readily amenable to statistical analysis. Indirect assays, which result in construction of dose–response curves, can be analyzed statistically. Indirect assays include graded ("quantitative") responses, such as the degree of contraction of a muscle strip, the amount of urinary excretion of sodium chloride, or a change in blood pressure, and an all-or-none ("quantal") response, such as sleep, death, or success of the test animal in finding its way through a maze within a stipulated period of time.

The basis for quantitative assays

The basis for quantitative assays is that, as the dose of drug administered to a single animal or to a discrete organ or tissue is increased, the magnitude of the pharmacological response increases gradually and smoothly, provided that the threshold dose has been exceeded. Figure 5.1 shows a typical dose–response curve for a quantitative assay based upon contraction of an isolated muscle strip. The degree of muscle contraction is plotted on the Y axis, against the dose on the X axis. Eventually a dosage level is reached where additional drug does not cause any further muscle contraction.

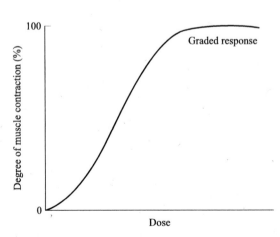

Figure 5.1 Graded response curve for the effect of a drug on muscle contraction

The middle of the dose–response curve is linear for a range of doses; in this portion of the curve the rate of change of response is directly related to the rate of change of dose. Because a linear relationship of dose to magnitude of pharmacological response offers convenience in both analytical and practical pharmacology, the pharmacologist often desires to extend the boundaries of linearity on the curve. This can be accomplished by mathematical manipulation of the dose parameter or of the response parameter, or both.

Experience has shown that in many graded-response assays linear relationships are obtained over a greater range of doses when the response (Y axis) is plotted against the logarithm of the dose (X axis). Such assays frequently give parallel curves for the standard drug and for the unknown drug, and they are referred to as *parallel line assays*. In the 2 + 2 protocol, two doses of the standard drug and two doses of the unknown are used. The doses are chosen to give responses lying on the linear part of the dose–response curves. The sequence of four doses (two of the standard and two of the unknown) is given as a series of randomized blocks. Administration of each dose several times permits measurement of the variability of the test system, and this information can be analyzed statistically to estimate the confidence limits. The 2 + 2 assay also reveals whether or not the two log dose–effect lines deviate significantly from parallelism. If the lines are not parallel, which may occur if the standard drug and the test drug have different mechanisms of action, then it is not possible to define their relative potencies in terms of a simple ratio. Low-ceiling diuretics produce only a modest diuretic effect, whereas high ceiling diuretics produce intense diuresis (see Chapter 16). A meaningful comparison of two such drug types requires measurement of the dose of each required to produce an equal low-level diuretic response, and also measurement of the relative heights of the ceilings attainable with each. In some rare instances, linear relationships are obtained when response is plotted against the arithmetic dose of the drug. These are termed *slope ratio assays*.

5.3 Tachyphylaxis: Drug Tolerance

It is sometimes observed that if the same biological object (the same animal or group of animals) is used in repeated assays for a given drug, the position of the dose–response curve from later assay runs may shift to the right along the X axis (dose parameter). The sensitivity of the test object to the drug is in a dynamic state. The term *tachyphylaxis* is applied to tolerance to a drug that develops rapidly after only a few doses of the drug, that is, as stated before, the dose–response curve shifts to the right. Now, a larger dose of the drug is required to produce the same magnitude of effect that was elicited by the first dose. Ephedrine exemplifies a drug that displays the tachyphylaxis phenomenon. An aqueous solution of the sulfate salt of this drug is used occasionally as a nasal decongestant by direct application into the nasal passages. For the initial dose of ephedrine sulfate solution, one drop in each nostril is usually sufficient, and the effect lasts for 2–3 h. However, after only a few additional doses, one drop is not sufficient to relieve the congestion, and three or four drops are required. In addition to the possible significance of tachyphylaxis in drug therapy in patients, it is recognized that if a test drug produces tachyphylaxis in laboratory animals, these same animals cannot be used repeatedly for assays of the drug. The data obtained from tachyphylaxis-induced animals may be misleading and/or invalid.

The mechanism(s) for initiating the tachyphylaxis phenomenon is (are) incompletely understood. It seems likely that there are multiple causative factors, including depletion of stores of neurotransmitters in nervous tissue, increased metabolic degradation of biochemical substances involved in the drug's action, loss of receptors, and/or physiological adaptation (activation of physiological compensatory mechanisms which counteract the drug's effect). It should be noted that many drugs do not produce tachyphylaxis-related responses. Pharmacologists distinguish between tachyphylaxis and *drug tolerance*, which is a more gradual decrease in responsiveness to a drug that requires days or weeks to develop. However, the physiological distinction between tachyphylaxis and drug tolerance is not clear.

5.4   Cumulation

When a drug is administered to an animal (or to a human patient) at such time intervals between doses that the body cannot remove one dose completely before the next dose is administered, *cumulation* results. This is encountered especially with drugs that have a long half-life in the body. However, it should be emphasized that cumulation is a function of dose size and the interval between doses as well as half-life. Figure 5.2 shows plots of the amounts of two drugs in the body (Y axis) versus the number of doses administered (X axis).

The X axis is calibrated as time, in days, and the drug is given once daily. Drug (a) (solid line) is completely destroyed or excreted from the body in less that 24 h. Thus, it does not accumulate, and the peak drug levels on days 1, 2, 3, 4, and 5 are essentially the same. The residual amount of drug in the body falls to zero by the end of the 24-h period. However, drug (b) (dashed line) requires more than 24 h for metabolic degradation or excretion, so that it accumulates when it is administered once every 24 h.

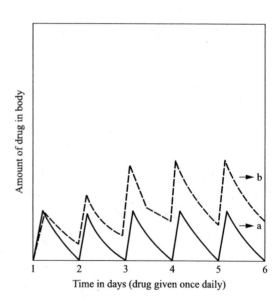

Figure 5.2 Drug cumulation   Time in days (drug given once daily)

The peak drug level on day 2 is greater than on day 1, and the day 3 peak level is greater than that of day 2. Cumulation may be a desirable factor in the pharmacology of some drugs, because it permits the attainment of high, constant blood levels for a prolonged pharmacological effect. In the case of anti-infective agents, such high, constant blood levels are frequently therapeutically useful. However, cumulation may represent an undesirable property of some drugs (for example, anticoagulants) which provide an effect greater than desired and/or toxic results.

5.5   Quantitative Expression of Effective Dose

A valuable result derived from dose–response curves is establishment of the magnitude of the effective dose of the test drug. What dose size can be expected to produce the desired pharmacological effect? There is a variety of methods for expressing the effective dose:
1. *Arithmetic mean dose* results when doses are plotted as their arithmetic value. It is simply an average dose: $X = \Sigma (x)/N$.
2. *Geometric mean dose* results when the logarithms of the doses are plotted: $G = $ antilog Σ (log dose)$/N$.
3. *Median dose* is the smallest dose that is effective in 50% of the individuals in the study. On the dose–response curve, the point for the median dose bisects the population of doses into equal halves. *Median effective dose*, expressed symbolically as ED_{50}, is most commonly used in the pharmacological literature, Similarly, LD_{50}, the *median lethal dose*, defines the smallest dose necessary to kill 50% of the population of animals.

The most important aspect of indirect pharmacological assays is that evidence for the validity and the error limits for each experiment can be obtained from the data of the experiment itself. Every statistic derived from experimental biological data is only an estimate of the "true" value of the statistic in a population of infinite size, and each estimate is associated with an error. A meaningful way of indicating the precision of a statistic is to use *confidence limits* (sometimes called *fiducial limits*). These are boundaries which are expected to contain the "true" value of the statistic at some selected level of probability. Confidence limits are calculated from the actual experimental data. When the 95% confidence limits are calculated for a median effective dose (ED_{50}), the assertion is made that the true ED_{50} for the drug in an infinitely large population of animals will be found within these numerical boundaries with a probability of 95% and will be outside these boundaries, by chance, only five times out of 100 repeated experiments. The median effective dose with its 95% confidence limits is expressed symbolically as

$$ED_{50} = 0.50 \text{ mg/kg } (0.28 - 0.69)$$

This expression states that 95% of the animals in a pharmacological assay of the drug show a response to doses between 0.28 and 0.69 mg per kg of their body weight and that the median dose is 0.50 mg$/$kg^{-1}. The numerical range of the confidence limits should be narrow, as in this example. For in vitro assays, the effective dose is usually expressed as the EC_{50}, the *median effective concentration* of the drug, rather than as an absolute amount. For assays of a drug's inhibitory effect on the catalytic activity of an enzyme, the median inhibitory concentration, IC_{50}, is reported. This is the concentration at which the inhibitor exerts its half-maximal effect.

5.6   Establishment of Pharmacological Antagonism as Competitive or Noncompetitive

Dose–response curves from quantitative assays provide information indicating whether an antagonist to a given drug's pharmacological effect acts by a competitive or a noncompetitive mechanism. Figure 5.3 is derived from a study of morphine-induced analgesia. In graph A, the solid curve results from an assay in which morphine was the only drug administered. The dashed curve represents a second experiment in which naive animals were predosed with a single dose level of a morphine-antagonist drug. Next, these predosed animals were administered graded dose levels of morphine, using the same protocol as in the study represented by the solid curve. Note that the dashed curve is shifted to the right along the X axis, demonstrating that for most dose levels of morphine, the antagonist-pretreated animals showed a lower analgesic response. The addition of the antagonist to the animal's body rendered the morphine less potent. However, this graph also demonstrates that the same maximum analgesic effect could be attained in the presence of the antagonist, but a larger dose of morphine was required. These results are characteristic of a competitive antagonism, a mass action effect.

In contrast, graph B in figure 5.3 shows a study involving a noncompetitive morphine antagonist. Again, the solid curve is the dose–response curve for morphine alone in naive animals. The dashed curve results from administering graded doses of morphine to animals which were predosed with a single dose level of the noncompetitive antagonist. It is seen that, regardless of the size dose of morphine administered to these predosed animals, a lower level of analgesia resulted. This antagonist drug has a stronger, longer lasting effect, and its interaction with the receptor apparently does not follow simple mass action law.

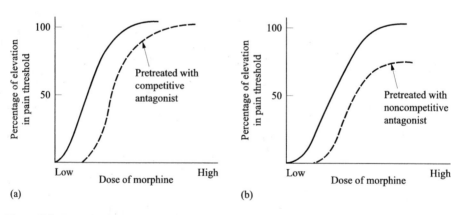

Figure 5.3 Competitive and noncompetitive drug antagonism. (Reproduced with permission from reference 1. Copyright 1984 Sinauer)

5.7   Quantal Assays

These describe the frequency with which any dose of a drug evokes a stated, fixed (all-or-none) response, and the dose–response curve describes the distribution of minimum doses that produce a given effect in a population of biological objects. Threshold doses for the effect can be determined, either by titrating a subject animal with the drug until the desired response is attained, or by giving a series of dose levels to different groups of animals and noting the number of animals in each group that responds to each dose level of the drug. The frequency of occurrence of threshold doses can be plotted against the actual dose used. Depending upon the way response is plotted on the Y axis, curves of two different shapes can be constructed (figure 5.4).

In one assay protocol, the animals are divided into several groups of 10. Each member of a given group receives the same size dose, and doses are graded between groups: 1 mg/kg^{-1}; 2 mg/kg^{-1}; 4 mg/kg^{-1}, etc. Figure 5.4A shows a plot of dose (X axis) versus frequency of response (Y axis); this is a bell-shaped curve, also termed a Gaussian or normal distribution curve. This curve describes the distribution of minimum doses that produce the biological end point in the population of test animals. It is seen (left side of the curve) that a small percentage of the 10 animals in a given dose group responds to a small dose of the drug. A maximum number of animals in a group of 10 responds to a median size dose of the drug, and only a small percentage of the animals in the group of 10 responds only to a very large dose of the drug. As it applies to drug doses that give a quantal (all-or-none) response, the bell-shaped curve suggests that the observed variation in doses needed to produce the response results from simple random variation.

A second format for plotting these test data is illustrated in figure 5.4B, in which the Y axis represents the *cumulative* frequency of responses of the animals to the range of doses; an S-shaped curve results. The central portion of the curve is linear, the advantage of which has been described previously. In addition to manipulating the X axis by plotting the logarithm of the dose, these S-shaped curves can be made linear over a wider dose range by simultaneously manipulating the plot of the Y axis. The "percent responding" parameter can be converted into *probits* (contraction of "probability units"). The probit designates the deviation from the median: a probit value of 5

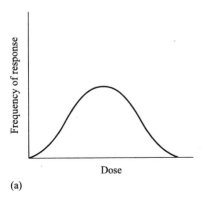

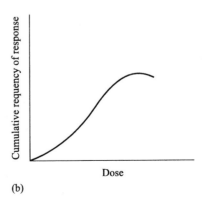

(a)   (b)

Figure 5.4 Formats for plotting quantal assay data

corresponds to a 50% response, and because each probit equals one standard deviation, a probit value of 4 equals 16% and a probit of 6 equals 84%. There are published tables which translate experimentally determined percentage responses into probit values. Contemporary pharmacologists utilize computer programs to convert raw experimental data into probit values.

Figure 5.5 illustrates departures from the ideal that are obtained in actual practice, and it shows some of the strategies utilized by the pharmacologist for plotting experimental numerical data so as to correct for these departures. For every drug, at least two dose–response curves can be constructed: one for the desired pharmacological response, and one for some unwanted toxic manifestation, such as death. The experimental data upon which the curves in figure 5.5 are based were obtained by injecting groups of 20 mice with graduated doses of the sedative/hypnotic drug phenobarbital, and observing the presence or absence of the *righting reflex*. In assaying a sedative/hypnotic drug, it is difficult to determine by visual observation whether the test animal is asleep. Accordingly, the tester picks the animal up and lays it on its back or side.

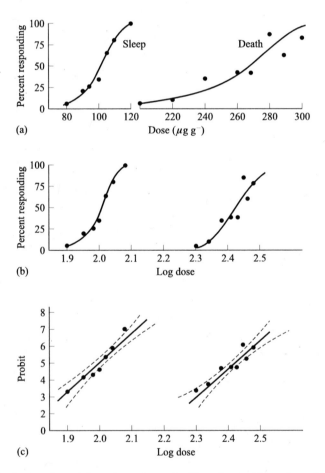

Figure 5.5 Plotting experimental data using various units. (Reproduced with permission from reference 2. Copyright 1981 McGraw-Hill)

If the animal immediately stands up, it is considered to be awake. If it remains on its back or side (that is, if it has lost the righting reflex), it is recorded as being asleep. In this assay protocol, if a test animal died within 24 h of administration of the drug, the dose was considered to be lethal.

The cumulative percentage of each group of mice responding with either sleep or death was plotted against the dose of the drug, as shown in the (a) set of curves of figure 5.5. The dose–response curve resulting from the plotted points for sleep is a moderately well-shaped S curve, and most of the points fall on or close to the S. However, the plotted points for death are scattered; they cannot be connected to produce a well-shaped S. Part of the difficulty with these "death" points is that a dosage level sufficient to elicit a 100% response (to kill all of the animals) was not utilized. Also, the response to the 280 μg dose is unexpectedly high compared to the other points. This amount of scattering of "death" points does not necessarily reflect any defect in the assay procedure or incompetence of the assayer. Great variation in response is common in pharmacological assays where the biological end point (in this case, death within 24 h) is affected by extraneous factors, such as development of a bacterial infection as a consequence of the prolonged depression of the central nervous system produced by the drug. In other words, some of the animals probably did not die as a direct result of the pharmacological effects of the drug. It would be anticipated that if the entire set of assays were repeated in a new group of animals in a different laboratory and by a different assayer, similar deviations of the points would result.

In the (b) set of curves in figure 5.5, the percent of animals responding (Y axis) is plotted against the logarithm of the dose (X axis). For the left (sleep) curve the plotted points describe an S rather well. In the case of the right (death) curve, there is less scattering of the points than in (a), but they still do not describe a well-shaped S.

The (c) set of curves in figure 5.5 shows the logarithm of the dose (X axis) plotted against probit values (Y axis). The sleep curve approaches a straight line, and even the death points show only moderate scattering, so that a straight line can be drawn which coincides with or comes close to the plotted points. The dashed lines on the (c) curves show the calculated 95% confidence limits. Note that the narrowest range of confidence limits is found in the middle portion of each curve in the region of the median dose. Statistically, the median dose (ED_{50} or LD_{50}) is expected to be the most accurate expression of the true potency of the drug.

5.8   Therapeutic Index

One of the most challenging aspects of experimental pharmacology is the extrapolation of pharmacological assay data (and, as previously discussed, pharmacokinetic and drug metabolism data) from animals to humans. Regardless of the scientific and statistical validity of a drug study in animals, an omnipresent question is: Do these animal results qualitatively and quantitatively reflect the effects to be seen in a population of humans?

The *therapeutic index* or *therapeutic ratio* of a drug is a crude quantitative attempt to indicate the safety of the drug, expressed as a ratio of its lethal dose to its therapeutic dose. Most often, the therapeutic index is based upon LD_{50}/ED_{50}. The larger the numerical value of the therapeutic index, the safer the drug.

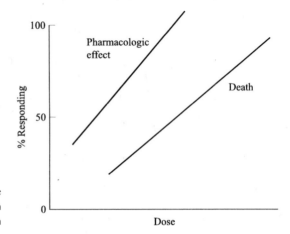

Figure 5.6 Parallel curves for effective dose and lethal dose: a desirable condition for therapeutic index calculation

5.8.1 Difficulties and Defects in Expressing the Therapeutic Index: Slopes of Dose–Response Curves

The above-cited method for expressing the therapeutic index has some disadvantage because median doses are merely points on the dose–response curves and they reveal nothing about the slopes of the curves. If the slopes of the linear parts of the two curves (for ED and for LD) are the same (that is, if the two dose–response curves are parallel), as shown in figure 5.6, then the ratio of lethal dose to therapeutic dose is the same at any level of response. However, if the therapeutic index were derived from two nonparallel dose–response curves (figure 5.7), then the magnitude of the therapeutic index would depend upon the level of response selected.

If the slope of the death curve were steeper than that of the pharmacological effect curve, in some individuals a large "therapeutic" dose could approach the level of a lethal dose.

In an attempt to avoid this potential problem, it has been suggested that the therapeutic index be defined as $LD_{0.1}/ED_{99.9}$: the lethal dose in 0.1% of the population

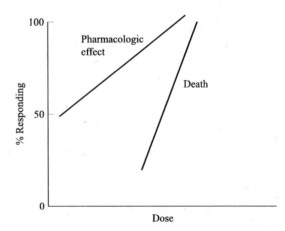

Figure 5.7 Nonparallel dose–response curves

divided by the effective dose in 99.9% of the population. However, there is a potentially dangerous flaw in this definition of therapeutic ratio. The $LD_{0.1}$ and $ED_{99.9}$ are at the far ends of the dose–response curves, and these values are not as reliable as the midpoints ED_{50} and LD_{50}. Refer to figure 5.5c which shows that the 95% confidence limits widen at both ends of the curves. The problem of finding an entirely satisfactory method for numerically expressing the safety of a drug has not been solved.

Regardless of the protocols used in calculating the therapeutic index, there remain added inherent difficulties. First, the lethal dose is based upon animal data and extrapolation to humans may not be appropriate. The LD does not address toxic side effects that are not lethal. Thus, MTD (minimum toxic dose) is often employed, especially in dose selection in clinical trials. Some significant forms of drug toxicity are idiosyncratic (they affect only a very small proportion of individuals who take the drug). Second the magnitude of the ED may depend upon the measure of effectiveness being used. The ED_{50} for aspirin to relieve a headache may be quite different from the ED_{50} for aspirin to relieve pain and symptoms of rheumatoid arthritis and different from the ED_{50} for aspirin's action in extending the clotting time of the blood. These shortcomings suggest that therapeutic index may have little value as a measure of the *clinical usefulness* of a drug. For example, the cardiac glycoside digoxin has a very small therapeutic index (approximately 3), but nevertheless, this drug is widely used to treat congestive heart failure. Some anticancer drugs have calculated therapeutic indices of less than unity. One authority has stated that the overall significance of the therapeutic index may be its application as a measure of the impunity with which an overdose can be given.

5.8.2 Clinical Significance

If a drug must bind to most of its receptors before a response is detected, the slope of the dose–response curve will be increased considerably. Although such a phenomenon is usually of little more than theoretical interest, occasionally there may be practical implications. For example, a steep dose–response curve for a central nervous system depressant implies that there is a small difference between a dose that causes mild sedation and one that produces coma. Therefore, an excessive response or an inadequate response may result if the dose of the drug is not carefully regulated for the specific patient.

5.9   Numerical Expression of Dose

Thus far, fortuitously, the dose data from assays have been expressed in milligrams or micrograms of drug per kilogram of body weight of the test animal. For comparison of doses of an unknown drug and a standard drug, this method is meaningful only if the test drug and the standard have the same or nearly the same molecular weights. Most pharmacologists prefer to express and to plot dose data on the basis of milli*moles* of drug per kilogram of body weight. In this system, the comparison reflects pharmacological effects produced by the same numbers of molecules of test drug and reference drug, and this is a more valid and meaningful comparison of intrinsic activities and potencies.

Bibliography

1. Feldman, R. S. and Quenzer, L. F. *Fundamentals of Neuropsychopharmacology*. Sinauer: Sunderland, Mass., 1984; p. 15.
2. DiPalma, J. R., Ed. *Drill's Pharmacology in Medicine*, 4th ed. McGraw-Hill: New York, 1981; p. 13.

Recommended Reading

1. Houston, J. G. and Banks, M. N. High-Throughput Screening for Lead Discovery. In *Burger's Medicinal Chemistry and Drug Discovery,* 6th ed. Abraham, D. J., Ed. Wiley Interscience: Hoboken, N.J., 2003; Vol. 2, pp. 37–69.
2. Ross, E. M. and Kenakin, T. P. Pharmacodynamics: Mechanisms of Drug Action and the Relationship Between Drug Concentration and Effect. In *Goodman and Gilman's The Pharmacological Basis of Therapeutics*, 10th ed. Hardman, J. G., Limbird, L. E. and Gilman, A. G., Eds. McGraw-Hill: New York, 2001; pp. 31–43.
3. Condouras, G. A. Natural Laws Concerning the Use of Drugs in Man and Animals. In *Drill's Pharmacology in Medicine*, 4th ed. DiPalma, J. R., Ed. McGraw-Hill: New York, 1971; pp. 10–20.
4. Modell, W., Schild, H. O. and Wilson, A. Nature and Measurement of Drug Responses. In *Applied Pharmacology*. W. B. Saunders: Philadelphia, Pa., 1976; pp. 21–48.
5. Goldstein, A., Aronow, L. and Kalam, S. M. *Principles of Drug Action*, 2nd ed. Wiley: New York, 1974; pp. 82–111.
6. Finney, D. J. *Statistical Method in Biological Assay*, 2nd ed. Griffin: London, 1964.
7. Snedecor, G. W. and Cochran, W. G. *Statistical Methods*. Iowa Press: Ames, Ia., 1967.
8. Finney, D. J. *Probit Analysis*, 2nd ed. Cambridge University Press: London, 1952.
9. Chappell, W. R. and Mordenti, J. Extrapolation of Toxicological and Pharmacological Data from Animals to Humans. In *Advances in Drug Research*. Testa, B., Ed. Academic Press: London, 1991; Vol. 20, pp. 1–116.
10. Williams, M. Receptor Binding in the Drug Discovery Process. In *Medicinal Research Reviews*. deStevens, G., Ed. Wiley: New York, 1991; Vol. 11, pp. 147–184.
11. Williams, M., Mehlin, C. and Triggle, D. J. Receptor Targets in Drug Discovery and Development. Recommended Readings ref. 1. Vol. 2, pp. 319–355.
12. Wallace, R. W. and Goldman, M. E. Bioassay Design and Implementation. In *High Throughput Screening: The Discovery of Bioactive Substances*. Devlin, J. P., Ed. Marcel Dekker: New York, 1997; pp. 279–305.
13. Brown, S. J., Clark, I. B. and Pandi, B. Rapid, High Content Pharmacology. Recommended Readings ref. 1, Vol. 2, pp. 71–80.
14. Williams, M., Deesher, D. C. and Sullivan, J. P. Receptor Binding Assays. In *Burger's Medicinal Chemistry and Drug Discovery*, 5th ed. Wolff, M. E., Ed. Wiley-Interscience: New York, 1995; Vol. 1, p. 390.

II

The Peripheral and Central Nervous Systems

6

Some Basic Concepts of the Anatomy and Physiology of the Nervous System

6.1   Aspects of Functional Anatomy of the Nervous System and Its Related Effector Organs

6.1.1 The Nerve Cell

The nervous system detects changes in the environment or within the body itself and controls the ability of the body to readjust itself to these changes. There are two sets of apparatus to control these tasks, one for detection and one for response. They are remarkably similar, both anatomically and biochemically/physiologically. Both involve special nerve cells (neurons), which consist of a cell body with a nucleus, and attached long fiber(s) that conduct the nerve impulse. The human nervous system is estimated to contain 10 billion neurons. Figure 6.1 illustrates the neuron, the functional unit of the nervous system. The dendrites carry nerve impulses toward the cell body, and the axon, the extremely long fiber emerging from the other side of the nerve cell, carries impulses away from the nerve cell body. Large numbers of individual nerve fibers (axons) are collected together, much like strands of wire in an electric cable. The resulting structure is termed a *nerve trunk,* and it is illustrated in cross section in figure 6.2. The individual aggregations of nerve fibers, as illustrated, are collected together into six separate nerves.

Functionally, there are two kinds of nerves: *afferent* or *sensory* nerves, which carry impulses toward the central nervous system, and *efferent* or *motor* nerves, which carry impulses away from the central nervous system. Sensory nerves are stimulated when some change in the environment occurs: a stimulus such as heat, cold, or pain. Motor nerves function oppositely; an electrical impulse travelling from the central nervous system out to the nerve endings produces a mechanical response such as a muscle contraction or relaxation, or a change in the activity of an organ. In low orders of animal life, for example some worms, a sensory nerve is connected directly to its appropriate motor nerve at an anatomic entity called a *synapse* (figure 6.3). If some point on the worm's body is touched with a sharp object, a sensory impulse is generated which travels along the nerve fiber to the synapse where the sensory impulse is translated into a motor nerve impulse. This resulting motor impulse travels along a separate motor nerve

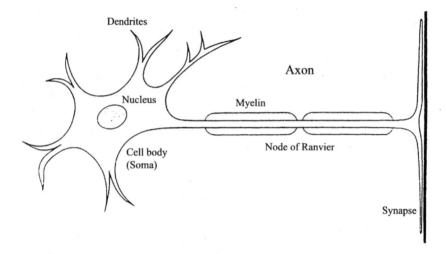

Figure 6.1 Schematic diagram of a nerve cell (neuron). Reproduced with permission from reference 1. Copyright 1986 Wiley-VCH

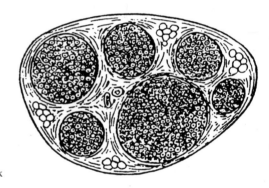

Figure 6.2 Cross section of a nerve trunk

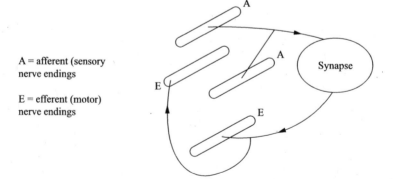

A = afferent (sensory nerve endings

E = efferent (motor) nerve endings

Figure 6.3 Diagrammatic representation of a synapse in lower forms of animal life

to the region of the muscles that were touched by the sharp object, and these muscles are caused to respond in the form of a twitch.

6.1.2 Central Nervous System: The Brain

In higher forms of life, the impulses that the body receives from an enormous number and variety of sensory nerves must be coordinated with the motor nerve system. The anatomic entity that accomplishes this task is the central nervous system (CNS) which consists anatomically of the brain and the spinal cord (figure 6.4). The highest region of the brain is the cerebrum, the command/control center of the body. The processes of thought, memory, and learning occur in the cerebrum. Other portions of the brain (in descending order of priority) are the diencephalon (including the thalamus, the

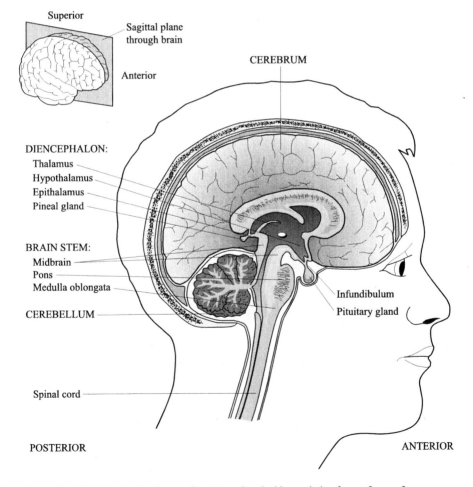

Figure 6.4 The human brain: sagittal section. (Reproduced with permission from reference 3. Copyright 1999 Benjamin/Cummings Science)

hypothalamus, and the pineal gland), the cerebellum, the pons, and the medulla oblongata. These "lower centers" of the brain are responsible for functions not normally performed by the higher centers in the cerebrum.

The cerebellum coordinates involuntary motor activity and it regulates posture and balance. The pons contains sensory and motor pathways connecting the cerebrum to the cerebellum. The medulla oblongata is the site of the respiratory center whose level of activity is governed by the levels of carbon dioxide and oxygen in the blood. Chemoreceptors in the aorta and the carotid artery continually monitor blood CO_2 and O_2 levels. Lowered O_2 blood content or elevated CO_2 causes generation of motor nerve signals in the respiratory center, traveling to the periphery of the body, instructing the respiratory apparatus (the diaphragm and the intercostal muscles) to increase the rate and depth of breathing. It is not necessary physiologically that the individual be conscious of or think about breathing. The process is initiated and regulated automatically. However, if one so desires, one can consciously stop breathing; the higher centers in the cerebrum can temporarily over-ride the automaticity of the respiratory center. The individual can also consciously will the respiratory rate to increase or decrease. However, it is not normally possible to commit suicide by holding one's breath. The willed cessation of breathing soon results in loss of consciousness, and at this point the cerebrum loses its control over breathing. The respiratory center in the medulla oblongata regains control, and the individual's spontaneous breathing resumes.

Centers that control the diameter of peripheral blood vessels (and therefore participate in regulating blood pressure) and centers that regulate heart rate are also found in the medulla. Less vital centers in the medulla control coughing, vomiting, salivation, and sneezing.

6.1.3 Central Nervous System: The Spinal Cord

The spinal cord lies inside the hollow part of the vertebrae (the bones) which form the spine (figure 6.5). The spinal cord is continuous with the brain, and the boundary between the brain and the spinal cord is purely arbitrary. Both are covered by the same membrane (the meninges), and both are surrounded by the same fluid, the *cerebrospinal fluid* (CSF) which also circulates and fills the cavities within the brain and the spinal cord. The brain and the spinal cord float in this aqueous solution, which serves three physiological purposes. First, it is a hydraulic cushion to protect the brain and spinal cord from possibly traumatic mechanical jars and jolts. Second, it serves as a chemical protection for the CNS, maintaining the proper ionic composition for the environment of the CNS. Even slight changes in its ionic composition could seriously disrupt nerve impulse transmission in the CNS. Third, it serves as a medium of exchange of nutrients and waste products between the blood and the nervous tissue of the CNS.

The spinal cord (figure 6.6) in cross section consists of an inner core of so-called *gray matter*. Anatomists designate the *dorsal* (back side) horns and the *ventral* (front side) horns of the gray matter. This nervous tissue receives and integrates all incoming and outgoing information. The gray matter consists of nerve cells with very short fibers, which are arranged to connect a sensory nerve with its appropriate motor nerve. This entire arrangement is termed a *reflex arc*, and it provides reflex (i.e., automatic) motor responses to sensory stimuli through synapses in the gray matter, analogous to the

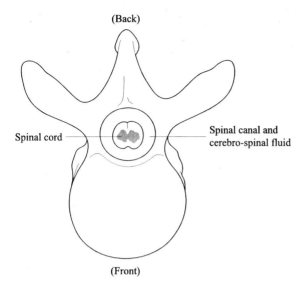

(Back)

Spinal cord

Spinal canal and
cerebro-spinal fluid

(Front)

Figure 6.5 Cross section of the spine

generation of a muscle twitch described in the case of touching a worm with a sharp object. The gray matter is surrounded by a dirty yellowish–white tissue, the *white matter*. This white matter contains nerve fibers, each of which carries messages up and down the cord, to and from the brain.

6.1.4 Peripheral Nerves and Spinal Cord

The spinal cord also makes possible direct control by the brain through nerve connections from the periphery into the white matter, thus linking sensory nerves with the brain.

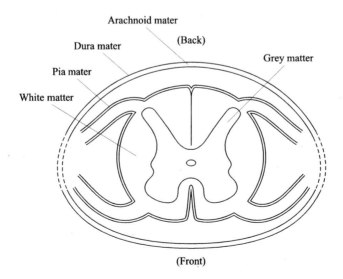

Arachnoid mater

Dura mater

(Back)

Grey matter

Pia mater

White matter

(Front)

Figure 6.6 Cross section of the spinal cord

As shown in figure 6.7, the sensory nerve entering the spinal cord diverges, one branch entering the gray matter to form a part of the reflex arc, and the other branch entering the white matter. A sensory nerve impulse received by the spinal cord travels to the brain via nerve fibers in the white matter, where it is received, integrated, and translated into a large number of motor nerve impulses which move back down appropriate motor nerve pathways in the white matter, arriving at locales in the cord where the nerve fibers leave the cord and travel thence to sites in the periphery where there is generated an appropriate multicomponent response to the original stimulus.

Associated with the neurons are *glial cells* which outnumber the neuron population 5- to 50-fold. Glial cells occupy more than half of the volume of the brain. There are six different anatomic/physiological types of glial cells and these perform a variety of tasks, some of which are associated with nerve impulse transmission, although the glial cells themselves do not conduct nerve impulses.

6.1.5 Physiological Types of Motor Nerves and Muscles

Motor nerves, physiologically, are of two types: *voluntary* and *autonomic* (involuntary, vegetative). Voluntary nerves control those muscles of the body (e.g., of the arms, legs, face) which are directly operated by will. Autonomic nerves control those physiological activities and organs of the body (e.g., heart, stomach, intestine, lung, blood vessels, pupil of the eye) which carry out the mechanical acts of life and over which the individual has little or no conscious control. Most of these involuntary organs are doubly innervated, one set of nerves being responsible for stimulation and the other set being responsible for depression of the activity of the organ. Thus, very delicate and exact control of the organ is possible. Its level of activity at any given time is the algebraic sum of the activities of the stimulant and the depressant nerves. The concept of double innervation of organs leads to further subdivision of the autonomic nervous system into *sympathetic* and *parasympathetic* branches, illustrated in figure 6.8.

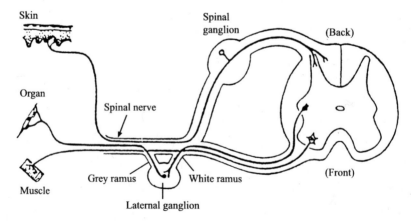

Figure 6.7 Nervous connection of the spinal cord with the periphery of the body

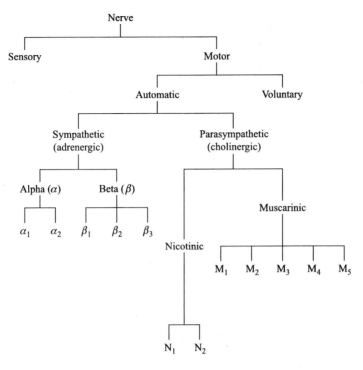

Figure 6.8 Functional organization of the nervous system

The anatomic site where an autonomic nerve fiber (the axon) meets the involuntary muscle fiber or the organ that it innervates is called the *nerve ending* or the *end plate*. The terms *myoneural junction* or *neuromuscular junction* are applied (synonymously) to the junction between a voluntary nerve and a voluntary muscle. Voluntary (skeletal) muscles are termed *striated muscles*. Under magnification, the individual voluntary muscle fibers are distinguished by striational markings on their surface (figure 6.9).

In contrast, involuntary muscles (those innervated by the autonomic system) lack these surface striations and they are termed *smooth muscles* (figure 6.10). As illustrated, there are two anatomic arrangements of smooth muscle fibers. *Multiunit* smooth muscle consists of discrete fibers, each operating independently, and each being activated by a single nerve ending. Fibers of this type are found in the eye and in many of the larger blood vessels. *Visceral* smooth muscle is arranged in sheets or bundles. The cell membranes contact each other at multiple points, and when one region of a visceral smooth muscle is stimulated, the stimulus is conducted to surrounding fibers. Visceral smooth muscle is found in most of the organs of the body, such as the walls of the intestine, bile ducts, and ureter.

Nerve fibers have only two responses to stimulation: they conduct fully or not at all. Partial nerve impulses do not occur. Anesthetic drugs do not lower the conducting efficiency of the individual sensory nerve fibers, but rather they reduce the proportion of fibers that are conducting impulses. Similarly, a muscle fiber (voluntary or involuntary) responds to a stimulus either by contracting maximally or by not contracting at all.

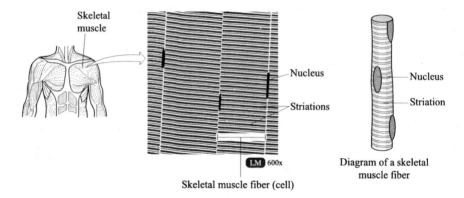

Figure 6.9 Voluntary (sketetal, striated) muscle. (Reproduced with permission from reference 3. Copyright 1999 Benjamin/Cummings Science)

The degree of response of a muscle (an arm, for example, completely extended, completely flexed, or in some intermediate position) depends upon the total numbers of individual muscle fibers that are contracted, not on partial contraction of all of the fibers.

6.1.6 Transmission of Nerve Impulses

A nerve impulse is an electric charge which moves along a nerve fiber, a membrane-based phenomenon. The membrane covering the nerve fiber is said to be *polarized* in

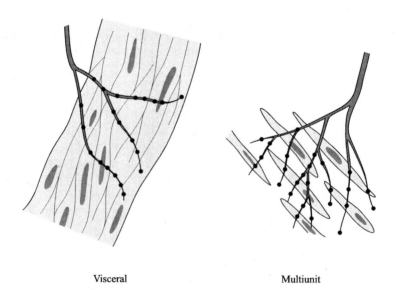

Figure 6.10 Smooth muscle: visceral and multiunit systems. (Reproduced with permission from reference 4. Copyright 1981 Saunders)

its resting, nonconducting state. In this resting state, a very small excess number of anions (chiefly organic, but some chloride plus some potassium cations) accumulates on the inner surface of the membrane, and an excess number of cations (chiefly sodium) accumulates on the outer surface of the membrane. The concentration of Na^+ inside the body cells is approximately 12 mM; that outside the cells, 145 mM. The intracellular concentration of K^+ is 155 mM, and the K^+ extracellular concentration is 4.0 m. In its resting state, the membrane is impermeable to these ions, and the net result of this ion separation is that an electrical potential is established across the membrane, the *resting potential*, having a magnitude of approximately −85 mV (figure 6.11).

A stimulus can suddenly increase the permeability of the nerve membrane to ions; channels in the membrane open and sodium ions migrate across the membrane from the outer surface to the inner surface. This is a spontaneous process, the ions moving from a region of high concentration (outside the membrane) to a region of relatively lower concentration. At the same time, large numbers of anions and smaller numbers of potassium cations migrate in the opposite direction, again from a region of high concentration (inside the membrane) to a region of relatively lower concentration (outside the membrane). This portion of the membrane is now said to be *depolarized*. The magnitude and sign of the electric potential across this part of the membrane change from −85 mV to +45 mV. This is the *action potential*. The original point of depolarization affects adjacent areas of the membrane, causing them to become permeable, transmitting ions in and out of the nerve fiber, and becoming depolarized. Thus, the region of depolarization moves along the nerve fiber.

6.1.7 The Recovery Process

At the original site of the depolarization, there is almost at once a tendency for the ions to sort themselves out—for the sodium ions to move back outside the membrane and

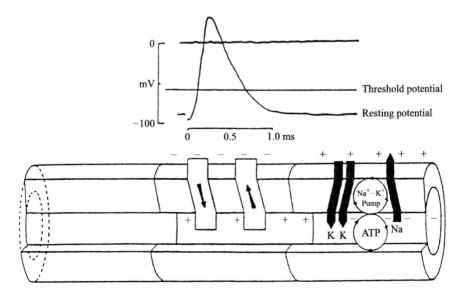

Figure 6.11 Relationship between action potential and ion flow across a nerve membrane

for the anions and potassium cations to move back inside the membrane. However, this recovery is not a spontaneous process because the sodium ions that migrated to the inside of the membrane must move back out against a concentration gradient. This same challenge is faced by the potassium cations that migrated to the outside of the membrane. Energy is required to accomplish these energetically uphill ion migrations, with the participation of the *sodium/potassium pump*. This protein complex is present in all cell membranes and it traverses the membrane structure. Initially it binds sodium ions on its intracellular region and potassium ions on its extracellular region, then the pump protein complex undergoes a conformational change, such that the bound potassium ions are extruded into the intracellular domain and the bound sodium ions are extruded into the extracellular domain. Biochemically, this cation-specific mechanism involves hydrolysis of adenosine triphosphate to adenosine diphosphate. The energy liberated by cleavage of the high energy phosphate bond triggers the conformational change in the pump protein complex. Thus the Na^+ and K^+ ions are transferred back to their original locations on each side of the membrane, and the original point of depolarization becomes repolarized. The membrane regains its impermeability to the ions (ion channels close), and the resting potential is restored. This process in which a point of depolarization moves along the membrane of the nerve fiber is the nerve impulse. The ion migration across the nerve membrane explains the observation that a nerve fiber which has been stimulated will not respond to a second stimulus for 3 to 5 ms. A finite period of time is required for the nerve membrane to become repolarized so that the depolarization process can occur again.

6.1.8 The Myelin Sheath

Many (but not all) mammalian motor nerve fibers possess an additional lipid-rich covering on the outermost surface, in addition to the nerve membrane which was described previously. This is the *myelin sheath* and it functions as a key participant in nerve impulse transmission. Postganglionic autonomic nerves are not myelinated. As shown in figure 6.12, the myelin sheath is not continuous along the nerve fiber, but there are gaps, the *nodes of Ranvier*, named for the French anatomist who first described them. The myelin sheath is found on those nerve fibers in which, physiologically, it is necessary that the nerve impulse travel extremely rapidly. The velocity of a nerve impulse on a small, nonmyelinated fiber is approximately $0.25~m\,s^{-1}$, and for a large, myelinated fiber, as great as $100~m\,s^{-1}$.

Because the nerve membrane is depolarized only at the nodes of Ranvier, much less energy is expended in operating the ion pumps to restore the resting potential than with the continuous movement characteristic of nonmyelinated fiber impulse transmission. As illustrated in figure 6.13, in a myelinated fiber the action potentials occur only at the nodes of Ranvier. Yet, the action potentials are conducted from node to node. This is *saltatory conduction*. Electrical current flows through the surrounding extracellular fluid outside the myelin sheath, as well as through the axoplasm inside the axon, from node to node, thus exciting successive nodes, one after another.

In the central nervous system, a subpopulation of glial cells (oligodendrocytes) forms a supporting network around the neurons to produce the myelin sheath. In the peripheral nervous system, a different type of glial cells (Schwann cells) produce the myelin sheath.

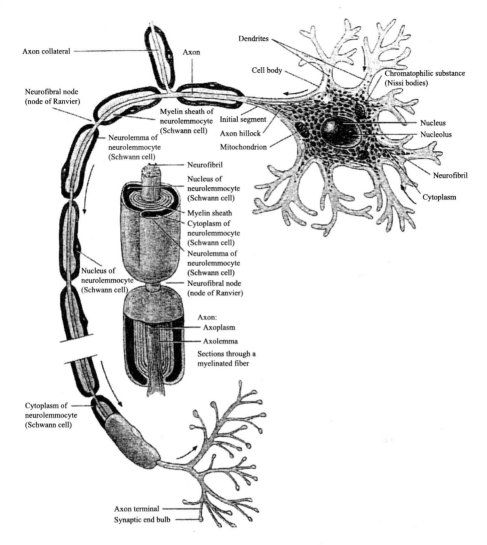

Figure 6.12 Myelinated motor (efferent) nerve fiber. (Reproduced with permission from reference 3. Copyright 1999 Benjamin/Cummings Science)

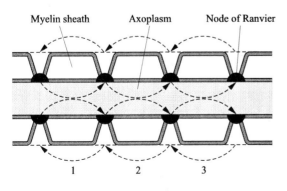

Figure 6.13 Saltatory conduction along a myelinated nerve fiber. (Reproduced with permission from reference 5. Copyright 2000 Saunders)

6.2   Trans-synaptic Nerve Impulse Transmission

6.2.1 Anatomy

A voluntary efferent (motor) nerve consists of a single, extremely long neuron fiber (axon) connecting the CNS with a voluntary muscle fiber. In contrast, as illustrated in figure 6.14, an autonomic efferent nerve extending from the spinal cord to a peripheral effector organ (smooth muscle or gland) is composed of two neurons, laid end-to-end. The branched endings of the axon of the first neuron closely approach the dendrites of the second neuron, but there is no anatomic connection between them. An impulse moving along the nerve fiber must traverse the *synaptic cleft* (more simply, the synapse), some 10 to 50 nm wide, to continue along the next nerve segment and reach its ultimate destination. Structurally, the branched axon endings (the *axon tree*) are not as simple as shown in figure 6.14, but rather they are covered with specialized terminal structures approximately 1 µm in diameter, projecting from their surfaces, as shown in figure 6.12. These are synaptic knobs or synaptic boutons, and there are several hundred of them on each axon terminal. Figure 6.15 illustrates one of these synaptic knobs, the synaptic cleft, and the membrane of the dendrite of the second neuron.

6.2.2 Chemical Mediation

The passage of a nerve impulse across the synaptic cleft is a chemically mediated process. Located within the synaptic knob are large numbers of *transmitter vesicles* or *synaptic storage vesicles* in which is stored a chemical substance, acetylcholine, bound in an inactive form to a "storage protein". It is estimated that there are approximately 3000 acetylcholine molecules in each storage vesicle and that there are enough storage vesicles in a synaptic knob to accommodate 10 000 nerve impulses. The nerve impulse moves along the nerve fiber to the region of the storage vesicles, and alters the presynaptic terminal's permeability to Ca^{2+}, permitting these ions to enter the synaptic knob and to trigger the release of some of the stored acetylcholine molecules into the synaptic cleft. Molecules of this *neurotransmitter substance* migrate across the cleft to the *postsynaptic membrane* (the membrane of the dendrite of the second neuron), where protein receptors specific for acetylcholine are embedded in the membrane lipid matrix. These receptors are ion channels which, in the resting stage, are closed.

Acetylcholine molecules chemically and reversibly bind to these receptors, and the net effect of this chemical interaction is that the ion channels open, permitting sodium cations, anions, and potassium cations to migrate in and out of the membrane as described previously. The postsynaptic membrane (on the dendrite) becomes depolarized. Thus, the nerve impulse, the action potential, is established on the far side of the synaptic cleft, and this impulse can now move on along the membrane of the second neuron. Subsequently, the acetylcholine molecules are desorbed very rapidly from their postsynaptic receptors on the dendrite membrane and they are enzymatically inactivated. The postsynaptic membrane receptor area re-establishes its resting potential by becoming repolarized and the postsynaptic membrane receptor is now ready and able to accept more acetylcholine and transmit another nerve impulse.

Botulinum toxin, one of the most potent poisons known (a lethal dose in humans is approximately 0.3 µg), blocks the release of acetylcholine from the storage vesicles.

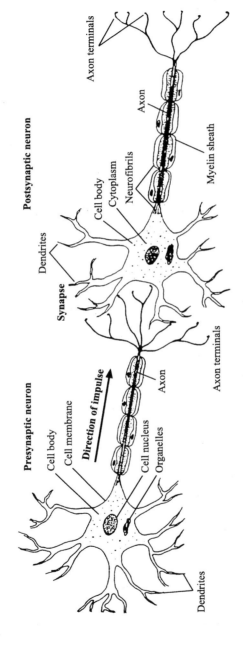

Figure 6.14 Chemical transmission across a synapse linking two associated autonomic neurons

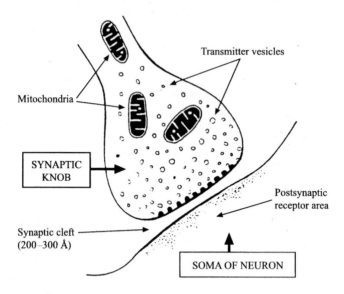

Figure 6.15 Functional autonomy of the autonomic synaptic knob and the synapse. (Reproduced with permission from reference 6. Copyright 1991 Saunders)

Hence, nerve impulse transmission across the synaptic clefts is blocked and flaccid (limp) paralysis of the muscles results. In contrast, the toxic effect of black widow spider venom is caused by its ability to trigger a massive release of acetylcholine into the synaptic clefts, even in the absence of a nerve impulse.

Figure 6.16 illustrates another cellular component, the *mitochondrion* (pl. *mitochondria*). These anatomic entities are found in the cytoplasm of the cell and, as illustrated, they are composed of two membranes, a smooth outer membrane and an inner folded membrane (*crista*). The folded cristae provide a very large surface area for a group of chemical reactions known as cellular respiration. Enzymes that catalyze these reactions are located on the cristae. Cellular respiration occurs only in the presence of oxygen, and the process results in the catabolism of nutrient materials, such as glucose,

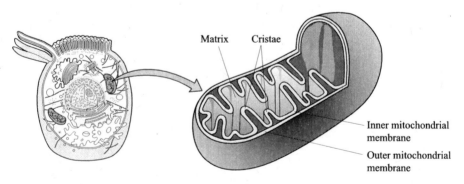

Figure 6.16 A mitochondrion. (Reproduced with permission from reference 3. Copyright 1999 Benjamin/Cummings Science)

forming carbon dioxide and water and liberating energy that is used in the production of adenosine triphosphate, a principal source of chemical energy for the body.

In addition to its role at the synapses between the two neuronal segments of an autonomic nerve pathway, acetylcholine is the mediator of nerve activity at the myoneural junction (between a voluntary nerve and the striated muscle fiber that it innervates). Unless and until acetylcholine is released at the myoneural junction or if for some reason it cannot interact with its receptors on the striated muscle membrane, the nerve impulse will not produce a muscle contraction. The muscle will be paralyzed.

Introduction to the Autonomic Nervous System

6.3.1 Physiological Aspects

It was previously stated that characteristically the organs of the body are innervated by both branches of the autonomic system, the sympathetic and the parasympathetic. However, specific and exclusive control over organ stimulation or organ depression cannot be ascribed to either branch; neither is universally "stimulating" or "depressing". Stimulation of the body's sympathetic system generally depresses all physiological functions except those which are necessary for maximal physical exertion to cope with the challenge of an emergency situation. A mnemonic suggests that the sympathetic system prepares the individual for "fight or flight". In contrast, the parasympathetic system controls those physiological functions that are necessary to build up or to conserve the body's energy store. With few exceptions, the effect of parasympathetic stimulation on a given gland or organ opposes the effect of sympathetic stimulation. Examination of table 6.1 confirms these statements. If nerve impulse transmission in a voluntary nerve

Table 6.1 Effects of automatic stimulation on selected body organs

Organ	Sympathetic (adrenergic)	Parasympathetic (cholinergic)
Heart muscle	Increase activity	Decrease activity
Coronary vessels	Dilate	Constrict
Systemic blood vessels		
Abdominal	Constrict	None
Skeletal (voluntary)	Dilate	Constrict
Eye (pupil)	Dilate	Constrict
Lungs		
Bronchi	Dilate	Constrict
Blood vessels	Constrict (weak effect)	Constrict
GI tract	Decrease peristalsis and tone	Increase peristalsis and tone
Kidney	Decrease urinary output	None
Liver	Decrease glucose	None
Blood glucose	Elevate	None
Mental activity	Increase	None
Intestinal glands	Depress activity	Increase secretion of digestive enzymes

to a striated muscle is disrupted for a period of time, the muscle becomes paralyzed and atrophies. In contrast, smooth muscles and glands, innervated by autonomic nerves, show some degree of spontaneous activity, independent of intact innervation.

Figure 6.14 showed two end-to-end neurons that comprise the autonomic nerve connection between the spinal cord and an effector organ. It should be noted that one presynaptic fiber may synapse with many postsynaptic fibers, thus multiplying the effects produced. In the autonomic system, a large number of individual synapses between the component neurons is collected together in one discrete mass, termed a *ganglion* (pl. *ganglia*). For clarity and convenience, the neuron originating in the spinal cord is termed a *preganglionic* fiber, and the subsequent neuron which leads to and terminates at an effector organ is termed a *postganglionic* fiber. At the synapses within the ganglion (both sympathetic and parasympathetic) the nerve impulse traverses the synaptic cleft because acetylcholine is released, as described previously. Heretofore, the term synapse has been used to designate the gap between the two neurons that comprise the autonomic pathway from the spinal cord to the effector organ. However, strictly speaking, any break in a nerve is a synapse. To avoid ambiguity, pharmacologists restrict use of the term synapse to that region between the nerve ending and the effector organ.

6.3.2 Post- and Presynaptic Receptors

As with the nerve synapses in ganglia, at the nerve terminus at an effector organ, there is no anatomic connection between the nerve terminal and the effector organ. A stored chemical transmitter is released from the nerve terminal by the effect of the nerve impulse, and freed neurotransmitter molecules migrate across the synaptic cleft to attach to *post-synaptic receptors* located on the surface of the membrane of the effector organ. This interaction triggers a sequence of biochemical events which results in the organ response.

In some nerve systems in the body there are neurotransmitter receptors located on the outer surface of the membrane of the synaptic knobs on postganglionic nerve termi-nals (figure 6.17). These are *presynaptic receptors* (sometimes called autoreceptors) and they are frequently specific for the neurotransmitter substance located in the storage vesicles in the nerve terminal. When molecules of the neurotransmitter substance, which have been released into the synaptic cleft, bind to the presynaptic receptors, this interaction triggers a response in the nerve terminal to depress the further release of neurotransmitter molecules from the storage vesicles into the synaptic cleft, regardless of the numbers or frequency of arrival of nerve impulses at the nerve terminal.

Generally, presynaptic receptors are much less sensitive to the neurotransmitter substance than are the postsynaptic receptors. A higher concentration of neurotransmitter is required to activate presynaptic receptors than to stimulate postsynaptic ones. If a large series of nerve impulses arrives rapidly at the postganglionic nerve terminal, this may cause the neurotransmitter molecules to be released more rapidly than they can be removed from the synaptic cleft. The accumulated excess neurotransmitter molecules will continue to interact with the postsynaptic receptors on the effector organ, which could result in a prolonged, continual, magnified (and perhaps physiologically undesirable) response of the effector. However, when the concentration of neurotransmitter molecules in the synaptic cleft reaches a certain threshold level, these molecules will interact with the presynaptic receptors, causing the nerve terminal to cease further release of neuro-transmitter molecules. Therefore, the release of neurotransmitter molecules is, to a certain extent, self-limiting because the terminal can turn itself off; the body can prevent

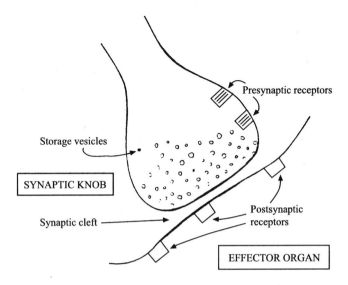

Figure 6.17 Pre- and postsynaptic receptors

a continual, uncontrolled stimulation of effector organs: a negative feedback mechanism. Not all of the nerves in the body are equipped with presynaptic receptors.

It is intriguing that some postganglionic nerve terminals bear presynaptic receptors which are highly specific for a neurotransmitter substance other than the one stored in the synaptic storage vesicles and utilized to activate the postsynaptic receptors. For example, some acetylcholine-mediated nerves in the central nervous system bear presynaptic receptors specific, not for acetylcholine, but for serotonin and insofar as can be determined, there are no stores of serotonin in the vicinity with which these presynaptic receptors can react. Whether these "foreign" presynaptic receptors represent an as-yet unidentified nerve mechanism or whether they are an evolutionary artifact is unknown. Regardless of their *raison d'être*, presynaptic receptors represent potentially useful targets for drugs which may selectively interact with them and thus produce a reverse selectivity compared with the neurotransmitter agonist.

6.3.3 Autonomic Nomenclature

The major biochemical difference between the sympathetic and the parasympathetic systems is that their postganglionic endings (nerve terminals at the effector organ) utilize different neurotransmitter substances (figure 6.18). The parasympathetic postganglionic fibers secrete acetylcholine, hence they are designated as *cholinergic* fibers. Sympathetic postganglionic fibers release norepinephrine (noradrenalin), and these are called *noradrenergic* or *adrenergic* fibers. In contemporary literature, the term *cholinergic nervous system* is used synonymously with the term *parasympathetic nervous system*. Similarly, *adrenergic* or *noradrenergic* nervous system is used synonymously and interchangeably with *sympathetic nervous system*. However, it should be borne in mind that the preganglionic fibers (in the synapses within ganglia) of both the sympathetic and the parasympathetic systems secrete and utilize acetylcholine, Hence, both

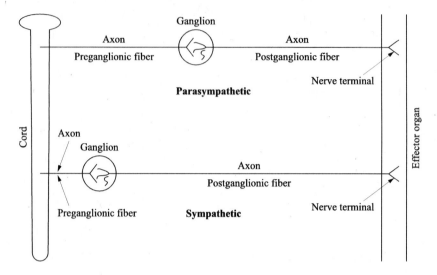

Figure 6.18 Sympathetic and parasympathetic nerves

are cholinergic fibers, even though in one case the entire nerve is designated physiologically as sympathetic (adrenergic).

Bibliography

1. Hucho, F. *Neurochemistry*. Wiley-VCH: Hoboken, N.J., 1986; p. 15.
2. Barlow, R. B. *Introduction to Chemical Pharmacology,* 2nd ed. Wiley: New York, 1962; p. 432, 434, 439.
3. Tortora, G. J. *Principles of Human Anatomy,* 8th ed. Benjamin/Cummings Science Publishing: Menlo Park, CAL., 1999; pp. 43, 82, 505, 545.
4. Guyton, A. C. *Textbook of Medical Physiology,* 6th ed. Saunders: Philadelphia, Pa., 1981; p. 122.
5. Guyton, A. C. *Textbook of Medical Physiology,* 10th ed. Saunders: Philadelphia, Pa., 2000; p. 63.
6. Guyton, A. C. *Textbook of Medical Physiology,* 8th ed. Saunders: Philadelphia, Pa., 1991; p. 483.

Recommended Reading

1. Hoffmann, B. B. and Taylor, P. Neurotransmission. The Autonomic and Somatic Motor Nervous Systems. *In Goodman and Gilman's The Pharmacological Basis of Therapeutics,* 10th ed. Hardman, J. G. and Limbird, L. E., Eds. McGraw-Hill: New York, 2001; pp. 115–154.
2. Guyton, A. C. and Hall, J. E. Membrane Potentials and Action Potentials. In *Textbook of Medical Physiology,* 9th ed. W. B. Saunders: Philadelphia, Pa., 1996; pp. 57–72.
3. Hucho, F. *Neurochemistry*; VCH: Deerfield Beach, Fla., 1986.
4. Webster, R. A., Ed. *Neurotransmitters, Drugs, and Brain Function.* John Wiley & Sons: Chichester, UK, 2001.

7

The Noradrenergic and Dopaminergic Nervous Systems

7.1 The Noradrenergic System

7.1.1 The Noradrenergic Postganglionic Nerve Terminal

Figure 7.1 represents a noradrenergic nerve terminal. Norepinephrine **7.1** is stored in the vesicles in the postganglionic terminal of a noradrenergic (sympathetic) nerve, chemically bound to a "storage protein". This norepinephrine is released from its protein-bound form and from the storage vesicles into the synaptic cleft by the action of a nerve impulse. Liberated norepinephrine molecules migrate across the synaptic cleft and interact chemically with specific postsynaptic receptors on the membrane of

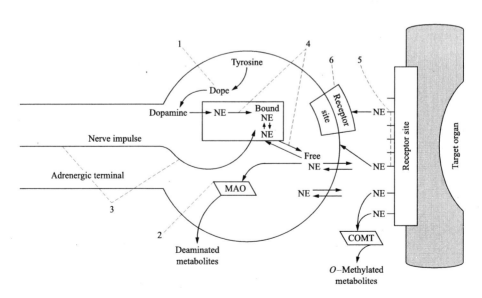

Figure 7.1 Adrenergic nerve terminal

109

the effector organ, and this agonist–receptor interaction translates the nerve impulse into an organ response. The receptor–norepinephrine interaction is a reversible process; norepinephrine molecules dissociate from the membrane receptor and are released back into the synaptic cleft. As long as these molecules remain in the cleft, they will continue to react reversibly with the postsynaptic receptors and stimulate the effector organ, which is usually a physiologically undesirable result. Norepinephrine is removed from the cleft by a re-uptake process (uptake-1), active transport of the norepinephrine molecules from the cleft across the nerve terminal membrane into the nerve terminal. Thereafter, a second active transport process (uptake-2) conveys the norepinephrine molecules across the membrane surrounding the synaptic storage vesicles. Once inside the vesicle, the norepinephrine recombines with the storage protein and remains there until another nerve impulse triggers its release. The details of the biochemical process involved in norepinephrine release into the synaptic cleft are not fully understood, but it has been established that the process requires the entry of calcium cations into the synaptic knob via calcium ion channels (figure 7.2). These ions trigger a sequence of events that causes some of the neurotransmitter-filled synaptic storage vesicles to migrate to the inner surface of the nerve terminal membrane where the storage vesicle membrane fuses with that of the nerve terminal. Then, the membrane bursts, releasing molecules of norepinephrine into the synaptic cleft.

7.1

7.1 R=H Norepinephrine

7.2 R=CH$_3$ Epinephrine

7.3 R=2-C$_3$H$_7$ Isoproterenol

7.1.2 The Noradrenergic Neurotransmitter: Biosynthesis

Figure 7.3 illustrates the biosynthesis of norepinephrine and of its *N*-methyl secondary amine homolog, epinephrine, from two dietary amino acids, phenylalanine and tyrosine. Norepinephrine is believed to be the body's sole neurotransmitter substance in the peripheral sympathetic nervous system. Epinephrine is not synthesized, stored, or used in the peripheral nervous tissue. It is synthesized peripherally in the medulla of the adrenal glands; is liberated into the blood stream (as with any hormone); and is carried in the blood to a multitude of sites in the periphery of the body where it is involved in a large number and variety of functions. Epinephrine's complex hormonal role in the body is physiologically separate from the peripheral autonomic nervous role of norepinephrine. At appropriate dose levels, exogenous epinephrine acts as a potent stimulant (direct agonist action) of peripheral sympathetic receptors. However, the human body physiology does not utilize it for this purpose.

Contrary to a belief held for many years, evidence now indicates that epinephrine is synthesized in the brain and that it is a true neurotransmitter in certain neural pathways there. One of the central mechanisms for regulation of blood pressure involves epinephrine-mediated pathways. Neither epinephrine nor norepinephrine

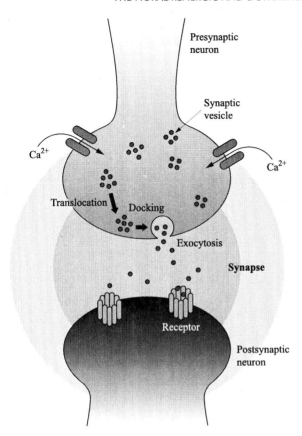

Figure 7.2 Release of neurotransmitter into the synaptic cleft. (Reproduced with permission from reference 1. Copyright 2001 American Chemical Society)

readily penetrates the blood–brain barrier. These substances which are so important in the physiology of the central nervous system must be synthesized there.

7.1.3 The Noradrenergic Neurotransmitter: Enzymatic Inactivation

The uptake-1 and uptake-2 mechanisms for norepinephrine are not 100% efficient. A relatively small number of norepinephrine molecules escapes one or the other of these two active transport systems, and physiological provision is made for the metabolic inactivation of these molecules (illustrated in figure 7.4). Located in the nerve terminals and also in may other parts of the body, including liver and kidney, is an enzyme system, monoamine oxidase (MAO) which catalyzes the oxidative deamination of norepinephrine and epinephrine. The aldehyde product of this reaction sequence is physiologically inert as a neurotransmitter. A second enzyme, catechol-*O*-methyltransferase (COMT), is found in high concentrations in the brain, skin, liver and kidney. The literature is inconsistent and conflicting concerning the presence of COMT in adrenergic nerve tissue. COMT converts epinephrine and norepinephrine into their 3-methyl ethers, metanephrine and normetanephrine, respectively. These ether metabolites

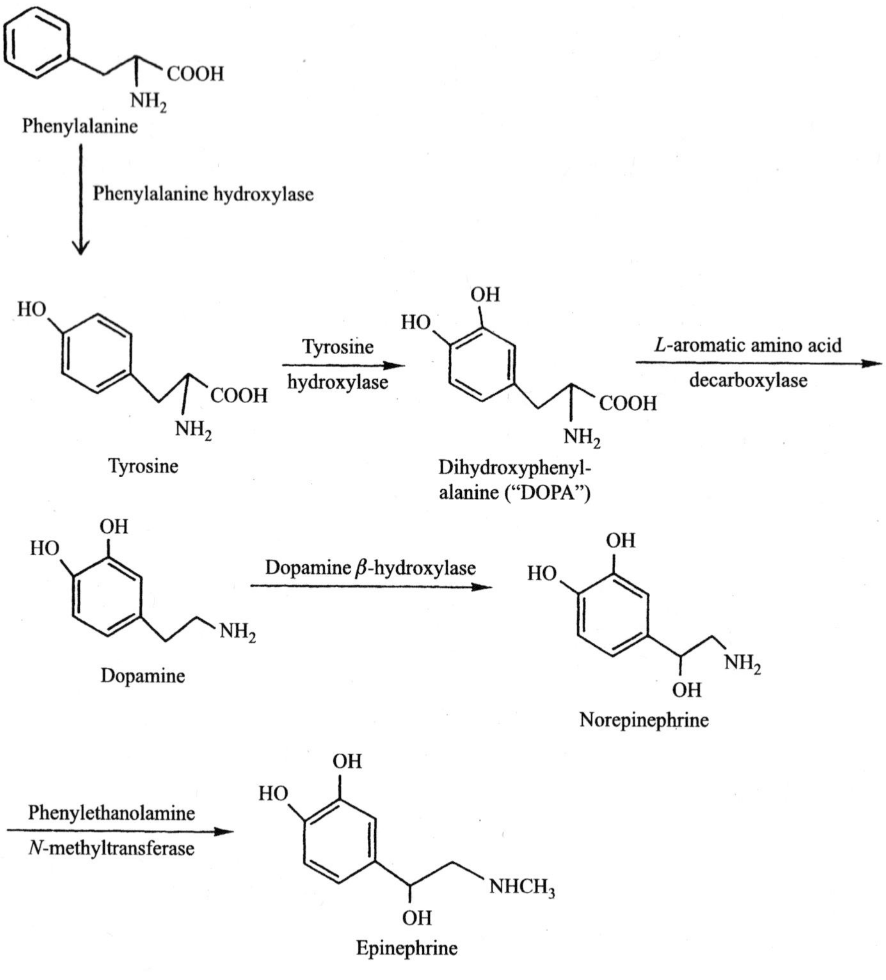

Figure 7.3 Biosynthesis of endrogenous catecholamines

do not stimulate noradrenergic receptors. As shown in figure 7.4, each of the first-formed metabolites of epinephrine and norepinephrine undergoes further metabolic change in the body, and the end products, 3-methoxy-4-hydroxymandelic acid and 3-methoxy-4-hydroxyphenylethylene glycol, are excreted in the urine.

Additionally, a small amount of norepinephrine diffuses out of the synaptic cleft and becomes bound to extraneuronal sites. However, it must be emphasized that the prime mechanism by which the effect of a noradrenergic nerve impulse at an effector organ is terminated is the active re-uptake of intact norepinephrine molecules into the synaptic storage vesicles. These enzyme-mediated inactivations described represent scavenger processes, to assure that no physiologically viable norepinephrine molecules remain in the synaptic cleft or in the nerve terminal outside the synaptic storage vesicles. It is noteworthy that both epinephrine and norepinephrine are carried in the blood without undergoing significant enzyme-mediated inactivation. However, passage of the

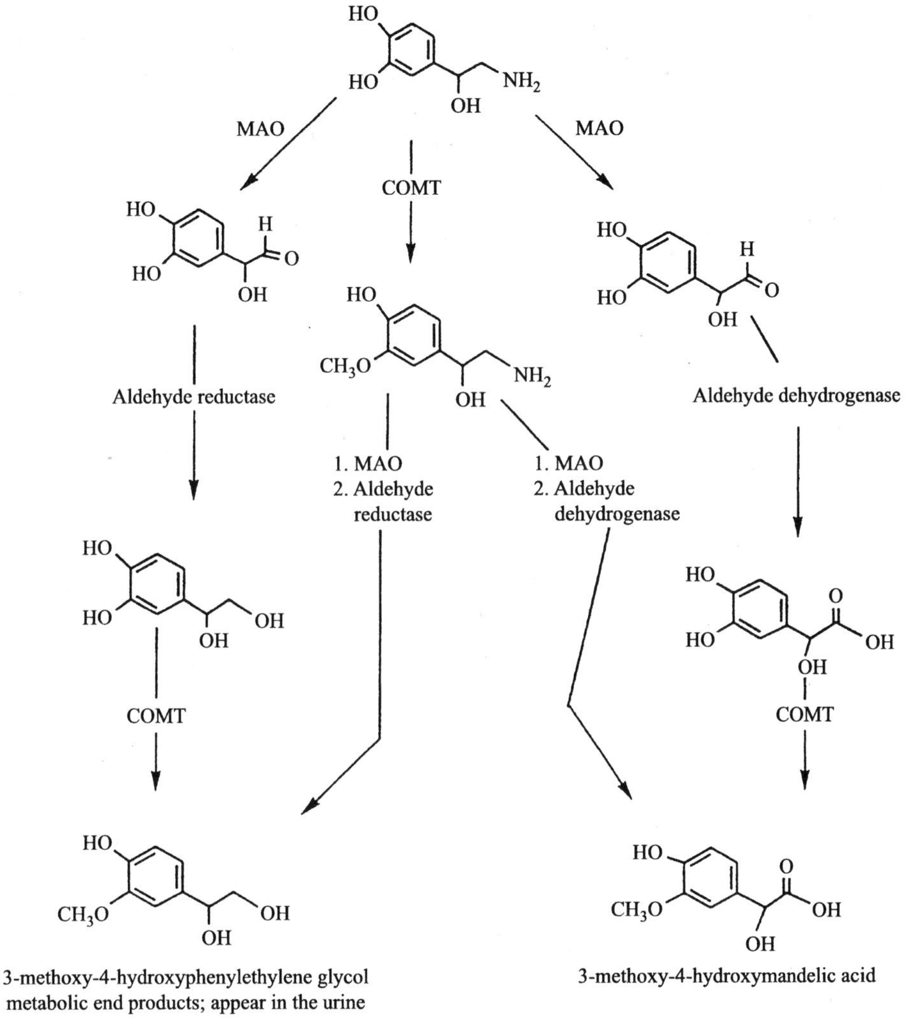

3-methoxy-4-hydroxyphenylethylene glycol
metabolic end products; appear in the urine

3-methoxy-4-hydroxymandelic acid

Figure 7.4 Metabolic fate of norepinephrine and epinephrine. Illustrated for norepinephrine

blood through the liver results in attack by monoamine oxidase and catechol-*O*-methyl-transferase there.

Catechol-O-methyltransferase

This enzyme requires magnesium cation as a cofactor, and *S*-adenosylmethionine as the source of the methyl group. It is highly substrate-specific for 1,2-dihydroxy aromatic systems and it converts only one of the aromatic OH groups into its methyl ether. Not only are endogenous catecholamines (epinephrine, norepinephrine, dopamine) substrates for COMT, but also many exogenous molecules bearing the 1,2-diphenolic moiety are converted into their monomethyl ether derivatives by the enzyme. In the case of the endogenous catecholamines, the predominant in vivo product is the 3-methyl ether. However, the reaction is not completely regiospecific, and small amounts of the isomeric 4-methyl ether are formed. In the case of drug molecules containing a catechol moiety, the structure(s) of the product(s) of COMT-mediated O-methylation are not easily predictable, and the reaction frequently gives mixtures of the two possible isomeric monomethyl ethers. In vitro, the ratio of isomeric monomethyl ether products can vary, depending upon experimental conditions for the enzyme incubation and the biological source of the enzyme. The catalytic effects of COMT can be blocked, both in vitro and in vivo, by drugs (e.g., tropolone or pyrogallol).

Monoamine oxidase

This enzyme is also found in many tissues of the body, in addition to nervous tissue. It exists in two isoforms, MAO-A and MAO-B. The generally applied distinction between the two forms is that MAO-A (but not MAO-B) is inactivated by low concentrations of clorgyline whereas MAO-B (but not MAO-A) is inactivated by low concentrations of deprenyl. The two forms exhibit some degree of substrate selectivity but not absolute specificity. β-Phenethylamine is said to be the preferred substrate for MAO-B; 5-hydroxytryptamine is preferred by MAO-A. It was once thought that there are substrates specific for MAO-A or MAO-B. However, current opinion holds that most substrates, even those that have been reported to be specific for one or the other form of MAO are indeed substrates for both forms of the enzyme. Monoamine oxidase is widely distributed throughout the animal kingdom; it has been found in all mammals in which it was sought. Moreover, although it is absent from red blood cells (a common site for many enzymes), it is found in almost every other tissue in which it was sought. The monoamine oxidases metabolize endogenous neurotransmitters and they also accept a variety of exogenous amines, including drugs, as substrates. For some drugs, monoamine oxidase is an important metabolizing enzyme. The recognition of substrates on the basis of their chemical structure by the two forms of MAO has been reviewed (see Recommended Readings at the end of this chapter).

7.1.4 Phenylketonuria

A serious condition can arise from a defect in the norepinephrine–epinephrine biosynthetic pathway (figure 7.3). The enzyme phenylalanine hydroxylase, which converts dietary phenylalanine into tyrosine, is found only in the liver. In some newborn infants, this enzyme is missing or is severely hypofunctional, and as a result dietary phenylalanine cannot be hydroxylated to form tyrosine. In this situation, metabolism of dietary phenylalanine follows an alternate pathway. As shown in scheme 7.1, the molecule is oxidatively deaminated to form phenylpyruvic acid, and this product is converted in part to phenyllactic acid.

These two products are trivially known as *ketone bodies*, and they are removed from the body in the urine, hence the name *phenylketonuria*. If the condition is not treated immediately, the long-term result will be that the infant will not grow and develop normally. It will be retarded mentally and physically and death will ensue within a few years. The accumulation of large amounts of phenylpyruvic acid in the infant's body causes degeneration of the myelin sheath of nerve fibers, and a broad spectrum of disruptions of nerve impulse conduction results. This effect has been likened to the destruction of insulation on an electric wire, which leaves the bare wire exposed. The degeneration of the myelin sheath is irreversible. However, prompt, early diagnosis of the condition (based upon a simple analysis for phenylpyruvic acid in the urine) can lead to successful preventive intervention. No later than the first month of life, the infant is placed on an artificial diet, low in phenylalanine, which precludes the accumulation of ketone bodies.

| Phenylalanine | Phenylpyruvic acid | Phenyllactic acid |

7.1.5 Classification of Noradrenergic Receptors

Some effector cells respond to noradrenergic impulses by being stimulated (e.g., a muscle contracts), whereas other effector cells respond by being depressed (e.g., a muscle relaxes). To rationalize these phenomena, it was postulated that there are two types of norepinephrine receptors: (1) α *receptors*, responsible for excitatory actions, and (2) β *receptors*, responsible for inhibitory actions. It is now recognized that these original definitions for α and β receptors are overly simplistic and that these definitions are no longer completely valid. Experimental distinction between the two receptor types was made on the basis of the response of the body's population of norepinephrine receptors to three compounds: norepinephrine **7.1**, epinephrine **7.2**, and isoproterenol **7.3**.

These compounds as well as other *o*-dihydroxybenzene derivatives of pharmacological interest are frequently called catecholamines. At α receptors, norepinephrine is approximately equipotent to epinephrine, and both are far more potent than isoproterenol. Norepinephrine is considered to be the prototypical sympathetic α-receptor agonist with which all other α -agonists are compared. At β receptors, isoproterenol is the most potent, followed by epinephrine, and finally by norepinephrine, which is the least potent of the three. Isoproterenol is the prototypical β-receptor agonist, with which all other β agonists are compared. Epinephrine has high potency and activity at both types of receptors. Although this system of classification of sympathetic receptors is useful, it is artificial and nonphysiological, in that isoproterenol is not a natural nerve impulse mediator and indeed it is not found anywhere in nature. It is strictly a product of the chemist's laboratory. Moreover, despite the potency comparisons resulting from these laboratory experiments, norepinephrine is the body's physiological neurotransmitter at all peripheral noradrenergic sites, even at β receptors.

Subtypes of β adrenoceptors

On the basis that some synthetic β-adrenergic agonist drugs produce only a portion of the multiple responses resulting from administration of isoproterenol, it was concluded that these drugs interact only with certain populations of the isoproterenol-responsive β receptors, not with all of them. Hence, it was proposed that there is more than one type of β receptor, defined as follows. β_1, located in the heart and small intestine, and β_2, located in the smooth muscle of the bronchi of the lungs, the vascular bed, and the uterus. Most β_1 and β_2 receptors are postsynaptic, although evidence suggests that there are some presynaptic β receptors. Cloning studies have established that there is also a β_3 receptor which is involved in lipolysis (the hydrolytic cleavage of body fat). The β_3 receptors are activated by isoproterenol, as with β_1 and β_2 types. They are coupled with adenylate cyclase and they promote production of cyclic AMP. It is unclear how or whether β_3 receptors are directly involved in nerve impulse transmission or in the translation of a nerve impulse into a response of an effector organ. Studies are in progress, evaluating the utility of agonists for β_3 receptors in treatment of obesity. The existence of β_4 receptors has been proposed, but as yet little is known of their physiological role(s).

Subtypes of α adrenoceptors

From the same kinds of experimental data that were invoked to demonstrate the existence of multiple types of β adrenoreceptors, α adrenoceptors are also concluded to be heterogeneous. α_1 Adrenoceptors were originally defined as being postsynaptic on the membrane of the effector organ, and α_2 adrenoceptors were defined as being presynaptic. However, evidence now points to the likelihood that there are some *postsynaptic* α_2 receptors located in smooth muscle of blood vessels and in the central nervous system. Several subtypes of both α_1 and α_2 adrenoceptors have been identified. Table 7.1 illustrates the effects of α and β adrenoceptor stimulation on a number of organs and physiological systems.

Table 7.1 Selected effects of α and β adrenoceptor stimulation

Organ	α Receptors	β Receptors
Heart	Excitation	Augmentation of contraction; acceleration of rate of beat β_1
Vascular system		
Muscular vessels	Vasoconstriction; decrease in blood flow	Great decrease in blood flow; dilation β_2 and some β_1
Brain vessels	Decrease in blood flow; vasoconstriction	Increase in blood flow; dilation β_2 and some β_1
Kidney	Great increase in blood flow	
Skin	Strong decrease in blood flow; vasoconstriction α_1	Slight increase in blood flow; β_2 and some β_1
Bronchi of lungs		Relaxation β_2
Intestine	Relax smooth muscle α_2	Relax smooth muscle β_1
Ureter	Contract	
Uterus	Excitation; uterine contractions	Inhibition of contractions β_2
Dilator muscle of iris of eye	Contraction (mydriasis)	
Carbodydrate metabolism	Increase blood sugar glucogenolysis in liver	Increase blood sugar glycogenolysis in muscle β_2
Fat metabolism	Mobilization of fat (shifts from deposit to liver)	

7.1.6 Biochemistry of β Adrenoceptors

The biochemical processes involved in the interaction of norepinephrine with its postsynaptic receptors to produce a response of the effector organ are extremely complex. The following discussion is intended to present only a superficial description. β-Adrenoceptor activation involves the neurotransmitter agonist norepinephrine as a cofactor with a membrane-bound enzyme, *adenylate cyclase* (also known as adenylyl cyclase), which catalyzes the conversion of adenosine triphosphate ("ATP") **7.4** (figure 7.5) into adenosine 3′, 5′-monophosphate ("cyclic AMP") **7.5**. Among the multiplicity of its complex biochemical functions, cyclic AMP participates with calcium cations and the enzyme *calmodulin* in regulating the conduction of ions across membranes, which in turn comprises one step in the biochemical process of muscle fiber contraction. Cyclic AMP is biochemically inactivated in the cell by *cyclic nucleotide phosphodiesterase* enzymes, which catalyze cleavage of the 3′ bond of the molecule to produce the biologically inactive 5′-AMP **7.6**. There are at least 40 different isoforms of the family of phosphodiesterases. Cyclic AMP is an example of a *second messenger* bioactive substance.

Figure 7.5 represents a cross section of an effector cell membrane containing the β adrenoceptor and attendant mechanistic components.

The "R_S" structure (right half of the diagram) is a transmembrane protein that contains a region on its outer surface that specifically binds and interacts with norepinephrine (represented by the triangle). The region labeled G_S, on the inner surface of the membrane, represents a trimeric *"G"-protein* (guanyl nucleotide-binding protein) or *transducer*

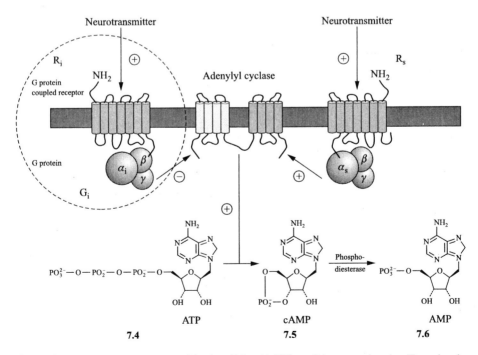

Figure 7.5 G-linked protein receptors. Stimulant (R_S) and inhibitory (R_i) receptor domains. (Reproduced with permission from reference 2. Copyright 1995)

protein complex which abuts the inner surface of the norepinephrine receptor protein R_S. The catalytic surface of adenylyl cyclase is on the inner side of the membrane, and in the resting state the conformation of the enzyme molecule is such that the catalytic surface is sequestered and is not available to substrate molecules. When norepinephrine interacts with its receptor on the outer surface of the membrane, the ligand–protein interaction induces a profound conformational change in the receptor protein. This norepinephrine-induced change in the conformation of the receptor protein induces a concomitant dissociation of one of the components of the trimeric G-protein complex. This component migrates in the plane of the membrane to the region of the sequestered adenylate cyclase catalytic surface where it interacts with and induces the enzyme molecule to undergo a conformational change such that the catalytic surface is exposed and now the enzyme can accept ATP substrate molecules and convert them into cyclic AMP which then exerts its appropriate intracellular effects. There are at least nine isoforms of adenylate cyclase.

The term *second messenger* derives from the concept that norepinephrine (the first messenger), released from the nerve terminal into the synaptic cleft, triggers a series of biochemical events leading to the formation of a chemical substance (cyclic AMP) inside the effector cell, which subsequently acts as a chemical messenger inside the cell. This remarkable sequence of events demonstrates a method of chemical communication in the body: norepinephrine triggers an extensive, profound sequence of intracellular biochemical events, without its ever entering the cell.

It is believed that there are two types of norepinephrine receptors and of the accompanying G-proteins: an R_S receptor and a G_S transducer protein, which act together to *stimulate* adenylate cyclase, as demonstrated previously, and also an R_I receptor and a G_I transducer protein (left half of the diagram) which, when activated, *inhibit* adenylyl cyclase by altering the conformation of the enzyme catalytic region such that the catalytic surface of the enzyme is sequestered and production of cyclic AMP ceases. This inhibitory phenomenon (a "negative coupling" effect) provides an explanation for the previously cited concept of inhibitory presynaptic receptors.

7.1.7 Biochemistry of α Adrenoceptors

α Adrenoceptors are also G-protein linked; stimulation of α_1 adrenoceptors leads to a rise in intracellular Ca^{2+} which initiates a cascade of events leading to the formation of inositol triphosphate and diacyl glycerol as the second messengers. α_2 Receptors (chiefly presynaptic) are negatively coupled to adenylate cyclase, and their stimulation results in depression of cyclic AMP production. The biochemical nature of *postsynaptic* α_2 receptors is uncertain.

7.1.8 Direct-, Indirect-, and Mixed-Acting Adrenergic Drugs

Drugs which mimic some or all of the effects of stimulation of the noradrenergic system can be categorized according to their mechanism(s) of action.

Direct-acting

This type of drug is an agonist. It binds to and reacts with the adrenoceptor, presumably similarly to norepinephrine, and produces a response in the effector organ.

Indirect-acting

This is a drug that does not itself interact with adrenoceptors. Rather, it stimulates the release of the body's own stored norepinephrine from the synaptic storage vesicles into the synaptic cleft, and this endogenous norepinephrine interacts with postsynaptic receptors to produce a response. Alternately or additionally, an indirect-acting drug may act by preventing the re-uptake of norepinephrine molecules from the synaptic cleft into the nerve terminal and subsequently into the synaptic storage vesicles. Thus, the adrenergic effect resulting from an indirect-acting drug is actually produced by the body's own norepinephrine.

Mixed-acting

This type of drug has some degree of direct agonist action (generally categorized as partial agonism), and it also triggers release of norepinephrine from storage vesicles and/or inhibits its re-uptake from the synaptic cleft into the nerve terminal.

Adrenergic drugs

In addition to norepinephrine, epinephrine, and isoproterenol, there is a multiplicity of adrenergic drugs available in the medical marketplace. Most are congeners of the β-phenethylamine skeleton of norepinephrine, bearing various substituents in various positions on the aromatic ring. Some exhibit selectivity for one or another of the subcategories of adrenergic receptors. Many of the clinically useful adrenergics are effective via the oral route. However, due to the presence of both MAO and COMT in the liver, orally administered norepinephrine and epinephrine do not survive first pass metabolic inactivation and they must be administered parenterally. The following discussion is limited to three representative compounds.

Phenylephrine **7.7** is a synthetic noradrenergic agonist. The receptors cannot distinguish between this molecule and norepinephrine. In the case of phenylephrine, removal of the ring position 4-hydroxyl group from the epinephrine molecule results in a drug which is a highly selective α_1-adrenoceptor agonist. Phenylephrine shows agonist effects at β adrenoceptors only at very high doses. *Ephedrine* **7.8** is a natural product, isolated from plants of the genus *Ephedra*. It is a mixed-acting agent (partly agonist, partly indirect-acting), and it produces stimulant effects nonselectively at adrenoceptor subpopulations. *Amphetamine* **7.9** is entirely indirect-acting. It triggers the release of norepinephrine into the synaptic cleft, and it also inhibits the uptake-1 process, so that the norepinephrine released remains in the synaptic cleft, continually to bombard the postsynaptic receptors. The summation of these amphetamine actions produces a profound adrenergic stimulant response.

7.7   **7.8**   **7.9**

These three drugs illustrate some other pharmacological principles. Phenylephrine, with its two hydroxyl groups, is a relatively hydrophilic molecule. In comparison, ephedrine, with its single alcoholic hydroxyl group, is considerably less hydrophilic, and its partition coefficient indicates a significant degree of lipophilic character. Amphetamine, which bears no hydroxyl groups and whose sole hydrophilic moiety is the amino group, has a markedly lipophilic partition coefficient. Because a high degree of lipophilicity is a common requirement for penetration of the blood–brain barrier by passive diffusion, it is possible to rationalize why phenylephrine's pharmacological effects are almost exclusively peripheral. Like epinephrine and norepinephrine, it cannot penetrate the blood-brain barrier. Ephedrine is sufficiently lipophilic that it penetrates the blood–brain barrier by passive diffusion and, in addition to its peripheral effects, it stimulates the central nervous system. Instillation of ephedrine sulfate nose drops at bedtime for the relief of nasal congestion frequently results in insomnia. The *Ephedra* plant contains small amounts of another alkaloid, *pseudoephedrine,* a diastereomer of ephedrine. Pseudoephedrine exhibits most of the pharmacological properties of ephedrine, and it is an ingredient of many over-the-counter cold products, used for its nasal decongestant effect. Amphetamine penetrates the blood–brain barrier with great ease, and its principal pharmacologic actions are those which originate in the brain, involving multiple adrenoceptor subtypes: central nervous system stimulation, decreased sense of fatigue, elevation of mood, wakefulness, and alertness.

7.1.9 Therapeutic Uses of Adrenergic Receptor Stimulants

The vasoconstrictor actions of α_1-adrenoceptor stimulants are exploited in treating shock (by intravenous administration) and relief of nasal congestion (by topical application). Epinephrine is the drug of choice in life-threatening allergic reactions (such as bee stings or drug allergy). It relieves edema of the lips, tongue, and glottis which may interfere with breathing by shutting off the airway. Epinephrine is not an antihistaminic agent; it does not directly block histamine H_1 receptors which are involved in the allergic response. However, epinephrine stimulates certain β adrenoceptors which initiate suppression of histamine release. Adrenergic effects on the heart are reflected by use of adrenoceptor stimulants in cardiac arrest and in relieving cardiac arrhythmias (irregular or abnormal heart rates or beats). β_2 Adrenoceptor stimulants are useful in dilating the bronchi of the lungs in bronchial asthma, which is characterized by constriction of the bronchi. Presynaptic α_2 receptor agonists are used in treatment of hypertension (see chapter 14).

Aqueous solutions of ephedrine salts have been widely used as nasal decongestants, and ephedrine has been used systemically in treatment of narcolepsy. Its use as a nasal decongestant has largely been replaced by pseudoephedrine which is reputed to be less potent than ephedrine in producing hypertension, tachycardia, and CNS stimulation. Ephedrine is no longer recommended for treatment of narcolepsy. Ephedrine-containing products that are available over the counter have been used as appetite-suppressing agents, but this use presents the same problems that will be described for amphetamines. Commercial distribution and sale of products containing powdered *Ephedra* (*ma huang*) or extracts thereof as an all-natural "nutritional supplement" are now being subjected to some degree of regulation and control by federal and some state authorities. There can be considerable variability in the alkaloidal content of these *Ephedra* products.

Anecdotal accounts of fatalities following inappropriate use of *Ephedra* are not uncommon in the public press.

Probably there are only two *bona fide* therapeutic uses for amphetamine drugs. First in the treatment of narcolepsy (an epilepsy-like condition in which the individual suddenly and uncontrollably falls asleep). The region of the brain involved in narcolepsy is believed to be the hypothalamus, where certain neurotransmitter cells that normally produce a peptide neurotransmitter which regulates the sleep-wake cycle are acting abnormally. However, the details of narcolepsy remain unclear, and amphetamine's mechanism of action here is also not well understood. Second, amphetamine is also used to treat the symptoms of hyperkinesia (attention deficit disorder, "ADD") in children, although this use is somewhat controversial.

Amphetamine (racemic) and dextroamphetamine ("dexedrine") have been used in treatment of obesity. Amphetamine has anorexigenic (appetite depressing) effects, but most authorities now discourage its use for this purpose. Patients rapidly become tolerant to the appetite-depressing effects of amphetamines and hence there is a tendency for the individual to increase his/her dosage. This is a dangerous practice; individuals do not become tolerant to the actions of amphetamines to raise the blood pressure, and a dangerous hypertension may be induced. The illicit use of amphetamine products by long-distance vehicle drivers to stay awake is to be deplored because of the accompanying distortion of distance estimation and judgment produced by the drug.

An amphetamine congener, *methylphenidate (ritalin)* **7.10** is probably more widely used than amphetamine to treat hyperkinesia in children. Methylphenidate is also useful in the relief of the symptoms of narcolepsy. The drug grossly mimics many of the CNS-stimulant effects of amphetamine, but some evidence suggests that the details of the mechanisms of action of amphetamine and of methylphenidate differ. It seems enigmatic that CNS-stimulant drugs like amphetamine and methylphenidate are useful in calming hyperkinetic children. A widely accepted, definitive mechanism for this property of these drugs has not appeared. However, recent research reveals that methylphenidate blocks the uptake-1 mechanism for dopamine in the CNS, resulting in elevated levels of this neurotransmitter in the synaptic clefts. This finding complements an earlier proposal that the cause of ADD/hyperkinesia seems to be linked to an abnormal increase in the CNS in the number of dopamine uptake-1 sites. In some regions of the United States, methylphenidate is employed by adults as an amphetamine-like recreational drug.

Atomoxetine (strattera) **7.10a** is a relatively new drug used to control attention deficit disorder. In contrast to the psychostimulant drugs used in ADD (amphetamine, ritalin) atomoxetine is believed to act by selectively blocking re-uptake of norepinephrine. It has little effect on dopamine or serotonin re-uptake. Atomoxetine is not a "schedule" drug and it seems to have a low tendency for abuse as a recreational drug. It is said to have significant clinical advantages over the older agents.

Desoxyephedrine **7.11**, better known as *methamphetamine*, is a very lipophilic molecule, and it was formerly employed as a *bona fide* therapeutic agent. It is now encountered chiefly as a recreational drug (street name: *speed*).

Methamphetamine produces CNS stimulation, similar to amphetamine: euphoria, heightened libido. Mechanistically, many of the effects of methamphetamine are due to its ability to inhibit re-uptake of dopamine and serotonin in the brain, in addition to its effect on norepinephrine re-uptake. When chronically smoked or taken intravenously, it produces an abstinence/abuse syndrome similar to that of cocaine. Methamphetamine is a widely and easily available drug of abuse, due in part to the ease with which it can be prepared chemically (reductive de-oxygenation of ephedrine or pseudoephedrine, both available over-the-counter in a variety of proprietary cold remedies). Intravenous overdosage with methamphetamine produces profound vasoconstriction, rise in blood pressure, and all-too-frequently, fatal cerebral hemorrhage.

7.10   7.10a   7.11   7.12

"Ecstasy", 3,4-methylenedioxymethamphetamine (MDMA) **7.12**, is another popular recreational drug. It is thought by users to enhance insight and self-knowledge. High doses of ecstasy induce visual hallucinations and panic attacks. It is believed that a major component of the CNS effects of ecstasy is its inhibition of re-uptake of serotonin. Recent and current research strongly suggests that even a few occasions of usage of ecstasy can cause severe damage to brain cells, potentially leading to memory loss or to other psycho-logical problems such as changes in mood, impulse control, and sleep cycles. Some authorities hold the opinion that some of the dire effects described as resulting from use of ecstasy may be exaggerated. Whether ecstasy-induced brain damage is permanent is not known.

7.1.10 Imidazoline-Derived Adrenergic Drugs

Most β-adrenoceptor agonists are structural variations of the β -phenethylamine system. However, some α adrenoceptors can accomodate other molecular systems, such as some imidazoline derivatives (illustrated by *clonidine* **7.13**, a central α_2 agonist with some peripheral α_1 agonist activity). Some years ago it was proposed that some of the pharmacological effects, including hypotensive responses, of synthetic imidazoline derivatives are mediated in part by a separate, unique group of autonomic "imidazoline receptors." Currently, the physiological/pharmacological status of these "receptors" seems uncertain. The contemporary literature reveals that there are imidazoline binding sites in the human body that recognize synthetic ligands. However, the signaling path-ways involved and the relative importance of these imidazoline-binding proteins are unclear. Whatever the physiological role of imidazoline binding sites may be, they seem to be chemically heterogeneous. Current thinking tends to the opinion that the hypoten-sive effects of the synthetic imidazoline-derived drugs such as clonidine can be explained on the basis of their actions at central α_2 adrenoceptors.

7.1.11 Adrenergic Blocking Drugs

Adrenergic blocking agents counteract the effects of norepinephrine, epinephrine, isopro-terenol, and other direct- and indirect-acting stimulants of noradrenergic receptors. Some blockers are selective for α adrenoceptors and some are selective for β adrenoceptors, but within these categories there may be little or no selectivity. Typical of agents that block both α_1 and α_2 receptors are *phenoxybenzamine* **7.14**, a 2-haloalkylamine that undergoes in vivo intramolecular cyclization to form an aziridinium ring system which subsequently opens to link covalently to α-adrenoceptors through some nucleophilic moiety at the receptor, and thereby produces long-lasting blockade; and *phentolamine* **7.15** and *tolazoline* **7.16** which are reversible, competitive, relatively short-acting inhibitors. These drugs produce a variety of effects on the cardiovascular system, but their side effects are so numerous and pronounced that their therapeutic value is minimal.

| 7.13 | 7.14 | 7.15 |

Prazosin **7.17** is typical of a modest number of drugs that are selective for blockade of α_1 adrenoceptors. These cause vasodilation and a resultant fall in blood pressure and there is less tachycardia (rapid heart rate) than with nonselective blockers. A detriment to the therapeutic value of drugs like prazosin is their short duration of action. Overall, α_1-adrenoceptor blocking drugs have been disappointing.

| 7.16 | 7.17 |

Propranolol **7.18** is a β-adrenoceptor blocker, with equal blocking effects at β_1 and β_2 receptors. The β-adrenoceptor blocking effect is found only in the (*S*)-enantiomer. Propranolol is quite lipophilic and it is efficiently absorbed across the intestinal wall, but it undergoes extensive first-pass metabolic inactivation. Only approximately 25% of an oral dose of the drug reaches the systemic circulation. Hence, an oral dose must be considerably larger than an injected dose. Propranolol and similar agents are useful in controlling hypertension. The mechanism(s) by which blockade of β-adrenoceptors produces a fall in blood pressure is (are) not well understood, but apparently β_1 receptors are involved. Further discussion of this therapeutic role of β-adrenoceptor blockers is found in chapter 14.

Stimulation of β_2 adrenoceptors causes dilation of the bronchi of the lungs (see table 7.1). Blockade of these receptors by a drug such as propranolol produces bronchoconstriction, which is a dangerous result in the individual suffering from bronchial asthma. This ability to precipitate an asthmatic attack precludes the use of propranolol and other nonselective β blockers in individuals suffering from bronchospastic-related conditions. Some newer agents, such as *metoprolol* **7.19**, which are selective for

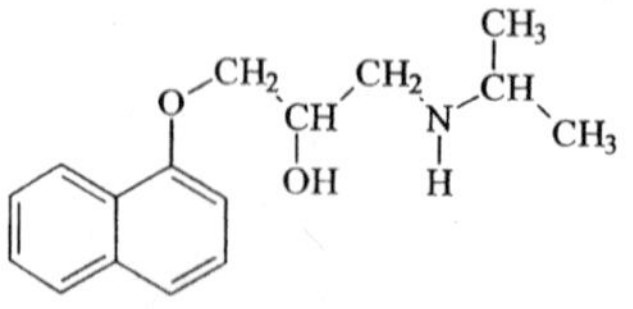

7.18

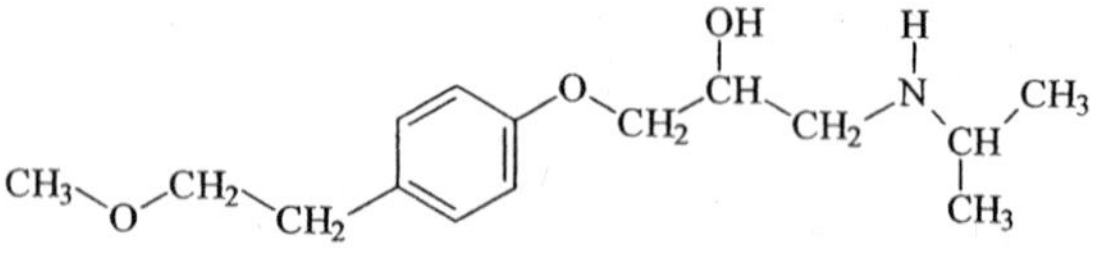

7.19

β_1 adrenoceptors, and have relatively little effect on β_2 receptors, present less danger in asthmatic patients. *Labetalol* **7.20** is a competitive antagonist at α_1 and at β_1 and β_2 receptors. It is also a partial agonist at β_2 adrenoceptors and it inhibits the uptake-1 active transport of norepinephrine into the nerve terminal. The hypotensive effect seen with labetalol results from its combined effects on α_1, β_1, and β_2 adrenoceptors. The labetalol molecule has two chiral centers, and the commercially available product is a mixture of the four possible optical isomers. These have qualitatively and quantitatively different pharmacological properties, all of which contribute to the observable effects of the drug. As with propranolol, this markedly lipophilic drug is efficiently absorbed across the intestinal wall, and it undergoes extensive first-pass metabolic inactivation ($\approx$20–40% bioavailability), necessitating large oral doses as compared with injected dose levels.

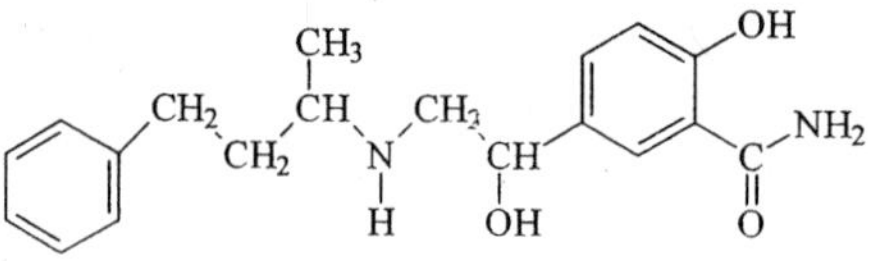

7.20

7.1.12 Sympatholytic Agents: Depletion of Norepinephrine in the Nerve Terminal

Representative of this category is the alkaloid *reserpine* **7.21**, a pentacyclic indole derivative that bears little structural similarity to any of the previously discussed adrenergic stimulants or blockers. This drug enters the synaptic storage vesicles at peripheral and central postganglionic noradrenergic termini, and it displaces the stored norepinephrine. The reserpine molecules remain tightly (and irreversibly) bound within the vesicles, and they disrupt the uptake-2 active transport mechanism. The displaced norepinephrine molecules tend to remain in the cytoplasm of the nerve terminal where

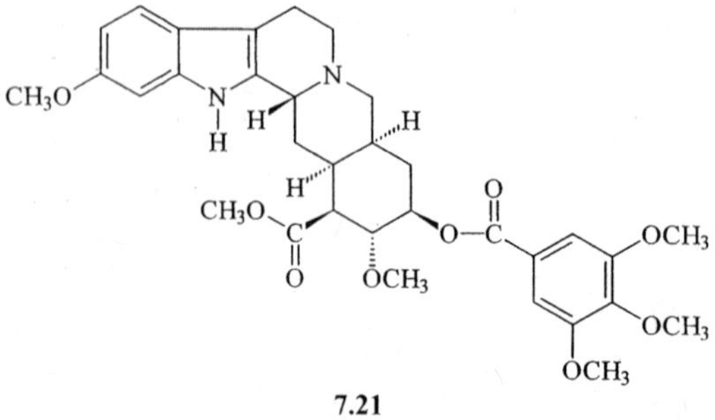

7.21

they are metabolically inactivated by the monoamine oxidase there. Thus, over a period of several days, the stores of norepinephrine are depleted and the body's ability to translate a noradrenergic nerve impulse into a response of an effector organ is diminished. This central and peripheral adrenolytic action is the pharmacological basis for reserpine's blood pressure lowering effect. Reserpine exerts a similar effect on the synaptic storage vesicles in the nerve terminals of dopaminergic and serotonergic (5-hydroxytryptamine) pathways, especially those in the central nervous system. Reserpine's inhibition of noradrenergic and serotonergic neurotransmission in the brain may produce symptoms of clinical mental depression, which is a serious side effect of the drug.

7.2   The Dopaminergic System

7.2.1 Physiology

For many years, dopamine was considered to be significant chiefly as an intermediate in the biosynthesis of norepinephrine and epinephrine (see figure 7.2). The pharmacological effects observed following administration of dopamine to laboratory animals were believed to reflect its bioconversion to epinephrine and/or norepinephrine. Now it is firmly established that dopamine itself is a neurotransmitter, and there is a dopaminergic nervous system. The anatomy/physiology of dopaminergic postganglionic nerve terminals is quite similar to that of noradrenergic terminals. As at norepinephrine-mediated sites, the effect of a dopaminergic nerve impulse at the postsynaptic receptor is terminated by a series of re-uptake processes, and scavenger enzymes (catechol-O-methyltransferase and monoamine oxidase) inactivate those relatively few dopamine molecules that escape re-uptake. The urine contains small amounts of the resulting metabolic end products, dihydroxyphenylacetic acid **7.22** and homovanillic acid **7.23**, formed in reaction pathways analogous to those shown for norepinephrine (figure 7.3).

| | 7.22 | 7.23 |

Dopamine functions in many parts of the body, but it is especially important in the brain. Nerve activity in the *nigrostriatal pathway*, that portion of the brain where the tone of the voluntary muscles is regulated, is mediated by dopamine. Neurons in the hypothalamic region of the brain secrete dopamine which is transported via a portal venous system to the anterior pituitary gland where interaction with dopamine receptors inhibits release of the hormone prolactin. The *chemoreceptor trigger zone* (CTZ) in the medulla oblongata, the region where the emetic response originates, is dopaminergically controlled. The CTZ is an example of a brain region which is functionally outside the blood-brain barrier. Dopamine is also involved in higher centers of the brain (limbic system) where mood and emotion are controlled, and an occasionally popular but unproven hypothesis stated that some types of schizophrenia are caused by a defect in the normal physiology of dopaminergic pathways in these higher centers.

Nicotine and heroin normally increase dopamine production in the brain from 150% to 200%. Methamphetamine (which the literature describes as "incredibly addictive") increases dopamine levels by as much as 2600%. Low doses of ethanol produce a moderate increase in dopamine production, but higher doses of ethanol seem to have no great effect on dopamine production.

Relatively few peripheral dopaminergic receptor sites have been identified. High blood concentrations of exogenous dopamine activate α_1 adrenoceptors in the blood vessel walls, causing vasoconstriction and elevation of blood pressure. Dopamine is used in emergency treatment of life-threatening states of shock. The short half-life of dopamine in the plasma makes it possible to titrate the shock patient with intravenous dopamine to attain and maintain the proper blood pressure range.

Dopamine receptors are heterogeneous. Subcategories of the two principal types, described later, have been cloned and evidence points to the existence of D_3, D_4, and D_5 receptors in the central nervous system and peripherally. The population of dopamine receptors has been subdivided into two major subfamilies: D_1-like (D_1, D_5) and D_2-like (D_2, D_3, D_4). The present discussion will address only the two best understood types. D_1 receptors are G-protein linked, and they activate adenylate cyclase to elevate intracellular levels of cyclic AMP. D_2 receptors are negatively coupled to adenylate cyclase, such that their interaction with dopamine inhibits intracellular production of the second messenger, cyclic AMP. Most D_1 receptors are postsynaptic, but D_2 receptors are found in a variety of locations, both postsynaptically and presynaptically.

7.2.2 The Parkinsonian Syndrome

Etiology

In Parkinson's disease, there is a deficiency of dopamine in the neurons of the nigrostriatal pathway in the brain, and in this condition the skeletal muscles of the arms, legs, neck, and face become rigid and spastic. Severe tremors of the limbs, arms, and hands develop. Walking becomes difficult, and the face assumes a characteristic blank expression because the spastic face muscles cannot relax. There may be undesirable personality changes.

The ultimate cause(s) of the Parkinsonian syndrome is/are, at best, poorly understood. The etiology of the disease may involve the biochemistry of dopamine precursors at some stage prior to the conversion of dihydroxyphenylalanine into dopamine. There are indications that exposure to environmental chemicals may be a significant factor in the onset of Parkinsonian symptoms, and several substances have been suggested as being involved. It has been reported that chronic long-term intravenous administration of the insecticide rotenone **7.24** to rats caused development of Parkinsonian symptoms and other biochemical/physiological abnormalities that accompany Parkinson's disease. These abnormalities involve the oxidative processes that occur in the mitochondria of cells. In these studies, the rats' brains developed protein aggregates ("Lewy bodies") which are also characteristic of human Parkinsonian brains.

The structure of rotenone suggests a high degree of lipophilic character, and it might be speculated that the molecule would have some ability to penetrate the blood–brain barrier. It should be emphasized that there seems to be, as yet, no definitive proof that rotenone causes Parkinson's disease in humans.

7.24

Drug therapy of Parkinson's disease

Although hypofunction of certain central dopaminergic pathways seems to be the gross cause of the symptoms of the Parkinsonian syndrome, attempts to treat Parkinsonian patients with dopamine have been completely unsuccessful. Orally administered dopamine is efficiently inactivated by first pass hepatic metabolism (by catechol-O-methyltransferase and/or monoamine oxidase) and, additionally, peripherally administered dopamine cannot penetrate the blood–brain barrier. However, it is found that the immediate biosynthetic precursor to dopamine, dihydroxyphenylalanine ("DOPA", levodopa) (figure 7.2) crosses the blood–brain barrier by an active transport system that does not accomodate dopamine itself. Once inside the brain, DOPA is enzymatically decarboxylated to generate dopamine. Most Parkinsonian patients seem to retain the ability to decarboxylate DOPA in the CNS.

In a high percentage of Parkinsonian patients, replacement therapy with DOPA produces dramatic relief of the symptoms. However, large oral doses (up to 8 g daily) of DOPA are required, due in large measure to extensive enzyme-mediated decarboxylation of DOPA in the periphery of the body. L-aromatic amino acid decarboxylase is found in a number of organs outside the CNS. Thus, a very large proportion of the dose of DOPA is therapeutically wasted. It has been estimated that less than 1% of an oral dose of DOPA enters the central nervous system. The enzyme-mediated decarboxylation of DOPA by L-aromatic amino acid decarboxylase requires the vitamin pyridoxine as a cofactor. Ingestion of doses of pyridoxine which are only slightly in excess of the recommended daily dietary allowance (such as are found in many over-the-counter multiple vitamin products) enhances the peripheral decarboxylation of DOPA and even less DOPA is available for transport into the central nervous system for conversion into dopamine there. Thus, if excess amounts of pyridoxine are ingested concurrently with DOPA, the beneficial effect of the DOPA therapy may be completely reversed.

Side effects of DOPA in Parkinsonian patients

These include nausea and vomiting (due to stimulation of the emetic mechanism in the medulla oblongata) which occurs soon after the dosage regimen is begun, but generally wears off after some days; involuntary head bobbing and shrugging of the shoulders and chewing motions of the jaws (so-called *extrapyramidal effects* caused by stimulation of other dopamine-mediated pathways in the brain that regulate other sets of peripheral voluntary muscles), which usually are manifested after the individual has been on DOPA therapy for some time and which persist; and changes in personality and mood (because of effects on dopaminergic pathways in higher brain centers). Past lurid and exaggerated popular press reports of DOPA-mediated, greatly increased libido in older Parkinsonian men reflect a rather rare side effect, which is believed to be caused by the ability of high levels of dopamine to stimulate certain subpopulations of adrenergic and serotonergic receptors in the central nervous system.

Clinical Deficiencies of DOPA therapy

DOPA is not a cure. It only ameliorates many of the symptoms of the disease, and relief from the symptoms lasts only as long as the individual maintains an appropriate dosage of the drug. Most frustrating is the fact that some 30% of Parkinsonian patients do not respond to DOPA therapy. In some instances, this is due to faulty intestinal absorption. In advanced cases of Parkinson's disease, the nerve tissue in the nigrostriatal pathway degenerates and becomes necrotic, such that the dopamine receptors themselves are destroyed. Postmortem examination of Parkinsonian brains reveals this necrosis. As the disease progresses, there are fewer and fewer living, functional dopamine receptors with which a drug can react. As the numbers of viable dopamine receptors decrease, the remaining viable receptors become hyperactive, a compensatory attempt to maintain the normal physiological activity of the nigrostriatal pathway. However, eventually and inevitably, as more and more dopamine receptors die, there will be no further beneficial therapeutic effect from DOPA or any similar drug. No known drug or other therapeutic strategy reverses or arrests the physical degeneration of the nigrostriatal pathways in the Parkinsonian patient. The cause of neuron death in the Parkinsonian syndrome is not well understood. However, the following explanation has been advanced. Dopamine metabolism by monoamine oxidase generates OH free radicals that damage cellular proteins, leading to cellular death (*apoptosis*). Cells can resist the effects of these free radicals with the help of a family of protein protective substances called 14-3-3. It has been suggested that a protein component of Lewy bodies, α-synuclein, binds to 14-3-3, thus inhibiting its ability to protect the dopaminergic neurons from death by OH free radicals. However, the complete, overall role of α-synuclein in Parkinson's disease is not known.

Carbidopa

The peripheral decarboxylation of DOPA has been offered previously as an explanation for the large doses of the drug required. This peripheral production of dopamine is likewise responsible for some of the unwanted side effects noted in DOPA therapy. *Carbidopa* **7.25**, an α-hydrazino acid, inhibits decarboxylation of DOPA peripherally but, like dopamine itself, carbidopa does not penetrate the blood-brain barrier. It cannot be accomodated on the active transport carrier for DOPA and it is insufficiently lipophilic to permit passive diffusion. In combination with carbidopa, much smaller doses of DOPA can be employed to attain the desired level of therapeutic effect. For the patient using combination therapy of DOPA and carbidopa, the antagonistic effect of pyridoxine on therapeutic efficacy is avoided. Nausea and vomiting from stimulation of the chemoreceptor trigger zone are largely prevented because this region of the medulla oblongata is functionally outside the blood–brain barrier. However, the extrapyramidal side effects and other central effects on mood and personality are not diminished. Extrapyramidal side effects may become so pronounced and so debilitating to the elderly Parkinsonian patient that DOPA therapy must be terminated. This will result in cessation of the extrapyramidal effects, but the Parkinsonian symptoms will return.

Ropirinole **7.26** and *pramipexole* **7.27** are newer dopaminergic agonists employed in treatment of Parkinson's disease. These drugs have selective activity at D_2-class sites (D_2 and D_3). They are effective orally, they have a longer duration of action than levodopa, and they are often better tolerated. They are being used increasingly as the initial treatment for Parkinson's disease rather than as adjunct therapy to levodopa.

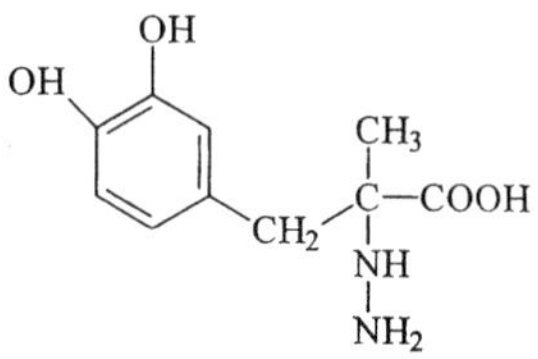

7.25 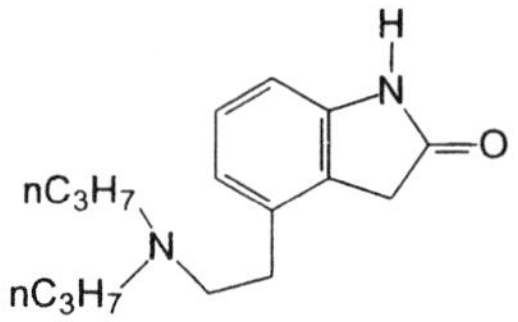   **7.26** 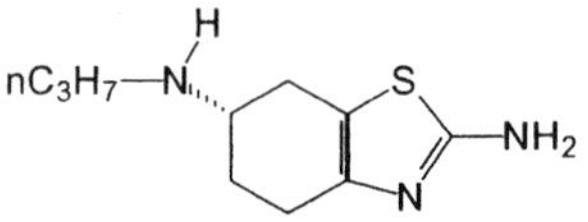   **7.27**

Selegiline **7.28** is a selective inhibitor of MAO-B, the form of monoamine oxidase that is responsible for the oxidative metabolic inactivation of dopamine in the nigrostriatal pathway. Selegiline has been used for symptomatic treatment of the Parkinsonian syndrome, although its benefit is said to be "fairly modest". Combination therapy of levodopa, carbidopa, and selegiline was found to be no more effective than a levodopa–carbidopa combination.

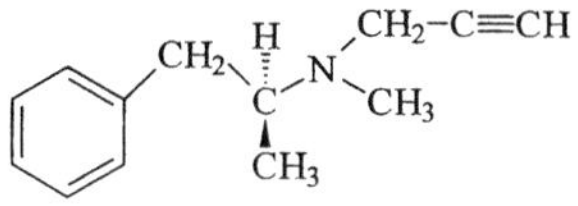

7.28

An alternate strategy for therapy of Parkinson's disease is the use of a catechol-*O*-methyltransferase (COMT) inhibitor. Two such drugs are entacapone **7.29** and tolcapone **7.30**. The mechanism of action of entacapone seems to involve mainly inhibition of peripheral COMT. The drug's beneficial effects are attributed to its protection of levodopa from peripheral inactivation by COMT. Thus, more levodopa survives for active transport into the brain. Severe hepatotoxicity in some patients has been associated with tolcapone, but this hazard has apparently not been reported for entacapone. These drugs are intended for concomitant administration with levodopa and carbidopa, to improve overall bioavailability of levodopa.

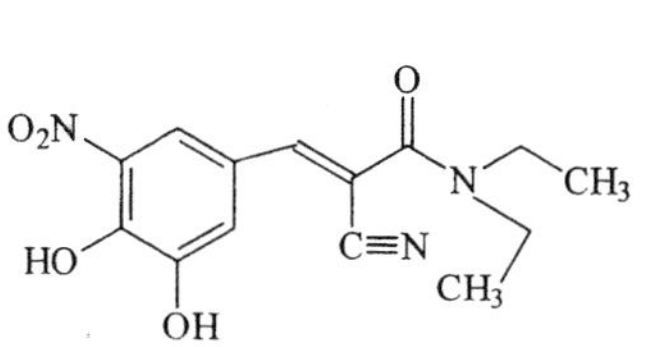

7.29 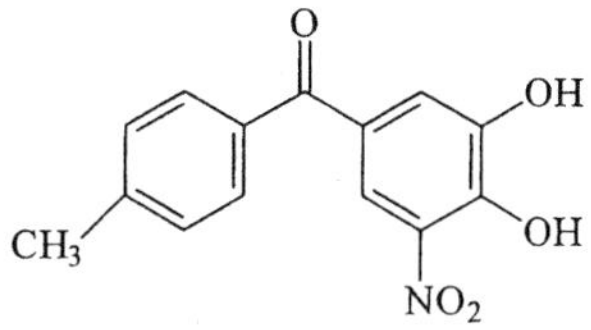   **7.30**

Bibliography

1. Grammer, S. *Modern Drug Discovery*. American Chemical Society: Washington, D.C., 2001; p. 39.
2. Broekamp, C. L. E., Leysen, D., Peeters, B. W. M. M. and Pinder, R. M. *J. Med. Chem.* **1995**, 38, 4624.

Recommended Reading

1. Watson, S. and Arkinstall, S. *The G-Protein Linked Facts Book*. Academic Press: New York, 1994.
2. Hoffmann, B. B. Catecholamines. Sympathomimetic Drugs and Adrenergic Receptor Antagonists. In *Goodman and Gilman's The Pharmacological Basis of Therapeutics*, 10th ed. Hardman, J. G. and Limbird, L. E., Eds. McGraw-Hill: New York, 2001; pp. 215–268.
3. Strolin Benedetti, M. and Dostert, P. Monoamine Oxidase. In *Advances in Drug Research*. Testa, B., Ed. Academic Press: London, 1992; Vol. 23, pp. 65–125.
4. Watling, K. J., Ed. Adrenoceptors. In *The Sigma–RBI Handbook of Receptor Classification and Signal Transduction*, 5th ed. Sigma–RBI: Natick, Mass., 2006; pp. 88–93.
5. Imidazoline Binding Sites. In Recommended Readings ref. 4, pp. 122–123.
6. Strosberg, A. D. and Pieri-Rouxel, F. Function and Regulation of the β_3-Adrenoceptor. *Trends Pharmacol. Sci.* **1996**, 17, 373–381.
7. Civelli, O., Bunzow, J. R. and Grandy, D. K. Molecular Diversity of the Dopamine Receptors. In *Annual Review of Pharmacology and Toxicology*. Cho, A. K., Blaschke, T. F., Loh, H. H. and Way, J. L., Eds. Annual Reviews, Inc.: Palo Alto, Calif., 1993; Vol. 33, pp. 281–307.
8. Dopamine Receptors. In Recommended Readings ref. 4, pp. 98–100.
9. Nichols, D. E. Physiology and Pharmacology of CNS Stimulants. In *Burger's Medicinal Chemistry and Drug Discovery*, 6th ed. Abraham, D. J., Ed. Wiley-Interscience: Hoboken, N. J., 2003; Vol. 6, pp. 179–184.
10. Neumeyer, J. L., Baldessarini, R. J. and Booth, R. G. Therapeutic and Diagnostic Agents for Parkinson's Disease. In Recommended Readings ref. 9, Vol. 6, pp. 711–741.
11. Small, K. M., McGraw, D. W. and Liggett, S. B. Pharmacology and Physiology of Human Adrenergic Receptor Polymorphisms. In *Annual Review of Pharmacology and Toxicology*. Cho, A. K., Blaschko, T. F., Insel, P. A. and Loh, H. H., Eds. Annual Reviews, Inc.: Palo Alto, Calif., 2003; Vol. 23, pp. 381–411.

8

The Cholinergic System

Acetylcholine-Inactivating Enzymes

On the basis of the multiplicity of nervous system phenomena in which acetylcholine plays a vital physiological role, it might be anticipated that an injection of acetylcholine would produce dramatic pharmacological responses. However, this is not the case. Exogenous acetylcholine elicits relatively few responses, and these are of short duration. These results are due in part to the strongly hydrophilic character of the acetylcholine molecule, which inhibits its transport to many sites of action in the nervous system, and to a significant extent to the rapid hydrolysis and biological inactivation of acetylcholine by cholinesterase (also known as pseudocholinesterase or butyrylcholinesterase) in the blood (equation 8.1).

$$CH_3-\overset{\overset{O}{\|}}{C}-O-CH_2-CH_2-\overset{\oplus}{N}Me_3 \longrightarrow CH_3-COO^{\ominus} + HO-CH_2-CH_2-\overset{\oplus}{N}Me_3 \qquad (8.1)$$

In addition to the blood esterase, there is a highly substrate-specific enzyme, acetylcholinesterase, in the nervous tissue in the immediate vicinity of the acetylcholine receptor sites. This enzyme differs chemically from blood cholinesterase which is not highly substrate-specific. Acetylcholinesterase is also found in certain other sites in the body, such as red blood cells, but its physiological role here is not established. The body uses this enzyme to terminate the effects of a cholinergic nerve impulse. The products of its action on acetylcholine, acetate anion and choline, are physiologically inert in nerve impulse transmission. Note that there is a fundamental physiological difference between the noradrenergic and dopaminergic systems and the cholinergic system. Effects of noradrenergic and dopaminergic impulses are terminated by active transport mechanisms (uptake-1 and uptake-2), and only a relatively minor proportion of the norepinephrine or dopamine molecules liberated into the synaptic cleft are enzymatically inactivated. There is no mechanism for active re-uptake (uptake-1) of acetylcholine from the synaptic cleft into the nerve terminal. However, there is an active transport mechanism for re-uptake of choline into the nerve terminal.

8.2   Anatomy/Physiology of the Cholinergic Nerve Terminal

Figure 8.1 illustrates and summarizes the nerve terminal's involvement in biosynthesis, storage, release, and the metabolic fate of acetylcholine. A nerve impulse moving down the postganglionic fiber causes the release of acetylcholine from the synaptic storage vesicles into the synaptic cleft. This release involves the fusion of the storage vesicle membrane with the nerve terminal membrane which then ruptures, releasing acetylcholine molecules into the synaptic cleft. The number of acetylcholine molecules liberated from a single storage vesicle is estimated to between 2000 and 200000, depending upon the specific nerve involved. The freed acetylcholine molecules migrate across the cleft to the postsynaptic receptor area, where they react reversibly with the receptor protein to initiate a response by the effector. Subsequently, the acetylcholine–receptor complex dissociates, and freed acetylcholine molecules are liberated back into the synaptic cleft where they are acted upon by acetylcholinesterase, which hydrolytically cleaves the ester linkage. The acetate ions migrate away, but the choline molecules are taken up into the nerve terminal by one of two different active transport systems, a *high-affinity sodium dependent system* (which requires sodium ions as a cofactor) and

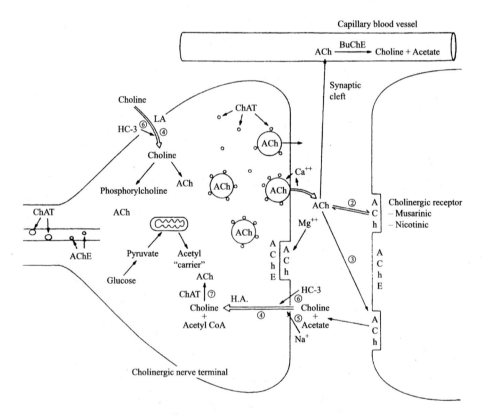

Figure 8.1 Functional anatomy of a cholinergic nerve terminal, synapse, and postsynaptic region. (Reproduced with permission from reference 1. Copyright 1982 Lippincott)

a *low-affinity system* which does not require sodium. The high-affinity system promotes the reconversion of choline into acetylcholine. Inside the nerve terminal an enzyme, *choline acetyltransferase*, catalyzes a reaction between choline and acetylcoenzyme A (the source of the acetyl group) to generate acetylcholine which is taken back up into the synaptic storage vesicle by an active transport mechanism, analogous to the uptake-2 process described for the adrenergic system. As shown in Figure 8.2 (a diagrammatic representation of an acetylcholine synaptic storage vesicle), adenosine triphosphate is hydrolytically cleaved to adenosine diphosphate, producing hydrogen ions. A protein complex traversing the membrane of the synaptic storage vesicle pumps these protons into the storage vesicle, and a second transport protein complex exchanges these protons for acetylcholine. Acetylcholine migrates into the storage vesicle as protons migrate out. This acetylcholine remains sequestered inside the synaptic storage vesicle until another nerve impulse triggers its release.

A relatively minor amount of the choline generated in the synaptic cleft is taken into the nerve terminal by the low-affinity uptake process. A portion of this choline is enzymatically esterified with orthophosphoric acid to form phosphorylcholine which functions in a variety of biochemical processes not directly related to nerve impulse transmission. A second portion of low-affinity uptake choline is converted into acetylcholine, but it appears that this acetylcholine is not taken up into the synaptic storage vesicles; it has a separate physiological function. Small amounts of acetylcholine are continuously and spontaneously released (without the involvement of a nerve impulse) from the nerve terminal out into the synaptic cleft. This small amount of acetylcholine constantly interacts with postsynaptic receptors on muscle cells to maintain a proper smooth muscle tone and prevent them from being completely relaxed and flaccid.

Acetylcholine cannot penetrate the blood–brain barrier. It must be synthesized in the central nervous system. Choline itself is not synthesized in the central nervous system.

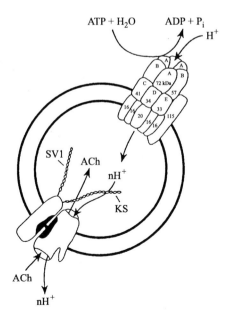

Figure 8.2 Uptake of acetylcholine into a synaptic storage vesicle in a nerve terminal. (Reproduced with permission from reference 2. Copyright 1993)

The body's choline supply comes in part from the diet and in part from biosynthesis in the periphery of the body. It is transported in the blood, and it penetrates the blood-brain barrier via a structure-specific process termed *facilitated diffusion*. This term describes a carrier-mediated process to which there is no input of energy (see active transport). Thus, the movement of choline across the blood–brain barrier cannot occur against a concentration gradient.

8.3   Cholinergic Receptors and Receptor Subtypes

8.3.1 Pharmacological Classification

The two principal types of acetylcholine receptors are distinguished on the basis of the response of the cholinergic nervous system to two alkaloids: muscarine **8.1**, a component of several species of mushrooms and nicotine **8.2**, the principal alkaloidal constituent of the tobacco plant. As with classification of noradrenergic receptors, this is an artificial, nonphysiological method of classsification, in that neither muscarine nor nicotine is an endogenous component of the body. Nevertheless, classification of acetylcholine receptors as *nicotinic* or *muscarinic* is helpful in understanding the physiology and pharmacology of the cholinergic nervous system. Nicotine exerts effects on acetylcholine receptors at the synapses in peripheral sympathetic and parasympathetic ganglia (N_2 receptors) and at the myoneural junctions between voluntary nerves and voluntary (striated) muscle fibers (N_1 receptors). There are also nicotinic receptors in the central nervous system. Peripherally, muscarine stimulates cholinergic receptors in smooth muscle tissue and glands.

8.1   **8.2**

8.3.2 Chemical Nature of Cholinergic Receptors

Nicotinic receptors are pentameric protein complex ion channels quite similar to the "generic" sodium ion channel described in chapter 1 (figure 1.3). The individual protein components are designated as α, β, γ, and δ, as illustrated by a typical nicotinic receptor (figure 8.3). Both N_1 and N_2 receptors are highly heterogeneous, comprising at least seven or eight different subtypes. These vary considerably in the nature of the protein subunit combinations. Stimulation of nicotinic receptors causes a rapid increase in cellular permeability to Na^+ and K^+, depolarization of the membrane, and excitation. Two molecules of acetylcholine are required to react simultaneously with the ion channel protein complex to open the channel; it is believed that these acetylcholine molecules bind specifically to each of the two α subunits.

Muscarinic receptors are located on the outer surface of the cell membrane and they are of the G-protein-coupled type, described in Chapter 7 (figure 7.5) for β adrenoceptors. However, some muscarinic receptors involve a different second messenger. On the basis of classical experimental pharmacological data and cloning studies, five subtypes of

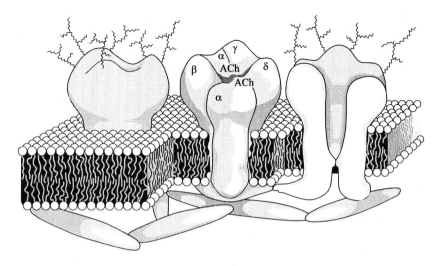

Figure 8.3 A nicotinic receptor (ion channel). (Reproduced with permission from reference 3. Copyright 1983)

muscarinic receptors have been identified: M_1–M_5. Some established locations of these receptor subtypes are as follows:

- M_1: in autonomic ganglia and various regions of the brain
- M_2: in the heart, ileal portion of the small intestine, and cerebellum
- M_3: in secretory glands and smooth muscle
- M_4: in brain
- M_5: in brain

M_1, M_3, and M_5 receptors are involved in producing inositol triphosphate as the second messenger, and these receptors indirectly modulate some K^+, Ca^{2+}, and Cl^- channels. M_2 and M_4 receptors are negatively coupled to adenylyl cyclase and thus they impede intracellular synthesis of cyclic AMP. This response indirectly affects K^+ and Ca^{2+} channels.

8.4   Muscarinic Agonists and Partial Agonists

Muscarinic receptors require that agonists bear a unit positive charge. An aspartic acid residue on the receptor macromolecule provides the negative charge for ligand binding. A dose of muscarine nonselectively stimulates all categories of muscarinic receptors, including those in the central nervous system, reflecting some ability of muscarine to penetrate the blood–brain barrier despite its hydrophilic quaternary ammonium moiety.

A number of simple congeners of acetylcholine lack some of its pharmacological disadvantages. *Carbamylcholine* (carbachol) **8.3**, *acetyl β-methylcholine* (methacholine) **8.4**, and *bethanechol* **8.5** retain the potent muscarinic stimulant effect of acetylcholine. Acetyl β-methylcholine is a poor substrate for acetylcholinesterase and it is almost totally resistant to hydrolysis by blood cholinesterase. Carbamylcholine and bethanechol are totally resistant to hydrolysis by both enzymes. Thus, all three drugs can be transported in the blood with little or no metabolic inactivation, and they demonstrate a long duration of action. The molecular modifications on the acetylcholine molecule represented by these drugs destroy nicotinic agonism. Acetyl β-methylcholine and bethanechol produce peripheral acetylcholine-like effects only on smooth muscles and glands. The relevant muscarinic receptors cannot discriminate between them and acetylcholine. However, carbachol produces some nicotine-like effects at N_2 receptors in peripheral autonomic ganglia; this is an indirect effect. Carbachol triggers the release of endogenous acetylcholine in the ganglia. Indeed, it is likely that the muscarinic effects of carbachol are also indirect rather than agonistic. Because of its prominent nicotinic side effects, carbachol is rarely employed therapeutically.

8.3   8.4

Two alkaloidal natural products, *arecoline* **8.6** and *pilocarpine* **8.7**, are classed as muscarinic partial agonists. Additionally, arecoline produces stimulant effects on nicotinic receptors. Once relegated to therapeutic limbo, both of these older drugs (as well as synthetic analogs and congeners of them) have been the subject of considerable contemporary investigation for possible therapeutic utility in such diverse conditions as glaucoma and cognitive dysfunction, for which long term drug therapy is indicated. It has been speculated that because these agents are only partial agonists, they may produce a lower incidence and milder side effects than full agonist drugs.

8.5   8.6   8.7

8.5   Nicotinic Agonists

8.5.1 Pharmacology of Nicotine

The free base of nicotine is a viscous, oily liquid which has remarkable solubility characteristics. It is miscible with water and is very soluble in alcohol, chloroform, ethereal solvents, hydrocarbon solvents (both aromatic and aliphatic), and fixed oils. These properties account for its rapid and efficient absorption from a large variety of body portals of entry, including mucous membranes and intact skin. Nicotine penetrates the blood–brain barrier with great ease. The average cigarette contains 6–11 mg. of nicotine, and it delivers 1–3 mg to the smoker. Nicotine is absorbed rapidly from inhaled cigarette smoke, and only 10–19 seconds are required for it to reach the brain. Nicotine produces initial stimulation of peripheral autonomic ganglia, which is followed by a persistent depression and blockade of nerve impulse transmission. A similar effect occurs at the other peripheral nicotinic sites, the myoneural junctions. Nicotine markedly stimulates the central nervous system, but with large doses stimulation is followed by profound depression, and death is due to respiratory failure. The depression of the respiratory center in the medulla oblongata is accompanied by paralysis (peripheral blockade of the myoneural junctions) of the respiratory muscles, the diaphragm and the intercostal muscles. Nicotine produces tachyphylaxis in the central nervous system, caused by desensitization of the receptors there (the ion channels which are the nicotinic receptors remain closed).

The liver metabolizes 70–80% of a dose of nicotine to the lactam derivative *cotinine* **8.8**, and this may be further metabolized to 3′-hydroxycotinine **8.9** which is pharmacologically inert. Cotinine has no effect on nicotinic receptors, but it is thought to affect the release of other neurotransmitters in the brain. It also affects a number of enzymes, including those which are involved in the biosynthesis of steroids. A small percentage of a nicotine dose is converted into the N′-oxide **8.10**, which is pharmacologically inert in the nervous system. The carbinolamine (aminal) moiety of another minor metabolite of nicotine, 2′-hydroxynicotine **8.11**, opens to generate the free ketonic structure **8.12**. This product, in vitro, reacts with nitric oxide (NO) to form the *N*-nitroso compound **8.13** which has been shown to be highly carcinogenic. It is known that levels of nitric oxide are elevated in the lungs of smokers. It has not been demonstrated that N-nitrosation of the open chain ketone metabolite occurs in the human body. However this sequence of chemical transformations may offer some explanation for the carcinogenicity of tobacco smoke. Nicotine metabolites are excreted in the urine via glomerular filtration.

R=H
8.8

R=OH
3′-hydroxycotinine
8.9

8.10

8.11

8.12

8.13

Nicotine is a truly addicting drug. The nicotine delivery system, which controls the route and rate of nicotine dosing, is a critical factor in developing and supporting nicotine addiction. Cigarette smoking and intravenous nicotine highly reinforce the addiction, whereas in one study smokers did not rate transdermally applied nicotine as reinforcing. Nicotine chewing gum and nasal spray were rated as reinforcing. These "reinforcing" delivery systems permit more rapid absorption than the transdermal route.

8.5.2 Possible Future Therapeutic Utility for Nicotinic Receptor Stimulants

For many years, no pathological conditions were linked to central or peripheral hypo-function of nicotinic receptors, and for this reason study of nicotinic agonists was largely neglected. Recent findings have suggested that central nicotinic receptors are involved in some disease states, indicating possible therapeutic role(s) for nicotinic agonists in Alzheimer's syndrome, Parkinson's disease, mental depression, schizophrenia, and Tourette's syndrome. Alzheimer's syndrome has been linked to a loss of nicotinic recep-tors in the brain. Nicotine itself enhances cognitive function in individuals with normal brains. Nicotine is under investigation as a treatment for ulcerative colitis, but a mecha-nism for this therapeutic use has not been established. In addition to nicotine, natural products for which nicotinic agonist action has been demonstrated are the plant alkaloid *lobeline* **8.14** and an analog of nicotine, *epibatidine* **8.15**, isolated from the skin of a poisonous frog. Remarkably, this latter compound shows potent analgesic activity; it is reported to be 200–400 times more potent than morphine. However, epibatidine's very small therapeutic index probably precludes its clinical use as an analgesic replacement for morphine. Some synthetic nicotinic agonists related structurally to epibatidine have been described in the literature, but the scope/spectrum of their nicotinic agonism at subpopulations of central and peripheral nicotinic receptors is incompletely defined.

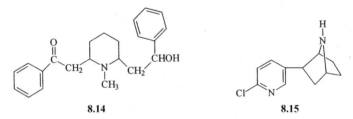

8.14   **8.15**

8.6   Indirect-Acting Cholinergic Agents

8.6.1 Inhibitors of Acetylcholinesterase and Cholinesterase

Cholinergic effects can be produced by administering a drug which inactivates acetyl-cholinesterase in the vicinity of the acetylcholine receptors. Thus, the metabolic inacti-vation of acetylcholine in the synaptic cleft is disrupted. Acetylcholine molecules will persist in the cleft and will continue to interact with their postsynaptic receptors, and the effects of the nerve impulse will be prolonged and magnified. Grossly, the pharma-cological effects observed with acetylcholinesterase inhibitor drugs are quite similar to those observed with cholinergic agonists such as acetyl β-methylcholine. Most of the drugs in this category inhibit both acetylcholinesterase and cholinesterase.

Enzyme-mediated hydrolysis of acetylcholine

Understanding acetylcholinesterase inhibitor drugs requires some understanding of the nature of the enzyme's catalytic site and of the chemical process involved in hydrolysis of the ester linkage by the enzyme. Figure 8.4 is a simplified representation of this catalytic region. Significant features of the enzyme molecule are a site that anchors the quaternary (trimethylammonium) head of the substrate, facilitating the proper interaction of the remainder of the acetylcholine molecule with the other catalytic subsites; a serine residue, the primary alcohol portion of which participates in a transesterification reaction with acetylcholine, resulting in acetylation of the enzyme; and an imidazole ring (part of a histidine residue) that, as illustrated, participates in hydrogen bonding to the carbonyl oxygen of acetylcholine, thus increasing the electrophilicity of the carbonyl carbon and facilitating transesterification. Once this process is complete, the choline product migrates away from the catalytic surface. The resulting acetylated serine moiety is extremely unstable, and it rapidly undergoes nonenzymatic hydrolytic cleavage to liberate acetate anion and to regenerate the active catalytic surface. The enzyme can now accept another substrate molecule. For many years it was believed that the cationic head of acetylcholine is anchored to the catalytic surface by a complementary anionic site on the enzyme. It is now established that the quaternary head of acetylcholine is attracted to and interacts with the π-electron cloud of an aromatic center in the enzyme molecule. Acetylcholinesterase is an extremely efficient, fast-acting catalyst. Modern biophysical methodology has revealed that the time required for hydrolysis of acetylcholine by the enzyme is less than a millisecond.

Acetylcholinesterase and cholinesterase inhibitors

These drugs are rather arbitrarily subdivided into three groups, according to their mode and duration of action:

• Short-acting reversible inhibitors
• Agents that have a carbamate ester linkage which is hydrolyzed by acetylcholinesterase, but at a much slower rate than acetylcholine
• Phosphorus compounds that are true hemisubstrates for the enzyme

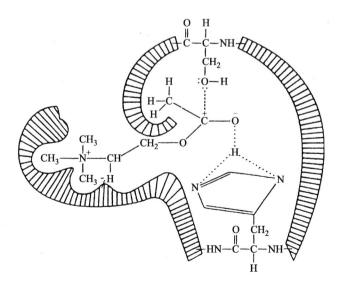

Figure 8.4 Hydrolytic cleavage of acetylcholine by acetylcholinesterase

Typical of the reversible inhibitors is tetramethylammonium cation, which interacts with the aromatic site on the enzyme that binds the quaternary head of acetylcholine, and denies access of acetylcholine to this anchoring site. Thus, the substrate molecule is unable to attach properly to the enzyme, and it is protected from metabolic inactivation. The interaction between tetramethylammonium cation and the enzyme subsite is reversible, and the drug–enzyme complex dissociates in a relatively short time; the tetramethylammonium cation migrates away and the enzyme catalytic surface again can accept acetylcholine molecules and cleave them. Hence, the duration of effect of this type of drug is short. *Edrophonium* **8.16**, having the same mechanism of action as tetramethylammonium cation and a short duration of effect, is administered intravenously as an antidote for acute D-tubocurarine intoxication (vide infra). Its pharmacological value lies in its ability to permit rapid acumulation of acetylcholine in peripheral nervous system synapses.

Another therapeutic disadvantage of quaternary ammonium enzyme inhibitors is their pronounced hydrophilic character and low degree of lipophilicity which generally are detrimental to absorption from the gastrointestinal tract and frequently limit transport in the body, even following intravenous administration. They are also rapidly excreted by glomerular filtration.

The prototypical carbamate-type acetylcholinesterase inhibitor is the alkaloid *physostigmine* **8.17**. The N^1 of the molecule is protonated at physiological pH, and the resulting cation presumably serves to anchor the molecule to the enzyme's quaternary ammonium binding subsite, facilitating transfer of the carbamate moiety to the hydroxyl group of the serine residue, analogous to the acetyl transfer of acetylcholine.

However, unlike acetylated serine, the carbamylated serine is resistant to hydrolytic cleavage. Rather than occurring within less than a millisecond as with the acetylated enzyme, in vivo ester cleavage of the carbamylated system requires hours, during which time the enzyme catalytic surface cannot hydrolyze acetylcholine. Simpler synthetic molecules bearing a carbamate moiety (e.g., *pyridostigmine* **8.18** and *neostigmine* **8.19**) show pharmacologic properties similar to physostigmine, and these are also clinically useful.

8.16

8.17

8.18

8.19

Phosphorus-containing compounds (most commonly derivatives of phosphonic acids or orthophosphoric acid) are among the most powerful acetylcholinesterase inhibitors.

Much of the developmental work in this area was done to prepare chemical warfare agents ("nerve gases"). Typical drugs are *diisopropylfluorophosphate* (DFP) **8.20** and *echothiophate* **8.21**. Typical "nerve gases" are *tabun* **8.22** and *sarin* **8.23**. The phosphorus compounds react at the esteratic site of the enzyme, and the phosphoric or phosphonic acid moiety is transferred to the serine hydroxyl group, again comparable to the reaction with acetylcholine and the carbamates. The resulting phosphorylated or phosphonylated serine is exquisitely stable. Depending upon the exact structure of the drug, regeneration of the catalytic surface of the enzyme may require several hours, or (in the case of the "nerve gases") the enzyme may be permanently inactivated, requiring de novo biosynthesis of new enzyme molecules to restore normal nerve activity.

8.20   8.21   8.22

8.6.2 Enhancers of Acetylcholine Release

No drugs in this category are presently available in the medical marketplace. However, experimental agents typified by *DuP-996* (*linopirdine*) **8.24** enhance *in vitro* acetylcholine release from various rat central nervous system tissues, but only when the release has already been triggered physiologically. Parallel effects have been reported in the human central nervous system. Additional acetylcholine release enhancers have been described in the literature, but their therapeutic utility remains to be established.

8.23   8.24

8.7   Therapeutic Uses of Cholinergic Receptor Stimulants

Many of the synthetic and natural cholinergic agonists and partial agonists show a preference for stimulation of peripheral muscarinic receptors rather than nicotinic receptors. The therapeutic utility of indirect-acting cholinergics is frequently (but not always) based on their ability to initiate stimulation of *peripheral* muscarinic receptors.

An important therapeutic role of peripheral muscarinic receptor stimulants is in relieving the symptoms of certain types of glaucoma. In this condition, the hydrostatic pressure of the intraocular fluid (the solution inside the eyeball, also called *aqueous humor*) is abnormally high. This elevated pressure is capable of producing severe pain and it can cause damage to the visual apparatus, resulting in permanent blindness. Physiologically, the intraocular fluid pressure is regulated in part by drainage of excess fluid from the eye via a duct, the *canal of Schlemm*. This anatomic entity is under cholinergic nerve regulation such that when appropriate muscarinic receptors in the eye are stimulated, drainage through the canal is promoted and the intraocular pressure is lowered. The muscarinic stimulants used for this purpose (e.g., physostigmine **8.17**, pilocarpine **8.7**, and phosphorus derivatives such as echothiophate **8.21**) are instilled directly into the eye. Systemic administration would produce widespread side effects due to nonselective stimulation of muscarinic receptors throughout the body. Even direct instillation of a drug into the eye can result in some occasionally severe systemic drug side effects due to the anatomic connections between the eye and the blood stream which have been cited previously.

Surgical procedures in the abdomen and adjacent areas can cause loss of smooth muscle tone and postoperative urinary bladder hypoactivity, resulting in the inability to void urine. Muscarinic agonists such as bethanechol **8.5** and acetyl β-methylcholine **8.4** are used to stimulate urinary bladder emptying by stimulating the smooth muscles involved.

Myasthenia gravis is a neuromuscular disease characterized by weakness and marked fatigue of the skeletal muscles. The disease involves nerve–muscle activity at the myoneural junction; it appears that the release of acetylcholine into the synaptic cleft is normal, but there is a decrease in the number of nicotinic N_1 receptors on the postsynaptic membrane (on the muscle fiber). As a result the voluntary muscles of the myasthenic victim cannot continually respond to efferent nerve impulses, and thus they cannot produce sustained contractions. In 90% of patients, the physiological cause of the disease is believed to be immunologically based. The serum of myasthenia patients contains an antibody against the nicotinic receptor protein. Systemically administered indirect acting cholinergic receptor stimulants, such as the acetylcholinesterase inhibitors pyridostigmine and neostigmine, relieve the symptoms of the condition in many sufferers, but these drugs do not effect a cure. These enzyme inhibitors are nonselective in their sites of action and for this reason they produce prominent side effects, especially those referrable to muscarinic receptor stimulation. These muscarinic effects can sometimes be controlled by administering a muscarinic blocking agent. In approximately 10% of myasthenia patients, the disease has a congenital rather than an autoimmune basis. These patients usually do not respond favorably to anti-acetylcholinesterase drug therapy.

8.8   Nicotinic Receptor Blockers

Peripheral nicotinic blockers block the effects of of acetylcholine at autonomic ganglia (N_2 receptors) and/or at the myoneural junctions (N_1 receptors). It is possible to design drugs that show a high degree of selectivity for their site of action: to block nerve impulse transmission at autonomic ganglia with minimal effects at myoneural junctions or to block nerve impulse effects at myoneural junctions with minimal effects at autonomic ganglia.

8.8.1 Ganglionic Blocking Agents

It will be recalled that all autonomic ganglia (both in parasympathetic and sympathetic nerves) are cholinergic. Ganglionic blocking agents impair nerve impulse transmission through both sympathetic and parasympathetic ganglia by blocking the nicotinic receptors there. But, unlike nicotine itself, they do not cause an initial stimulation. Typical of this category of drugs are *trimethaphan* **8.25** and *mecamylamine* **8.26**. Historically, ganglionic blockers were the first genuinely effective antihypertensive agents. However, they produce a wide variety of severe side effects due to their nonselective blockade of ganglionic neurotransmission, and they have been largely supplanted by other drugs (see chapter 14).

8.25   **8.26**

8.8.2 Myoneural Blocking Agents

Figure 8.5 diagramatically illustrates a nicotinic receptor, an ion channel. If blockade of nicotinic N_1 receptors on the membrane of a voluntary muscle occurs, the acetylcholine which is released into the synaptic cleft cannot interact with the receptor to open the ion channel and permit ion migration with the resultant depolarization of the

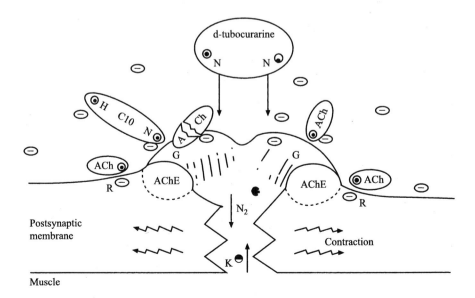

Figure 8.5 Myoneural blocking agents at the nicotinic N_1 receptor

membrane, which is an early stage in the biochemical process of constriction of the muscle fiber. The muscle is paralyzed and it cannot constrict; this paralysis is of the *flaccid* (limp) type. A bulky molecule typified by D-*tubocurarine* **8.27** straddles the closed ion channel and, because of its physical bulk, prevents acetylcholine molecules from approaching the receptors and triggering the depolarization process. D-Tubocurarine does not depolarize the postsynaptic membrane. While it denies access of acetylcholine to the nicotinic receptors, it cannot interact appropriately with these receptors to induce the conformational changes in the ion channel protein complex to open the ion channel. Drugs of this type are described as *nondepolarizing myoneural blockers*. The paralyzant effects of D-tubocurarine on voluntary muscles can be reversed by administering large doses of an acetylcholinesterase inhibitor such as physostigmine, which results in accumulation of larger than normal amounts of acetylcholine in the synaptic cleft. These high levels of acetylcholine displace the blocker molecules by mass action phenomena and the flaccid paralysis is terminated.

In contrast, *succinyl dicholine* **8.28** interacts with the nicotinic receptors on the muscle fiber membrane and depolarizes them, similar to the action of acetylcholine itself. However, unlike acetylcholine, succinyl dicholine is not rapidly desorbed from the nicotinic receptors, and the depolarization persists. The overall observable effect of succinyl dicholine on voluntary muscles is an initial (and transient) contraction of the muscle, a result of the initial depolarization of the membrane. This is followed by a persistent relaxation of the muscle, flaccid paralysis. It might have been anticipated that these agents which produce a persistent depolarization of the muscle fiber membrane would produce a persistent rigid paralysis of the muscle rather than persistent flaccidity; the muscle would constrict and remain constricted. However, this is not the case and grossly succinyl dicholine produces the same kind of flaccid paralysis as does D-tubocurarine. No universally accepted explanation for this peculiar biphasic action of depolarizing myoneural blockers has been advanced. Some authorities suggest that this is an example of tachyphylaxis. Acetylcholinesterase inhibiting drugs like physostigmine are ineffective in reversing the paralysis induced by the depolarizing myoneural blockers. The resulting accumulation of acetylcholine in the synaptic cleft has no effect, because the postsynaptic membrane has already been depolarized and it remains depolarized. Succinyl dicholine resists hydrolysis by acetylcholinesterase, but it is a substrate for plasma cholinesterase.

8.27

8.28

Uses of myoneural blocking agents

These drugs are used principally as surgical adjuvants to relax spastic voluntary muscles, especially those of the abdominal wall, which facilitates surgical procedures in that region.

They also find use in preventing trauma during electroshock therapy. They are almost always administered intravenously. D-Tubocurarine and, to a lesser extent, succinyl dicholine act directly on mast cells to induce the release of histamine which can produce bronchospasm, hypotension, and excessive secretion by the bronchial and salivary glands. This side effect can represent a serious clinical problem. The improper administration of the depolarizing muscle relaxants can produce life-threatening side effects. It will be recalled that the process of membrane depolarization involves migration of sodium cations into the intracellular regions and concomitant migration of potassium cations into the extracellular region. If a dose of a membrane depolarizing drug such as succinyl dicholine is administered very rapidly, the depolarization effect on the membranes will occur rapidly, resulting in release of a large amount of potassium cations into the extracellular environment. This sudden *hyperkalemia* can trigger a prolonged apnea (cessation of breathing) and it can also cause cardiac arrhythmias or even cardiac arrest in some individuals.

In vivo fate of myoneural blocking agents

The major part of a dose of D-tubocurarine is excreted chemically unchanged in the urine, with only a minor portion undergoing metabolic change. The duration of the effect of D-tubocurarine, to a large measure, reflects its rate of clearance from the body by glomerular filtration. As was mentioned previously, succinyl dicholine is a substrate for blood and liver cholinesterase, and for this reason it has a short duration of effect. This is a desirable property because the skilled anesthesiologist can literally titrate the patient with succinyl dicholine to attain the desired level of flaccid paralysis. Because of their pronounced hydrophilicity these drugs cannot penetrate the blood–brain barrier in sufficient amounts to attain an effective concentration in the CNS.

8.9   Muscarinic Receptor Blockers

8.9.1 Peripheral Effects and Therapeutic Uses

Muscarinic receptor blockers are frequently called parasympatholytic or anticholinergic drugs, but these terms are inaccurate because their effects are limited to muscarinic receptors. They are competitive inhibitors of acetylcholine; one molecule of antimuscarinic agent blocks one acetylcholine receptor. Many of the antimuscarinics have a greater affinity for the muscarinic receptor than does acetylcholine; nevertheless, their blockade is surmountable by inducing very high concentrations of acetylcholine in the synaptic cleft (use of an acetylcholinesterase inhibitor). Muscarinic receptor blockers have little or no action at nicotinic receptors.

The accumulation of acetylcholine at the nerve terminal on a smooth muscle produces a constant stimulus to the muscle. The muscle contracts and it will remain contracted; it is spastic. Antimuscarinic agents combat the spasm of smooth muscles, and they are sometimes called *antispasmodics*. This ability to relax spastic smooth muscles is exploited in treatment of such conditions as gastrointestinal spasm, biliary spasm (spasm of the smooth muscle of the bile duct), arterial spasm, and urinary duct spasm. Hyper-release of histamine and serotonin also can cause gastrointestinal spasms.

Antimuscarinics are usually ineffective against spastic effects produced by other than acetylcholine-derived mechanisms. Because they relax gastric smooth muscle and inhibit secretion of hydrochloric acid and the proteolytic enzyme pepsin in the stomach, antimuscarinic drugs have been used to treat gastric ulcer, However, due to the high incidence of side effects, they have been largely supplanted for this purpose by histamine H_2 receptor antagonists (see chapter 9) and proton pump inhibitors (see chapter 17).

Drug combinations for treating simple diarrhea frequently contain an antimuscarinic agent to counteract the hypermotility of the gut. The ability of antimuscarinic agents to inhibit exocrine glandular secretions is exploited in their use in over-the-counter rhinitis (common cold) mixtures and as an adjunct to general inhalation anesthetics to dry secretions. In ophthalmology, antimuscarinic drug eye drops are used to produce *mydriasis* (dilation of the pupil of the eye) for examination of the retina and the optic disc. Drug-induced dilation of the pupils of the eye was employed cosmetically by women in Renaissance Italy to produce "sparkling eyes". Indeed, the species designation "belladonna" in one botanical source of alkaloidal antimuscarinic agents, *Atropa belladonna,* is Italian for "pretty lady".

Typical antimuscarinic drugs are the alkaloids *hyoscyamine* **8.29**, *atropine* **8.30** (racemic hyoscyamine; the optically active natural product hyoscyamine is racemized in the extraction process from its natural source), and *hyoscine* (scopolamine) **8.31**. Because these alkaloids are derived from plants of the botanical family Solanaceae, they are sometimes called *solanaceous alkaloids*. A large number of synthetic agents, esters of a variety of lipophilic carboxylic acids and heterocyclic or acyclic amino alcohols, have antimuscarinic properties. These agents (natural products and synthetics) are all competitive antagonists of acetylcholine at muscarinic receptors. A major defect of almost all of the myriad of antimuscarinic drugs available in the marketplace is that they exert blocking effects on all subtypes of muscarinic receptors and they produce many pronounced side effects in addition to their desired therapeutic response, including dryness of the mouth, flushing of the blush areas of the face and neck, mydriasis, and cycloplegia (inability of the pupil of the eye to accomodate to changes in light or dark). Although these side effects are not life-threatening, they are annoying and they frequently become intolerable if the drug is administered chronically. A small number of antimuscarinic agents, typified by *pirenzepine* **8.32**, are selective for M_1 receptors. Pirenzepine is said to produce desirable gastrointestinal effects at dose levels that do not produce blurred vision and dry mouth. It has low lipid solubility and it exhibits only limited ability to penetrate the blood brain barrier. It does not cause the CNS side effects characteristic of older muscarinic receptor blockers (vide infra).

8.9.2 *Central Effects of Muscarinic Receptor Blockers*

Because most of the antimuscarinic drugs are lipophilic, they penetrate the blood–brain barrier, and in relatively large doses they produce a variety of central nervous system effects. Atropine and hyoscyamine cause central nervous system stimulation, and large doses result in hallucinations and delerium. In contrast, low doses of hyoscine produce central depression marked by drowsiness, fatigue, and a dreamless sleep. Hyoscine also can produce euphoria. In high doses, hyoscine produces atropine-like stimulation. It has been speculated that these differences in central nervous system response to atropine and to hyoscine may reflect dose-related differences in their abilities to block different subpopulations of central muscarinic receptors.

8.29 Hyoscamine (levo-)
8.30 Atropine (dl-)

8.31 Hyoscine

8.32

Hyoscine relieves nausea and vomiting of labyrinthine (involving disruption of inner ear balance) origin, as in motion sickness and vomiting caused by local stimuli in the stomach. Hyoscine is superior to atropine and hyoscyamine in this property. The antiemetic effect of hyoscine probably reflects its actions in the cerebral cortex. Hyoscine is without effect in controlling vomiting arising from stimulation of the chemoreceptor trigger zone in the medulla oblongata, which is under dopaminergic control. Hyoscine's marked lipophilic character permits its absorption across intact skin, and a dermal patch containing hyoscine is commonly employed to prevent motion sickness.

Central effects of tertiary amine antimuscarinics, such as atropine or hyoscine, are eliminated by quaternization of the amino group. The resulting permanent cation is much less lipophilic than the tertiary amine, and its ability to penetrate the blood–brain barrier by passive diffusion is greatly diminished. These quaternary derivatives generally retain peripheral antimuscarinic activity, although potency may be lowered somewhat and efficiency of absorption across the intestinal wall is diminished. However, oral dosage forms of the quaternary derivatives are available.

Stimulation of muscarinic receptors in the nigrostriatal pathway in the brain exerts an excitatory effect on skeletal muscle, opposite to the effect of dopamine in this region. Muscarinic excitation complements the dopaminergic hypoactivity in the Parkinsonian syndrome in production of rigid paralysis and tremors. Before the introduction of DOPA, antimuscarinic drugs were used to relieve Parkinsonian symptoms, and selective M_1 receptor blockers are still occasionally employed as adjuncts to DOPA therapy.

8.10   Cognitive Dysfunction: Alzheimer's Syndrome

Cognitive dysfunction is a prominent symptom accompanying aging, stroke, and such neurodegenerative diseases as Alzheimer's syndrome. Many authorities believe that a deficit in central cholinergic function is a significant contributor to cognitive dysfunction, both in aging and in Alzheimer's syndrome. Postmortem examination of Alzheimer victims' brains reveals shrinkage of brain tissue and a selective loss of cholinergic neurons in the basal forebrain nuclei. Postsynaptic M_1 receptors are largely intact, but presynaptic M_2 receptors degenerate. Also, the number of nicotinic receptors is reduced considerably. Lesions in the forebrain region in experimental animals produce deficits in memory and learning. Choline acetyltransferase activity is markedly lowered in the cortex of Alzheimer patients.

It has been postulated that direct- or indirect-acting cholinergic agents may provide symptomatic relief of the condition. Muscarinic partial agonists, such as arecoline, demonstrate some efficacy in animal models, but their therapeutic value is greatly limited by peripheral side effects due to broadly based cholinergic stimulation. Many muscarinic full agonists do not penetrate the blood–brain barrier and hence are of no value. Inhibitors of acetylcholinesterase such as physostigmine provide some beneficial effects in Alzheimer patients. However, the resulting increased cholinergic activity is nonselective and such drugs produce a variety of intolerable peripheral side effects.

Focus on novel inhibitors of acetylcholinesterase led to the study and exploitation of *9-aminotetrahydroacridine* (*THA*; *tacrine*) **8.33**, which inhibits the enzyme by reversibly binding to the enzyme's active catalytic site located in a deep chasm in the protein molecule, termed the *aromatic gorge*. Binding occurs between two specific aromatic groups in the aromatic gorge, one of which changes its conformation better to accomodate a complimentary region on the drug molecule. THA is an even better inhibitor of plasma cholinesterase. This drug's reported ability to improve cognition in Alzheimer patients is compromised by reports of hepatotoxicity.

It is unclear why therapeutic doses of tacrine display fewer and milder side effects than do the more traditional acetylcholinesterase inhibitors, such as physostigmine. Is it possible that tacrine is not transported to many sites in the body where acetyl-cholinesterase is located? Or, is it possible that tacrine (unlike physostigmine) inhibits only certain subtypes of the multiple forms of acetylcholinesterase? Or, is it possible that tacrine's anti-acetylcholinesterase activity is only one component of its overall effect on the Alzheimer patient? THA has been reported to block neuronal potassium channels; to inhibit monoamine oxidase; to act as a competitive antagonist to GABA at GABA-A receptors; and to inhibit neuronal monoamine uptake. Some of these actions may contribute to the beneficial effects of the drug. It must be remembered that tacrine is not curative; it only ameliorates the symptoms of Alzheimer's disease, and many Alzheimer victims experience little or no benefit from the drug.

bis-(7)-Tacrine **8.34** has superior anti-acetylcholinesterase activity to tacrine itself. This drug is said to have superior memory-enhancing activity compared to tacrine, and it is a more potent competitive antagonist to GABA at GABA-A receptors. This finding suggests that GABA-A receptor antagonism may contribute to the memory-enhancing properties of *bis-(7)-tacrine*, and it might be speculated that the sole cause of the Alzheimer syndrome is not a defect in the CNS cholinergic system, but that other neurotransmitter systems may also be involved.

8.33   8.34

Donepezil **8.35** is a selective inhibitor of acetylcholinesterase in the CNS, but it has little effect on the enzyme in peripheral tissues. It is said to produce less frequent and less severe cholinergic-related side effects than does tacrine, and it is not associated with hepatotoxicity. Several other acetylcholinesterase inhibitors are now available for treatment of the Parkinsonian syndrome.

8.35

Extensive studies implicate a peptide, β-*amyloid*, as an etiological factor in Alzheimer's syndrome. Plaques of this substance are deposited in the nerve pathways and in the walls of cerebral blood vessels of Alzheimer patients' brains. β-Amyloid is a 39–42 amino acid peptide; it is formed from the cleavage of a larger protein, catalyzed by the sequential action of two proteases, β- and γ-secretases. Secretase inhibitors are being evaluated for their clinical value in Alzheimer patients. β-Amyloid is prone to self-aggregation and another approach to blocking the formation of β-amyloid plaques is to inhibit the polymerization process. The inhibition of β-amyloid aggregation might be expected to slow or to arrest the progress of the Alzheimer syndrome, rather than merely treating the symptoms of the disease. However, it is uncertain how β-amyloid accumulation causes neurodegeneration.

Bibliography

1. Doerge, R. F., Ed. *Wilson and Gisvold's Textbook of Organic Medicinal and Pharmaceutical Chemistry*, 8th ed. Lippincott: Philadelphia, Pa., 1982; p. 434.
2. Parsons, S. M. and Rogers, G. A. *Annu. Rep. Med. Chem.* **1993**, 28, 247.
3. Lindstrom, J., Tzartos, S., Gullick, W., Hochschwender, S., Swanson, L., Sargent, P., et al. *Cold Spring Harbor Symposium on Quantitative Biology*. **1983**, 48, 93.
4. Foye, W. O., Lemke, T. L. and Williams, D. A., Eds. *Principles of Medicinal Chemistrty*, 4th ed. Williams & Wilkins: Baltimore, Md., 1995; p. 333.

Recommended Reading

1. Cannon, J. G. Cholinergics. In *Burger's Medicinal Chemistry and Drug Discovery,* 6th ed. Abraham, D. J., Ed. Wiley-Interscience: Hoboken, N.J., 2003; Vol. 6, pp. 39–108.
2. Watson, S. and Arkinstall, S. *The G-Protein Receptor Facts Book*. Academic Press: New York, 1994.
3. Dostert, P. I., Strolin Benedetti, M. and Tipton, K. F. Interactions of Monoamine Oxidase with Substrates and Inhibitors. *Med. Res. Rev.* **1989**, 9, 45–89.
4. Ember, L. R. The Nicotine Connection. *Chem. Eng. News* **1994**, 72(48), 8–18.
5. Arneric, S. P., Holladay, M. W. and Sullivan, J. P. Cholinergic Channel Modulators as a Novel Strategy in Alzheimer's Disease. *Exp. Opinion Invest. Drugs* **1996**, 5, 79–100.
6. McDonald, I. A., Cosford, N. and Vernier, J.-M. Nicotinic Acetylcholine Receptors: Molecular Biology, Chemistry, and Pharmacology. *Annu. Rep. Med. Chem.* **1995**, 30, 41–50.

7. Arneric, S. P., Sullivan, J. P. and Williams, M. Neuronal Nicotinic Acetylcholine Receptors: Novel Targets for Central Nervous System Therapeutics. In *Psychopharmacology: The Fourth Generation of Progress*. Bloom, F. E. and Kupfer, D. J., Eds. Raven Press: New York, 1995; pp. 95–110.

8. Benowitz, N. L. Pharmacology of Nicotine: Addiction and Therapeutics. *Annu. Rev. Pharmacol. Toxicol.* **1996**, *36*, 597–613.

9. Eglen, R. M., Hegde, S. S. and Watson, N. Muscarinic Receptor Subtypes and Smooth Muscle Function. *Pharmacol. Rev.* **1996**, *48*, 561–565.

10. Rang, H. P., Dale, M. M., Ritter, J. M. and Gardner, P. Pathogenesis of Alzheimer's Disease. In *Pharmacology*, 4th ed. Churchill Livingstone: New York, 2001; pp. 504–505.

11. Lucas, R. J., Changeux, J.-P., Le K. J., Lindstrom, J. M., Marks, M. J., Quik, M., *et al.* Review of nicotinic receptors. *Pharmacol. Rev.* **1999**, *51*, 397–401.

12. Corringer, P.-J., Le Novère, N. and Changeux, J.-P. Review of nicotinic receptors. *Annu. Rev. Pharmacol. Toxicol.* **2000**, *40*, 431–458.

13. Nicotinic Receptors: Muscarinic Receptors. In *The Sigma-RBI Handbook of Receptor Classification and Signal Transduction*, 5th ed. Watling, K. J., Ed. Sigma–RBI: Natick, Mass., 2006; pp. 78–82.

14. Rama Sastry, B. V. "Anticholinergic Drugs" in Recommended Reading ref. 1, pp. 109–165.

15. Rzeszotarski, W. J. "Alzheimer's Disease: Search for Therapeutics". In Recommended Reading ref. 1, pp. 743–777.

16. Eid, Jr., C. N., Wu, Y.-J. and Kinney, G. G. Cognition Enhancers. In Recommended Reading ref. 1, pp. 779–835.

17. Colquhoun, D., Shelley, C., Hatton, C., Unwin, N. and Silvilotti, L. Nicotinic Acetylcholine Receptors. In Burger's Medicinal Chemistry and Drug Discovery, 6th ed. Wiley–Interscience: Hoboken, N. J., 2003; Vol. 2, pp. 357–405.

18. Selkoe, D. J. and Schenk, D. Alzheimer's Disease. Molecular Understanding Predicts Amyloid-Based Therapeutics. In *Annual Review of Pharmacology and Toxicology*. Cho, A. K., Blaschke, T. F., Insel, P. A. and Loh, H. H., Eds. Annual Reviews: Palo Alto, Calif., 2003; Vol. 43, pp. 545–584.

19. Kornilova, A. Y. and Wolfe, M. S. Secretase Inhibitors for Alzheimer's Disease. Annu. Rep. Med. Chem. **2003**, *38*, 41–50.

9

The Central Nervous System.
I: Psychotropic Agents

Commonly Used Terms in Central Nervous System Pharmacology

Various authorities in the field present somewhat different definitions of many commonly used terms in psychopharmacology, and even optimally the terms and their definitions are inexact. No single system of categorization and definition of mental illnesses or of mental drugs is universally accepted. Many of these terms have entered into lay usage and are employed carelessly or improperly in everyday conversation. Discussions in this chapter are consistent with the following definitions.

Psychosis. In psychosis, the most severe of the psychiatric disorders, there is marked impairment of the individual's behavior and also serious inability to think coherently, to comprehend reality, or to gain insight into the abnormality. These conditions often include delusions and hallucinations. Causes of psychoses may be referrable to biochemical or metabolic defects or to neuropathological changes, or they may be *idiopathic* (having no detectable cause). *Depressive psychosis* is characterized by states of mental depression, melancholia, despondency, inadequacy, and feelings of guilt. *Manic depressive psychosis* is marked by emotional instability, dramatically excessive mood swings, and a tendency to recurrence.

Dementia praecox. The chief characteristics of dementia praecox, a usually idiopathic condition, are disorientation, loss of contact with reality, and splitting of the personality (*schizophrenia*). There may be paranoia (delusions of persecution) and auditory hallucinations. Chronic schizophrenia accounts for the majority of patients in long-stay mental hospitals. Some authorities use the terms schizophrenia and dementia praecox synonymously.

Neurosis. Neurosis is a less incapacitating disorder than psychosis. The ability to comprehend reality is retained, and the personality remains more or less intact. Nevertheless, the individual's suffering and disability may be extremely severe. One authority has described neurosis as "an abnormal reaction to external circumstances." Neuroses may be acute and transient or, more commonly, persistent or recurrent. There may be mood changes (anxiety, panic, depression), limited abnormalities of thought (obsessions, irrational fears), or limited abnormalities of behavior (rituals or compulsions).

Psychotropic drug. The term includes any chemical substance that affects psychic functions and/or behavior. It includes drugs used to treat mental disorders as well as drugs that cause mental abnormalities.

Tranquilizer. A tranquilizer affects the emotional state and quiets or calms the individual without affecting clarity of consciousness.

Ataractic. An ataractic agent induces calmness; the term is frequently used synonymously with the term tranquilizer.

Neuroleptic. The term is usually taken as a synonym for *antipsychotic*, a drug used to treat a psychosis. Contemporary usage attempts to restrict the term neuroleptic to emphasize more neurological aspects of drugs (e.g., the effects on movement and posture produced by many psychotropic drugs), and antipsychotic is the preferred general term.

For convenience, this discussion of psychotropic drugs is divided as follows:

• Antidepressants (thymoleptics, mood-elevating drugs)
• Mood-stabilizing drugs
• Antianxiety–sedative drugs
• Antipsychotic drugs

Psychotropic drugs, regardless of their specific pharmacological effects, are characterized by distinctly lipophilic oil/water partition coefficients, which permit efficient penetration of the blood–brain barrier by passive diffusion. Active transport of an exogenous psychoactive organic molecule into the CNS is unusual.

9.2   Biochemistry and Physiology of Neurotransmitters in the Central Nervous System

Previous chapters have described nerve impulse transmission, phenomena occurring at synapses and at peripheral effector organs, and the roles of neurotransmitters in the peripheral nervous system. Similar mechanisms and phenomena are also operative in the brain. However, the brain is much more complex biochemically in that, in addition to the modest number of neurotransmitter substances (norepinephrine, acetylcholine, dopamine, inter alia) described for the peripheral system, there are several other nonpeptide, low molecular weight chemical entities involved in central nervous system transmission (figure 9.1), and also a number of peptide neurotransmitters. The development of spectrophotofluorometry, among other modern techniques, has permitted quantitative measurement of these substances in nerve tissue at picogram (10^{-12}) levels. Thereby, it has been possible to "map out" the nerve pathways of the brain to show where the various substances shown in figure 9.1 are localized.

9.2.1 Serotonin (5-Hydroxytryptamine, 5-HT), Melatonin

Serotonin

This is one of the most important of the central neurotransmitters. It is involved in the physiological processes of sleep, thermoregulation, appetite control, sexual behavior, cardiovascular function, endocrine regulation, and muscle contraction. Serotonin stimulates some central neurons and inhibits others. Presynaptic serotonin receptors inhibit release of neurotransmitters (acetylcholine as well as serotonin itself) from some nerve

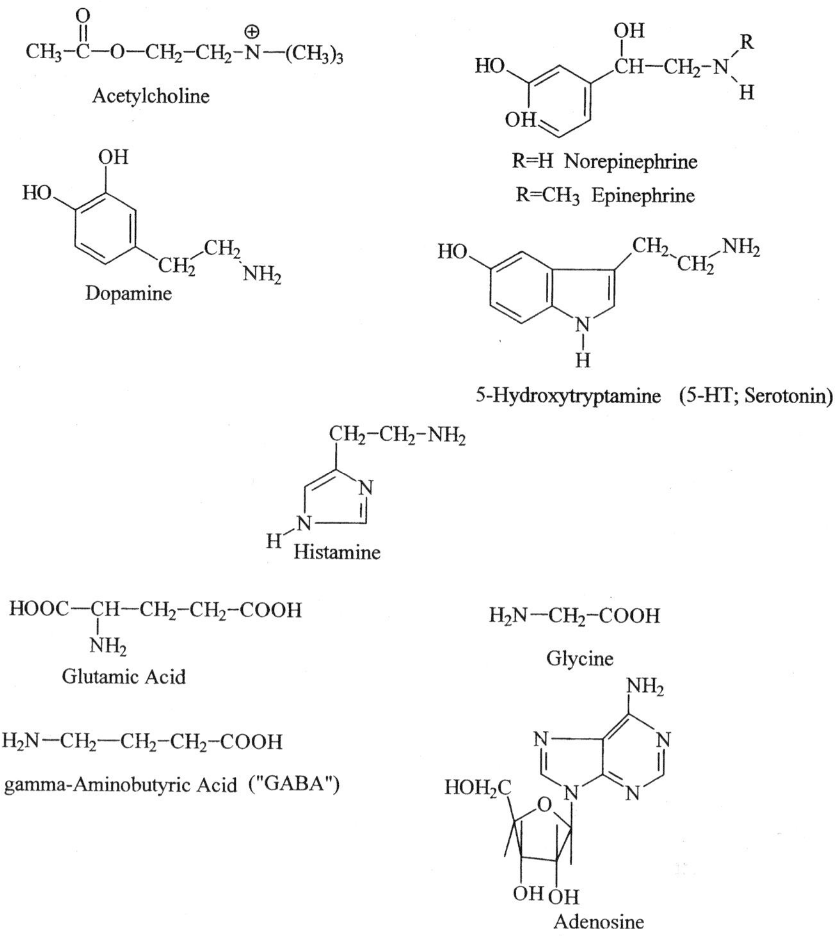

$$CH_3-\overset{\overset{\displaystyle O}{\|}}{C}-O-CH_2-CH_2-\overset{\oplus}{N}-(CH_3)_3$$

Acetylcholine

R=H Norepinephrine
R=CH$_3$ Epinephrine

Dopamine

5-Hydroxytryptamine   (5-HT; Serotonin)

$$CH_2-CH_2-NH_2$$

Histamine

$$HOOC-\underset{\underset{\displaystyle NH_2}{|}}{CH}-CH_2-CH_2-COOH$$

Glutamic Acid

$$H_2N-CH_2-COOH$$

Glycine

$$H_2N-CH_2-CH_2-CH_2-COOH$$

gamma-Aminobutyric Acid ("GABA")

Adenosine

Figure 9.1 Central nervous system neurotransmitters

terminals. Grossly, it may be considered that norepinephrine and serotonin have opposing physiological effects in the central nervous system. Generally the psychic role of
norepinephrine is arousal, stimulation, and increased activity in accord with the "fight
or flight" mnemonic. In contrast, the role of serotonergic nerve pathways in the brain is
to calm the individual and provide for restorative, recuperative actions. It has been
proposed that norepinephrine is involved in REM (rapid eye movement) sleep and that
non-REM sleep is controlled by serotonin (see chapter 10).

Serotonin is synthesized peripherally in the intestine where it is co-stored with norepinephrine. However, authorities disagree concerning the details of the mechanism(s) of
the peripheral effects of serotonin. Serotonin is biosynthesized in nervous tissue as
indicated in scheme 9.1.

Effects of serotonergic nerve impulses at postsynaptic receptor sites are terminated by
uptake (active transport) mechanisms analogous to those described previously for norepinephrine and dopamine. Scavenger enzymes metabolically inactivate those relatively few

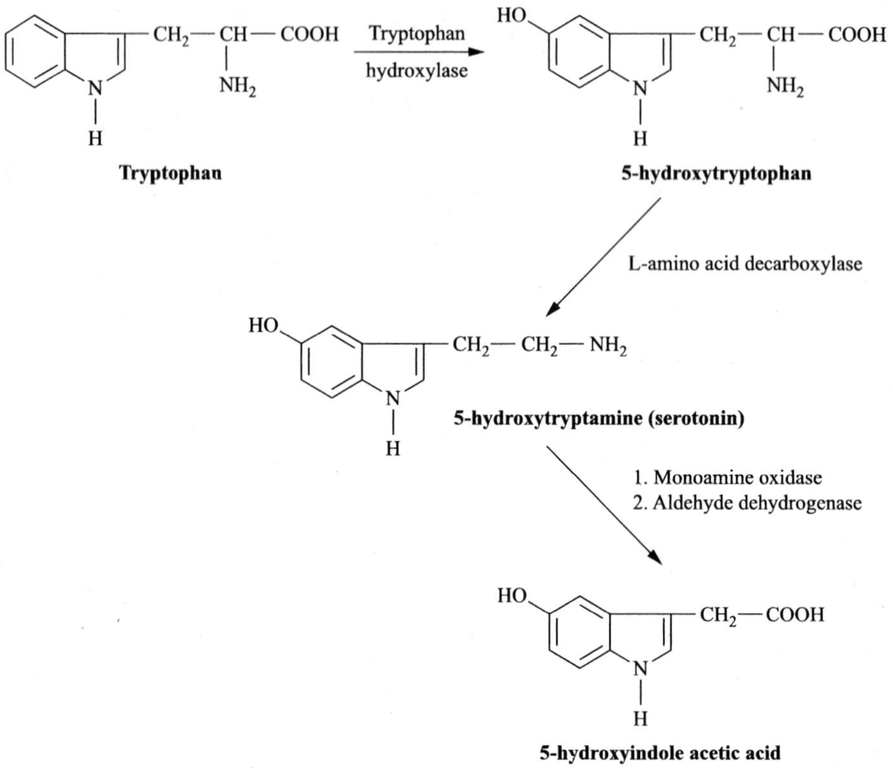

Scheme 9.1 Biosynthesis and metabolic disposition of 5-hydroxytryptamine

serotonin molecules that escape the reuptake process (see scheme 9.1). Monoamine oxidases and aldehyde dehydrogenases are some of the same types of metabolizing enzymes that are involved in norepinephrine and dopamine metabolism. The metabolic end product of serotonin, 5-hydroxyindoleacetic acid, is excreted in the urine.

Serotonin and/or its metabolites have been implicated in the craving syndrome for ethanol. An isoflavone glycoside, *daidzin*, has been reported to diminish the intake of ethanol by ethanol-preferring laboratory animals. Daidzin interferes with serotonin metabolism: it inhibits mitochondrial aldehyde dehydrogenase, the enzyme that converts 5-hydroxyindole-3-acetaldehyde into 5-hydroxyindole-3-acetic acid (scheme 9.1). Thus, the aldehyde accumulates. Daidzin has no effect on monoamine oxidase enzymes. Indeed, some congeners of daidzin have been found which do inhibit mitochondrial MAO, and these have no effect in suppressing ethanol intake by laboratory animals. This finding may be rationalized on the basis that inhibition of MAO prevents accumulation of 5-hydroxyindole-3-acetaldehyde.

Melatonin

In the pineal gland (which is anatomically but not physiologically a part of the thalamus gland) serotonin is N-acetylated and O-methylated to produce *melatonin* (equation 9.1).

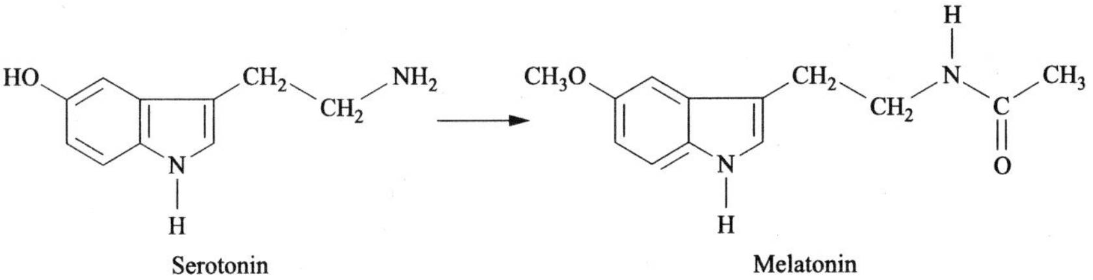

$$(9.1)$$

Serotonin   Melatonin

A complex triggering mechanism for melatonin synthesis is initiated in and proceeds from the so-called *master biological clock* located in the hypothalamus, thence to the pineal gland. The biochemical mechanism involves increased pineal gland production of adenosine 3′, 5′-monophosphate ("cyclic AMP") and subsequent activation of an acetyltransferase system. Melatonin is not a neurotransmitter, but rather it is a circulating hormone. It appears in highest concentrations in the blood during the dark period of the 24-h day. Environmental light inhibits its biosynthesis, and darkness promotes biosynthesis. Release of melatonin is controlled by input from the retina of the eye. Melatonin is involved in lightening of pigmentation in skin cells, and it suppresses ovarian functions. It is involved in the regulation of *circadian rhythms* in vertebrates, including humans. ("Circadian" means rhythmic repetition of certain phenomena in living organisms at about the same time each 24-h period.) Melatonin induces sleep in humans, and sleep disorders in the elderly have been ascribed to low melatonin levels. Drugs affecting the melatonin system have been investigated for their value in treating jet lag and other sleep disorders. There are three subtypes of melatonin receptors, all of which are G protein-linked; they are found mainly in the brain and in the retina of the eye.

Serotonin receptors

Serotonin receptors are highly heterogeneous, and subcategorization of them has been difficult. Several of the serotonin receptor subtypes listed below are known by different names in the literature, which adds to the confusion. In all, some 13 different subtypes have been identified. Definition of seven major subtypes ($5\text{-}HT_{1-7}$) has been widely agreed upon, with further subdivision of some of these as follows:

$5\text{-}HT_{1A}$, $5\text{-}HT_{1B}$, $5\text{-}HT_{1D\alpha}$, $5\text{-}HT_{1E\alpha}$, $5\text{-}HT_{1f}$ receptors

These are found in various regions of the central nervous system. They are all presynaptic, G protein-linked, and are negatively coupled to adenylate cyclase. $5\text{-}HT_{1A}$ receptors are found in high concentrations in portions of the brain where mood and emotion are controlled.

$5\text{-}HT_{2A-C}$ receptors

$5\text{-}HT_{2A}$ receptors are widely distributed through the central nervous system and also peripherally in the gastrointestinal tract; $5\text{-}HT_{2B}$ receptors have not been found in the central nervous system, but they are found in the stomach; 5-HT$_{2C}$ receptors are found in the central nervous system. All three of these receptor subtypes are involved with formation of second messengers, *diacyl glycerol* **9.1** and *inositol-1,4,5-triphosphate* **9.2**.

9.1   9.2

5-HT$_3$ receptors

These are located peripherally in the gastrointestinal tract and in the central nervous system where they are involved in the anxiety syndrome and in the emetic response. This is the only serotonin receptor subtype that is an ion channel.

5-HT$_{4-7}$ receptors

These have been cloned. All except 5-HT$_5$ are G protein-linked, increasing production of cyclic-AMP. The signal transduction mechanism involving 5-HT$_5$ receptors is unknown.

9.2.2 γ-Aminobutyric Acid (GABA)

GABA is found in the brain and spinal cord but only in trace amounts peripherally; it cannot penetrate the blood–brain barrier. GABA is a major postsynaptic inhibitory neurotransmitter in the central nervous system; it dampens dopamine production in the brain. It has been estimated that GABA serves as a neurotransmitter at 30% of all synapses there. It plays a key role in mechanisms involved in anxiety, and its receptors are heterogeneous. *GABA$_A$* receptors are postsynaptic and are involved with pentameric protein chloride ion channels. Some evidence suggests that there are multiple subtypes of GABA$_A$ receptors in the brain.

Certain naturally occurring steroids are potent allosteric modulators of GABA$_A$ receptors, but the physiological significance of this finding is not clear. Some central *GABA$_B$* receptors are presynaptic and some are postsynaptic. There are at least two different subtypes, and they are G protein-linked. One of the functions of central GABA$_B$ receptors is indirect regulation of potassium and calcium ion channels. Stimulation of GABA$_B$ receptors in certain areas of the spinal cord produces skeletal muscle relaxation. Some GABA$_B$ agonist drugs are used in treating skeletal muscle spasticity. *GABA$_C$* receptors are chloride ion channels and they are found in several regions of the brain and also in the retina of the eye. The physiology of these receptors is incompletely understood.

GABA is synthesized in the body by enzyme-mediated decarboxylation of glutamic acid (scheme 9.2). As with noradrenergic and dopaminergic nerves, GABAergic nerve impulses are terminated by re-uptake of the neurotransmitter from the synaptic cleft into the nerve terminal and thence back into the synaptic storage vesicles. GABA is destroyed by a widely distibuted enzyme, GABA-transaminase, which catalyzes transfer of the

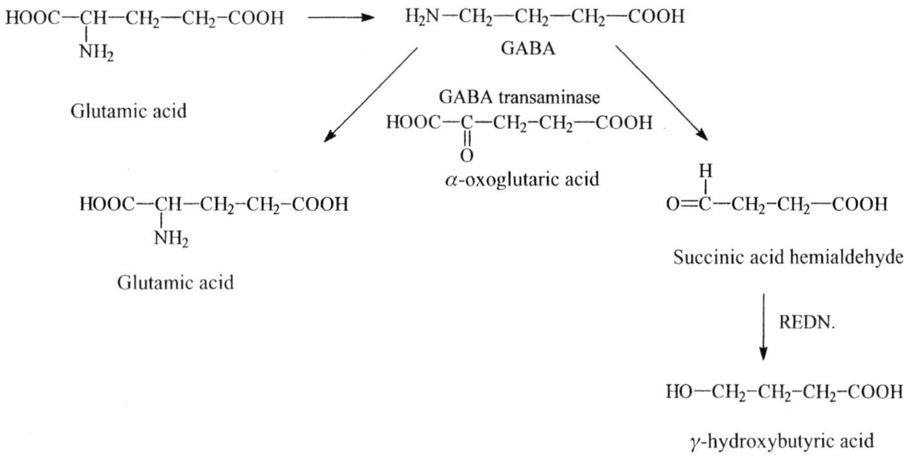

$$HOOC-\underset{\underset{NH_2}{|}}{CH}-CH_2-CH_2\text{-}COOH \longrightarrow H_2N-CH_2-CH_2-CH_2-COOH$$

Glutamic acid

GABA

GABA transaminase

$$HOOC-\underset{\underset{O}{\|}}{C}-CH_2\text{-}CH_2-COOH$$

α-oxoglutaric acid

$$HOOC-\underset{\underset{NH_2}{|}}{CH}-CH_2\text{-}CH_2\text{-}COOH$$

Glutamic acid

$$O=\overset{\overset{H}{|}}{C}-CH_2\text{-}CH_2-COOH$$

Succinic acid hemialdehyde

REDN.

$$HO-CH_2\text{-}CH_2\text{-}CH_2\text{-}COOH$$

γ-hydroxybutyric acid

Scheme 9.2 Biosynthesis and metabolic fate of gamma-aminobutyric acid (GABA)

amino group of GABA to α-oxoglutaric acid to form glutamic acid, thus converting GABA into succinic acid semialdehyde, which is subsequently in part oxidized to succinic acid.

γ-Hydroxybutyric acid (GHB) (scheme 9.2) is an endogenous component of mammalian (including human) brains. The physiological role of endogenous GHB is not well understood. It may be a so-called neuromodulator. γ-Hydroxybutyric acid in the blood stream penetrates the blood–brain barrier. It interacts with binding sites in the cerebral cortex, the limbic areas, and the thalamus, in addition to other brain regions innervated by dopaminergic neurons. The compound is recognized and is officially classed as a drug of abuse. Ingested orally, it induces euphoria and it has sedative properties. Low doses produce anxiolytic and muscle relaxant effects, while larger doses increase REM sleep time. Chronic use can lead to physical dependence. An overdose of γ-hydroxybutyric acid can result in dizziness, nausea, vomiting, confusion, epileptic seizures, coma, and respiratory depression. In rare instances, death can result. The toxic effects of the drug are potentiated by ethanol, opioids, barbiturates, and benzodiazepines. GHB is efficient in preventing memory storage in the brain, and it has been implicated in cases of "date rape".

9.2.3 Aspartic Acid and Glutamic Acid

The latest edition of *Goodman and Gilman's Pharmacological Basis of Therapeutics* states that aspartic acid (α-aminosuccinic acid) is "possibly" a neurotransmitter in the CNS. A complete definition of a unique physiological role for aspartic acid has not appeared.

Glutamic acid (the biosynthetic precursor to GABA) is described as the major excitatory neurotransmitter in the brain and spinal cord. It contributes input to the majority of excitatory synapses. One type of glutamate receptor, the *NMDA receptor* (*N*-methyl-D-aspartate) has been identified pharmacologically and cloned. This receptor for glutamic acid is so-called because *N*-methyl-D-aspartic acid is one of the principal agonists that activate it. Despite its name, this CNS receptor serves physiologically as a receptor for glutamic acid; the normal human body neither contains nor utilizes *N*-methyl-D-aspartic acid. NMDA receptors are ion channels, involving Na^+, K^+, and Ca^{2+}. Physiologically, NMDA receptor channels can be blocked by Mg^{2+}. The literature frequently refers to NMDA receptors as *ionotropic* glutamate receptors. The NMDA receptor plays a significant role in such diverse neurotransmitter-related disorders as epilepsy, stroke, and Huntington's chorea, a hereditary disease characterized by progressively worsening dementia and uncontrolled, involuntary jerking of the skeletal muscles.

When a stroke occurs, nerve cells deprived of a blood supply expend their energy reserves within a few minutes, and they can no longer maintain their resting potential. In response to this crisis, large amounts of glutamic acid are released into the synaptic cleft and they activate NMDA receptors. This activation of NMDA receptors causes a massive influx of Ca^{2+} into the cells and the resulting profound rise in levels of cytoplasmic Ca^{2+} stimulates Ca^{2+}-dependent intracellular enzymes to become hyperactive. The affected nerve cells are described as burning out from this "metabolic frenzy", and this phenomenon is postulated to be the cause of the irreversible skeletal muscle paralysis that frequently results from a stroke. Accordingly, drug therapy to prevent stroke-derived paralysis might involve administering an NMDA receptor-blocking agent. A problem that has arisen in the search for useful NMDA receptor blockers for treatment of stroke is the tendency of these agents to produce psychotropic side effects (hallucinations) similar to those resulting from the use of phencyclidine (vide infra), as well as some other side effects. The use of NMDA receptor blockers in treatment of stroke has not been successful.

Two additional central glutamic acid receptors have been identified: the *AMPA* (α-amino-3-hydroxy-5-methylisoxazole) **9.3** receptor and the *kainate* (2-carboxy-4-[1-methylvinyl]-pyrrole-3-acetic acid) **9.4** receptor, both named for the exogenous, non-physiological agonists by which they were first identified. Despite their names, as for the NMDA receptor, the AMPA and kainate receptors are both physiological receptors for glutamic acid. The physiological role(s) of these latter two glutamic acid receptor subtypes is/are imperfectly understood. Kainate and AMPA receptors are ligand-gated ion channels (Na^+, K^+, Ca^{2+}), and they are included in the ionotropic category of glutamate receptors.

9.3   9.4

There is another series of functional types of glutamic acid receptors, known as *metabotropic receptors*. These are G protein-linked, and there are several subtypes. Their physiological role is as yet imperfectly understood. They indirectly modulate K^+ and Ca^{2+} channels and they have some presynaptic functions to inhibit glutamate release from the nerve terminal. The action of glutamic acid as a neurotransmitter at all of its neural sites is terminated by an active transport uptake mechanism.

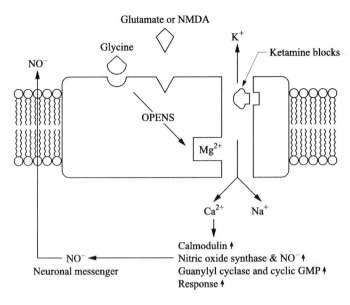

Figure 9.2 Glutamic acid NMDA receptor and associated glycine receptor

9.2.4 Glycine

The amino acid glycine is a major inhibitory postsynaptic neurotransmitter in the interneurons in the gray matter of the spinal cord. When exogenous glycine is applied to this region, it produces an inhibitory effect on nerve impulse transmission. However, ion channel opening by activation of the glutamic acid NMDA receptor requires simultaneous interaction of glycine and of glutamic acid with their respective receptors. Glycine is said to be a co-agonist with glutamic acid. Thus, in its physiological involvement with NMDA receptors, glycine is an excitatory transmitter. Figure 9.2 shows a diagrammatic perspective of the NMDA/glycine receptor complex.

Tetanus toxin blocks the release of glycine from its storage sites in the spinal inhibitory interneurons. The lethal convulsant effect of the alkaloid strychnine superficially resembles that of tetanus toxin and reflects competitive antagonist action on the inhibitory effects of glycine at the interneurons in the gray matter of the cord to produce violent muscular spasms. The widely abused hallucinatory, possibly addicting drug *phencyclidine* ("PCP", "angel dust") **9.5** blocks the ion channels at the NMDA/glycine receptor region. However, this effect does not represent the total mechanism of action of the drug. Phencyclidine also acts upon σ receptors, and some authorities believe this to be the drug's site of hallucinatory action (see chapter 12).

9.5

9.2.5 Adenosine and Adenosine Phosphate Esters

There are receptors in the central nervous system that are specific for adenosine (figure 9.1). These are also called purinergic receptors (The nonsugar portion of the adenosine molecule is a derivative of the purine ring, a structure closely related to those of caffeine **9.6** and theophylline **9.7**). Adenosine receptors are widely distributed peripherally, and stimulation of central and peripheral adenosine receptors produces a wide variety of cellular responses. The physiological role of adenosine in the nervous system is profound and complex and is as yet imperfectly understood. Many authorities do not classify adenosine as a true neurotransmitter, but rather they term it a *neuromodulator*. Adenosine inhibits the release of true neurotransmitters from presynaptic storage sites in the central nervous system, and in some instances adenosine modifies the postsynaptic effects of neurotransmitters. There are at least four subpopulations of adenosine receptors, each with some degree of agonist selectivity: A_1, A_{2A}, A_{2B}, and A_3. Adenosine receptors are G protein-linked, involving production of cyclic AMP and of some other second messengers. A_1 adenosine receptors inhibit adenylate cyclase and reduce cyclic AMP levels, and A_2 receptors stimulate the enzyme and increase cyclic AMP levels.

9.6 Caffeine R=R′=R″=CH$_3$
9.7 Theophylline R=R′=CH$_3$, R″=H

Some of the pharmacological effects of the xanthine alkaloids (caffeine, theophylline, theobromine) result from their antagonism of adenosine receptors, both centrally and peripherally. Also, caffeine and theophylline inhibit some subtypes of *phosphodiesterases*, the enzyme family that metabolically inactivates cyclic AMP. This action probably contributes to the overall pharmacological effects of the xanthine alkaloids. A complication to the therapeutic use of adenosine receptor ligands (agonists, antagonists) is that the effects of acute administration of a particular ligand can be diametrically opposite to the chronic effects of the same ligand. A explanation for this observation has not been provided.

Adenosine triphosphate (ATP) is, a neurotransmitter (or co-transmitter) both centrally and peripherally. Its heterogeneous receptors are designated as P_{2X} and P_{2Y}, and there are several subtypes. P_{2X} receptors are ligand-gated ion channels and P_{2Y} receptors are G protein-linked. The physiology of all of these ATP receptors is incompletely understood.

9.2.6 Histamine

Histamine is formed in the body by decarboxylation of the amino acid histidine (equation 9.2).

$$\text{(structure)} \quad \longrightarrow \quad \text{(structure)} \qquad (9.2)$$

Histamine plays a role as a chemical messenger in a variety of complex biological processes. It is stored in many body tissues in an inactive form from which it is released by a variety of stimuli and mechanisms. There are three well-defined histamine receptor types: The classic peripheral H_1 receptors are involved in the allergic response. H_2 receptors in the stomach promote the release of hydrochloric acid, and H_2 receptor antagonists are used therapeutically to treat peptic ulcer and acid reflux disease. H_2 receptors in the heart (like β-adrenoceptors there) elevate cyclic AMP levels resulting in increased heart rate and increased contractile force of the heart muscle. Postsynaptic H_1 and H_2 receptors are found in the central nervous system where histamine is a true neurotransmitter substance. The side effect of drowsiness produced by many of the antiallergy "antihistaminic" drugs reflects inhibitory actions on central histamine receptors. Central H_1 receptors are G protein-linked, and their activation causes elevation of inositol triphosphate and diacylglycerol second messengers. Activation of these receptors produces wakefulness. H_2 receptors are also G protein-linked and their activation causes elevation of cyclic AMP levels. H_2 receptors play an important role in the cerebral cortex and in the hippocampus where mood and emotion are controlled. H_3 receptors are found primarily in the central nervous system, and they are presynaptic. They control the synthesis and release of histamine in the cortex. These H_3 receptors have been described, together with α_2 adrenoceptors, as principal inhibitory neurotransmitter systems in the brain, modulating serotonergic, cholinergic, noradrenergic, dopaminergic, and peptidergic pathways. Histaminergic pathways in the brain play a role in regulating cerebral blood flow and in states of wakefulness. An H_4 receptor is found in white blood cells and intestinal tissue. It is negatively coupled to a G protein, but its physiology has not been competely described, and its role (if any) in nerve impulse transmission is not known.

The effects of histaminergic nerve impulses are terminated by a reuptake process. There are scavenger enzyme systems which metabolically inactivate those histamine molecules that escape reuptake (scheme 9.3).

9.2.7 Nitric Oxide

Nitric oxide, NO, is a physiological component of the human body; it is biosynthesized from the amino acid arginine, as shown in scheme 9.4. *Nitric oxide synthase*, the critical enzyme system in this biosynthetic sequence, exists as a family of closely related enzymes. They are activated by Ca^{2+}. Moment-to-moment control over the rate of NO synthesis depends upon the concentration of intracellular Ca^{2+}, which in turn is involved in activation of glutamic acid receptors. Inhibitors of these enzymes are subjects of investigation as potential therapeutic agents. It has been stated that there is no physiological storage mechanism for nitric oxide, and that it must be synthesized on demand. However, experimental data reveal that the amino acid cysteine can be converted in the

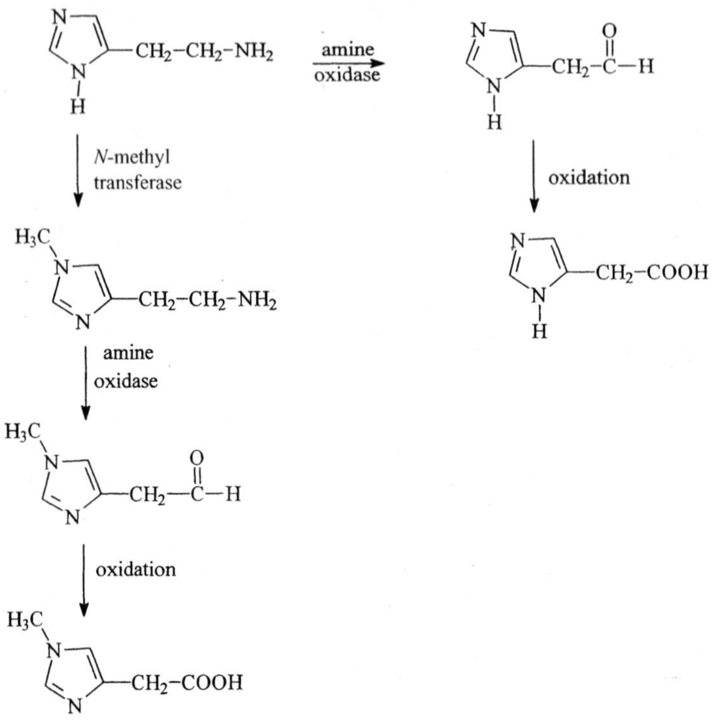

Scheme 9.3 Metabolic fate of histamine

body into *S*-nitrosocysteine, and that two of these molecules can undergo a reaction releasing two equivalents of nitric oxide (scheme 9.5). It has not been established that these reactions represent a physiological storage/release mechanism for nitric oxide.

Once released, nitric oxide diffuses into nearby cells where it interacts with appropriate target molecules. Nitric oxide is involved in peripheral and central neurotransmission and among other physiological actions, it participates in regulating blood flow and arterial pressure. The brain contains more nitric oxide than any other organ, thus reflecting nitric oxide's vital role as a central neurotransmitter. Nitric oxide exists in

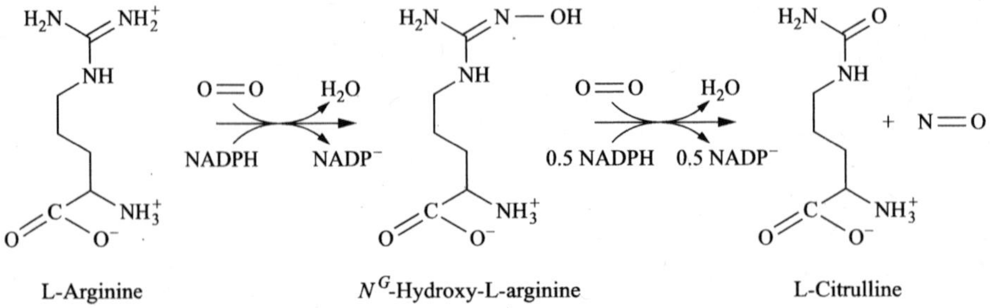

Scheme 9.4 Biosynthesis of nitric oxide

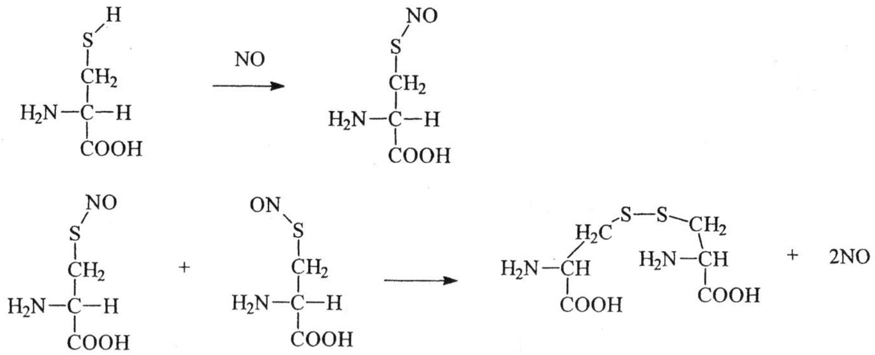

Scheme 9.5 Fixation and release of nitric oxide

three oxidation states: the electrically neutral NO· which is a free radical; a *nitrosonium cation*, NO^+; and a negatively charged *nitroxyl anion*, NO^-, each of which seems to have its own physiological role. It appears that the NO^+ (nitrosonium) cation attaches covalently to a sulfur atom of a disulfide linkage in the NMDA glutamate receptor domain to block neurotransmission, producing a neuroprotective effect. The nitrosonium ion forms an iron–nitrosyl moiety with the heme–Fe^{2+} portion of the G protein-linked enzyme guanylate cyclase and activates it. This promotes formation of a second messenger, guanosine-3′,5′-monophosphate (cyclic GMP) that exerts a multiplicity of physiological effects, among which is reduction of of levels of free intracellular Ca^{2+}, which leads to muscle relaxation. The mechanism of action of *sildenafil (Viagra®)* **9.8**, the drug used in treatment of erectile dysfunction, is related to one of the roles of nitric oxide in nervous tissue. As shown in figure 9.3, nerve cells in the penis manufacture nitric oxide which diffuses into the blood vessels of the penis where it binds to the G protein-linked enzyme guanylate cyclase, activating it and signaling it to synthesize more of the second messenger guanosine-3′,5′-monophosphate ("cyclic GMP"). As described previously, the effect of cyclic GMP on Ca^{2+} transport causes the smooth muscle fibers of the penile blood vessel walls to relax, and the vessels become engorged with blood, which is the erection. Sildenafil inhibits a phosphodiesterase ("PDE5") which normally inactivates cyclic GMP, protecting the second messenger from metabolic destruction, and prolonging and magnifying its action in producing the erection. Sildenafil has no aphrodisiac activity.

Nitric oxide has some gastrointestinal effects, in maintenance of the integrity of the gastric mucosa. It stimulates mucus secretion and participates in maintenance of mucosal blood flow.

The free radical form NO· can react with superoxide anion ($O\text{-}O^-$) to form peroxynitrite anion ($ONOO^-$) which is said to be responsible for some of the cytotoxic effects produced by nitric oxide. Nitric oxide produces generalized peripheral vasodilation, resulting in a fall in blood pressure (see chapter 15) and it may be a key participant in migraine and other vascular headaches.

Nitric oxide reacts with oxygen to form N_2O_4 which subsequently combines with water to form nitrate and nitrite anions. The nitrite anions are oxidized to nitrate by oxyhemoglobin.

9.8

9.2.8 Peptide Neurotransmitters

In addition to the small molecule neurotransmitters described previously, there is a significant number of low molecular weight peptides in the brain and in other parts of the central nervous system which probably act as neurotransmitters. This subject is incompletely understood, and it has been proposed that some of these peptides modify the effects of authentic nonpeptide neurotransmitter substances. Vida (Recommended Reading, ref. 31) has listed 40 peptides that have been detected in neurons and nerve terminals within the mammalian central nervous system, excluding those related to endocrine or neuroendocrine functions (Table 9.1).

One member of a family of putative neurotransmitter peptides, the *tachykinins*, is an undecapeptide, *substance P*. This peptide is involved in pain perception, both peripherally and centrally. Nerve fibers in peripheral pain-sensitive neurons use substance P as

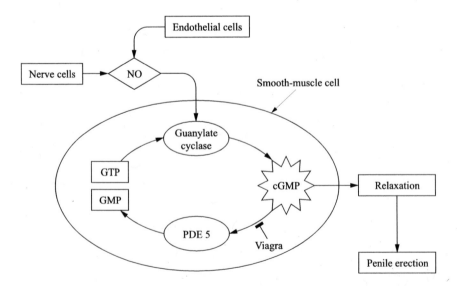

Figure 9.3 Site and mode of action of sildenafil (Viagra®)

Table 9.1 Neuropeptides

Pituitary peptides	Neuropeptide Y
Corticotropin (ACTH)	Neurolinin A and B
Growth hormone (GH)	Pancreatic polypeptide
β-Lipotropin	Secretin
α-Melanocyte-stimulating hormone	Substance P
Oxytocin	Vasoactive intestinal polypeptide
Thyroid-stimulating hormone	**Opioid peptides**
Vasopressin	Dynorphin
Circulating hormones	β-Endorphin
Angiotensin	Met-enkephalin
Glucagon	Leu-enkephalin
Gastric inhibitory peptide	Kyotorphin
Insulin	**Hypothalamic-releasing hormones**
	Bradykinin
Gut hormones	Corticotropin-releasing factor
Avian pancreatic polypeptide	Gonadotropin-releasing hormone
Bombesin	Growth hormone-releasing factor
Calcitonin gene-related hormone	Luteinizing hormone-releasing hormone
Cholecystokinin	**Miscellaneous peptides**
Gastrin	Bradykinin
Physalaemin	Camosine
Eleidosin	Neuropeptide γ
Cerulein	Neurotensin
Motilin	Proctolin
Neurotensin	

the excitatory neurotransmitter. Morphine-like analgesics inhibit the release of substance P in the spinal cord. Substance P acts in the central nervous system to regulate the release and re-uptake of catecholamines: norepinephrine, dopamine, and epinephrine.

Other central nervous system peptides are the families of *endorphins*, *enkephalins*, and *dynorphins*, molecules which have true analgesic effects and for which there are well-defined central nervous system receptors. It was originally believed that all of these receptors are postsynaptic, but it is now concluded that some presynaptic receptors exist.

Neuropeptide Y, a 35 amino acid molecule, interacts with several G protein-linked receptor systems. It is released as a co-transmitter with norepinephrine at many noradrenergic nerve terminals and it enhances the vasoconstrictor effect of norepinephrine. The mechanism for this process is unknown. Among the multiplicity of its other putative physiological effects are enhanced cognitive function, increased food intake, and shifts of circadian rhythms, in addition to effects on the gastrointestinal tract. There are at least three subtypes of neuropeptide Y receptors.

9.2.9 Physiological Inter-relationships of Neurotransmitters

From the preceding discussions, it is apparent that the physiological activities and the biochemistries of neurotransmitter substances are intimately and extensively interconnected and interdependent. The relationship of glutamic acid NMDA receptors and glycine has been cited. A noradrenergic or serotonergic nerve terminal may bear presynaptic cholinergic receptors. Some noradrenergic nerve terminals bear presynaptic dopamine receptors. Thus, neurotransmitters frequently regulate each other's activity. Although it is convenient to refer, for example, to cholinergic pathways, serotonergic pathways, or dopaminergic pathways, the nervous system and its multitude of neurotransmitter substances are actually integrated into a cohesive whole. Strictly speaking, it seems improper (and misleading) to designate cholinergic, noradrenergic, gabanergic, or other "nervous systems". The human body has only one unitary, indivisible nervous system.

9.3   Functions of Some Brain Regions

The *limbic system* (hippocampus, amygdaloid body, septal nuclei, olfactory bulb) (figure 9.4) is an assembly of brain regions involved in control of mood, emotion, memory, and motivational activities. Artificial stimulation of the limbic region of laboratory animals can cause patterns of rage, escape, and other emotional reactions.

One of the major pathways in the brain by which motor nerve signals are transmitted from motor areas of the cerebral cortex to the spinal cord is the *pyramidal tract* (so named for the conical shape of the cells) (figure 9.5). The *extrapyramidal tracts* collectively are all other tracts that transmit motor signals from the cortex to the cord. This latter system is involved with coordination and integration of fine muscular movements,

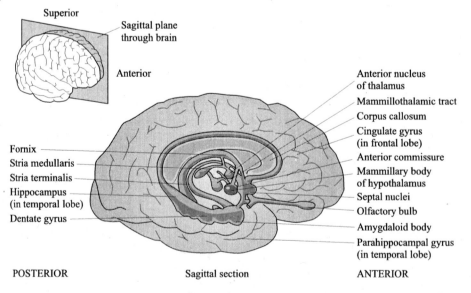

Figure 9.4 Components of the limbic system and surrounding structures. (Reproduced with permission from reference 3. Copyright 1999 Benjamin/Cummings Science)

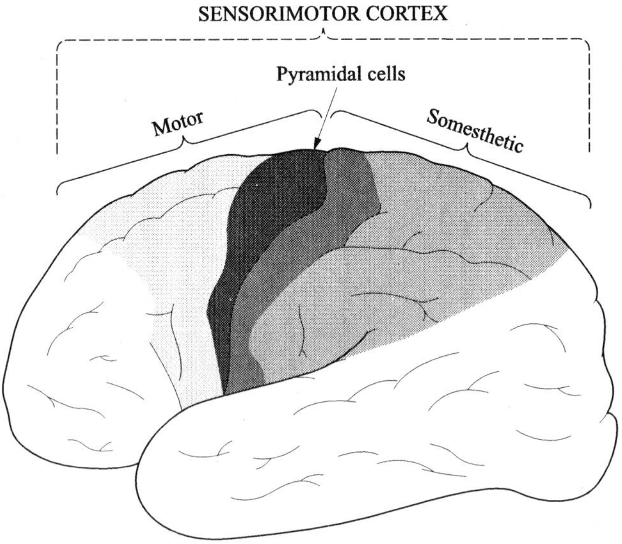

SENSORIMOTOR CORTEX

Figure 9.5 Pyramidal and extrapyramidal tracts. (Reproduced with permission from reference 4. Copyright 1981 Saunders)

such as picking up small objects. The extrapyramidal system also participates in controlling body movement and posture. Damage to the extrapyramidal region depresses the ability to initiate voluntary movements. The muscular tremors and spasticity of Parkinson's disease and the uncontrollable limb movements of Huntington's chorea originate in malfunction within the extrapyramidal region.

Figure 9.6 illustrates schematically that stimulation of the brain at any point along the arrows results in an upward movement of the stimulus reponse, ultimately to excite large portions of the cerebral cortex. This is the *reticular activating system* (RAS) that projects widely into the forebrain, midbrain, and cerebellum. The RAS is essential for regulating sleep, wakefulness, and arousal level, as well as coordination of eye movements. Stimulation of the RAS produces wakefulness, and depression results in sedation.

9.4   Animal Screening of Psychotropic Drug Candidates

It is difficult to develop animal tests to screen drugs that alter purely cognitive or other psychic aspects of human activity. Nevertheless, psychopharmacology has evolved to the point where distinct classes of psychotropic drugs have been found that alter predictably one or several types of measurable animal behavior. In those many instances where the animal response does not resemble the human condition, a correlated, quantifiable measure in the animal is substituted for a more psychic or cognitive behavior. These assays are useful and they have predictive value, but they are empirical and their applicability is frequently narrowly limited to a specific category of

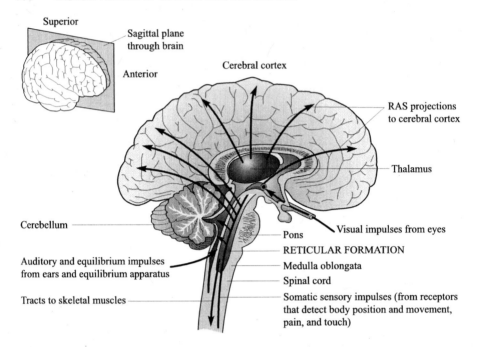

Figure 9.6 The reticular activating system. (Reproduced with permission from reference 3. Copyright 1999 Benjamin/Cummings Science)

chemical structures and mental ailments and to specific animal species. The following are representative of the kinds of whole animal screening strategies employed by the psychopharmacologist:

9.4.1 Avoidance

Animals can learn to control unpleasant stimuli, such as electric shock, by escape or avoidance responses. If a rat is given an opportunity to control the shock by pressing a lever, it has two options. (1) If it presses the lever while the shock is being delivered, it can shut it off. This is escape behavior. (2) If it presses the lever before the shock is delivered, it can prevent the shock. This is avoidance behavior. In a *signaled avoidance* a warning, such as a tone, signals the impending shock. In a *Sidman avoidance* the animal receives no warning, and the shocks occur at a fixed interval. Pressing the lever during the interval postpones the shock, and if the animal responds regularly, it can avoid the shock indefinitely. This type of test is used to screen and assess drug candidates for their effect on human memory and learning. The ability of the drug candidate to alter the animal's response/reaction to the shock stimulus is observed and quantified.

9.4.2 Rotating Rod

Rodents are placed on a slowly rotating rod, and the time before the animal falls off is measured. The difference between pre- and postdrug scores gives an indication of muscle coordination and muscular strength. Some aspects of antipsychotic drug activity can be detected in this assay.

9.4.3 Cataleptic Effects

Catalepsy is the abnormal maintenance of distorted postures, often termed "waxy flexibility." Animals that demonstrate catalepsy remain in bizarre postures when positioned by the experimenter. The time for restoring normal posture may indicate the extent of catalepsy. The relative potency of some drugs in producing catalepsy in rats parallels their relative potency in clinical treatment of schizophrenia and some other psychoses.

9.4.4 Swimming Maze

The ability of rodents to negotiate their way by swimming through a water-filled maze before and after candidate drug administration correlates with some human psychic responses.

9.4.5 Control of Aggressive Behavior in Rats or Mice

The aggressive tendencies of populations of animals can be greatly magnified by various experimental techniques: chronic mild electric foot shock; creation of brain lesions; isolation of single animals; administration of drugs such as levodopa. Potential human anxiolytic drugs are sometimes screened for their ability to diminish aggressive tendencies in these animals.

9.5 Antidepressants

Compounds **9.9–9.12** are typical of the so-called *tricyclic* antidepressants. For many years a widely held and popular theory of the biochemical cause of mental depression was centered on a deficiency of excitatory influences of biogenic amines (particularly norepinephrine and perhaps dopamine) in the brain. It was proposed that the tricyclic antidepressants increase the availability of these biogenic amines at their postsynaptic receptor sites, and thus they reverse the depression. These drugs do indeed inhibit re-uptake of norepinephrine, dopamine, and serotonin into the nerve terminals, and considerable

9.9 Imipramine   **9.10** Amitriptyline   **9.11** Doxepin

9.12 Protryptyline

experimental data suggest that inhibition of the uptake-1 process for norepinephrine parallels the antidepressant effect of several drugs. Inhibition of serotonin re-uptake also has antidepressant activity, but the increased serotonin level in the synaptic clefts also produces sedation.

Metabolic inactivation of the tricyclics involves two principal routes: hydroxylation (e.g., at position 2 of imipramine and at a benzylic position of amitriptyline) and N-demethylation of the *N,N*-dimethylamino moiety to produce the secondary amine. Both of the metabolite types retain some degree of antidepressant activity and some of the secondary amine metabolites are utilized clinically. A small amount of these tertiary amine drugs is converted into the *N*-oxide derivative of the *N, N*-dimethylamino moiety.

One of the most widely used antidepressant drugs, *fluoxetine* **9.13** is not a tricyclic. This drug inhibits re-uptake of serotonin at the presynaptic neuronal membrane. The resulting elevated concentrations of serotonin in the synaptic clefts in the central nervous system enhance serotonergic neurotransmission. Fluoxetine has little or no effect on re-uptake of other central nervous system neurotransmitters. In addition to its antidepressant action, fluoxetine has clinical value in relief of the symptoms of anxiety and of some cognitive disturbances.

9.13 Fluoxetine

Inhibition of dopamine re-uptake seems to be associated with central nervous system stimulant actions rather than with true antidepressant effects. There is increasing doubt that inhibition of norepinephrine and/or serotonin uptake adequately explains the antidepressant effects of the tricyclic drugs. Although those tricyclic antidepressants that inhibit re-uptake of biogenic amines produce this effect very soon after the first dose, the appearance of a clinical antidepressant response requires administration of the drug for several weeks. Thus, potentiation of noradrenergic or serotonergic transmission (if indeed this is significant mechanistically) may be only a very early event in a complex cascade of events that results in antidepressant action.

Another theory of the underlying cause of clinical depression suggests a supersensitivity of adenylate cyclase-involved norepinephrine receptors in the central nervous system, and these receptors may be improperly regulated in depressed patients. According to this hypothesis, clinical depression is not caused by a lack of neurotransmitters but rather by overstimulation of certain norepinephrine receptors. These receptors are desensitized ("down-regulated") during chronic treatment with tricyclic antidepressant drugs. This theory provides an explanation for the observation that there is a long induction period between the initiation of drug therapy and the appearance of a clinical effect. There is a strong implication that *presynaptic* receptors are involved in clinical depression: that is, those receptors on the nerve terminal for which stimulation results in inhibition of release of norepinephrine into the synaptic cleft for subsequent interaction with the post-synaptic receptors to produce an appropriate response to the nerve impulse. However, it seems likely that this explanation also is simplistic, and that clinical depression is a biochemically heterogeneous syndrome, resulting from a combination of several (many?) biochemical factors. It is reasonable to assume that not all depressed states represent exactly the same combination of contributing malfunctions.

It might be expected that an effective antidepressant drug would have a stimulant or mood elevating effect in normal individuals, but this is not the case with the tricyclics. For example, a normal therapeutic dose of imipramine produces sedation and light-headedness in nondepressed individuals. Most of the tricyclics display more or less prominent antimuscarinic side effects (dry mouth, blurred vision), quite similar to those of atropine, and these effects are manifested in both depressed and nondepressed individuals. Overdosage with tricyclics can be life-threatening, producing potentially dangerous changes in conduction of nerve impulses through the heart. *Postural hypotension* can occur, which can be quite severe. While the margin of safety of the tricyclics in humans is not known for certain, deaths have been reported from total doses of 2000 mg. The tricyclics potentiate the depressant effects of ethanol and probably of other sedatives also.

9.6   Cocaine

The alkaloid cocaine **9.14** blocks re-uptake of dopamine, norepinephrine, and serotonin into nerve terminals (uptake-1) in the central nervous system as well as at peripheral sites. Cocaine produces hyperactivity in central noradrenergic, dopaminergic, and serotonergic pathways. Dopaminergic hyperactivity has been invoked to explain the CNS excitement and euphoria produced by cocaine. Qualitatively, cocaine's CNS effects are very similar to those of amphetamine, whose mechanism of action also involves inhibiting uptake-1 mechanisms. It has been stated that experienced human subjects cannot distinguish between the subjective effects of cocaine and amphetamine. There is a cocaine receptor site at the region of the membrane protein complex that is involved in the active transport of dopamine from the synaptic cleft into the nerve terminal. When cocaine binds to this site, it denies access of dopamine to the transport mechanism and the concentration of dopamine in the synaptic cleft is elevated. The likely role of dopamine in those regions of the brain where mood and emotion are controlled has been cited previously. Probably this effect on dopamine re-uptake is the most important component of cocaine's psychic effects. Cocaine produces a dose-dependent increase in heart rate and blood pressure. There is CNS arousal and a sense of self-confidence and well-being. Higher doses produce euphoria of brief duration and a desire for more drug. The half–life of cocaine in plasma is about 50 min, but users by inhalation desire more drug after 10–30 min. Cocaine is truly addicting. The major route for its metabolism is hydrolysis of one or both of the ester groups. The major urinary metabolite is benzoyl ecgonine, the product of hydrolysis of the methyl ester moiety. Among the risks of cocaine use are cardiac arrhythmias, myocardial ischemia, cerebral vasoconstriction, and seizures. Symptoms of cocaine withdrawal are generally mild. The major challenge in treatment of the addiction is not detoxication, but rather helping the user to resist the compulsive use of the drug (psychological dependence).

9.14

Cocaine is readily absorbed by many routes and it readily penetrates the blood–brain barrier. The hydrochloride salt has been injected intravenously to produce an immediate and intense euphoria. Nasal inhalation of cocaine hydrochloride produces a somewhat less intense response, although nasal inhalation is popular due to its ease and convenience of administration. Chronic nasal administration can cause atrophy and necrosis of the nasal mucosa and septum, due at least in part to the potent vasoconstrictor effect on the small blood vessels of the nose (a result of elevated norepinephrine levels in these vessels which produces profound vasoconstriction and resultant deficiency of blood supply to these tissues). "Crack", the free base form of cocaine, can be smoked to produce a rapid, intense effect. Cocaine can severely impair brain development in utero and the incidence of neurological and limb malformations is increased.

Cocaine is not useful in treating clinical depression. Its qualitative effects in depressed patients differ from those of the tricyclics and the other antidepression drugs. This fact provides added support to the argument that elevation of neurotransmitter levels in synaptic clefts is an inadequate explanation for the antidepressant effects of the entire series of antidepressant drugs. Cocaine has local anesthetic properties. (see chapter 13).

9.7 Cannabis

Although a discussion of cannabis (marijuana) is not pharmacologically consistent with the topic of antidepressants, it can reasonably follow a discussion of cocaine as a drug of abuse. The crude drug *Cannabis sativa* and its smoke contain a multiplicity of compounds, collectively termed *cannabinoids*, but the most important of these is Δ^1-tetrahydrocannibinol (also named in the literature Δ^9-tetrahydrocannabinol) **9.15**.

Probably other pharmacologically active derivatives arise from thermolysis reactions accompanying the smoking of marijuana and from metabolism of marijuana constituents.

A *cannabinoid receptor* (CB$_1$) has been found (among other sites) in the cerebral cortex and the cerebellum: those regions of the brain involved in mood, emotion, and memory. Two endogenous ligands for this receptor have been identified: an arachidonic acid derivative, *anandamide* **9.16** and *2-arachidonylglycerol* **9.17**. There is a second cannabinoid receptor (CB$_2$) that is apparently found only peripherally. Little is known of its physiological role. Anandamide is a ligand for both types of receptors. Both CB$_1$ and CB$_2$ receptors are negatively coupled G protein-linked, leading to inhibition of the catalytic activity of adenylate cyclase. The physiological role(s) of cannabinoid receptors and of anandamide is/are incompletely understood. However, it seems established that anandamide is produced on demand when the body feels pain. When it interacts with the cannabinoid receptors, it blunts the pain sensation. An endogenous enzyme, *anandamide amidohydrolase*, metabolically inactivates anandamide. An emerging analgesic strategy is directed at blocking the catalytic effect of this enzyme, thus protecting anandamide from metabolic destruction. However, the physiological role(s) of anandamide and of 2-arachidonylglycerol is/are incompletely understood.

9.15

9.16

9.17

Marijuana smoking produces a series of central and peripheral responses. The most obvious subjective effects are a feeling of relaxation and well-being, not unlike the response to moderate-size doses of ethanol. There is a sensation of sharpened sensory awareness, with sights and sounds seeming more intense and fantastic. Subjects report that time seems to pass extremely slowly. Cannabis increases the appetite. Contrary to popular belief, increased sexual activity has not been demonstrated in controlled experiments with cannabis. Psychomotor performance is impaired: simple learning and memory tasks are not well performed, and complex motor coordination (such as is required in driving an automobile) is impaired. Subjective feelings of confidence and heightened creativity are not reflected in actual performance. Unpleasant sensations such as panic or hallucinations have been reported. A considerable percentage of chronic marijuana users has reported at least one anxiety experience.

Marijuana has demonstrated analgesic effects, and it is an impressive anti-emetic. A recent (1999) study suggested that the analgesic mechanisms of the opioids and of cannabis are "separate but related". Cannabis reduces intraocular pressure in the eye, and hence it has value in treatment of certain types of glaucoma. It produces bronchodilation.

It produces tachycardia (speeded heart rate) which is prevented by administration of drugs that block sympathetic transmission. For almost all of these effects, no mechanism has been determined.

There seems to be no convincing evidence that marijuana damages brain cells or that it causes permanent functional damage. The literature describes withdrawal symptoms of marijuana, but these are mild. Psychological dependence does not seem to occur with cannabis, and many authorities conclude that it should not be considered to be addictive. Chronic users of marijuana may experience residual drug effects for several weeks after cessation of use of the drug.

9.8   Mood-Stabilizing Agent

Lithium cation, utilized as its carbonate or citrate salt, stands alone in this category. This drug is used in treatment of manic depressive illness, also termed *bipolar disorder*. The condition is characterized by an expanded emotional state, elation, hyper-irritability, and over-talkativeness alternating with periods of severe depression. Lithium cation is remarkable in that therapeutic doses have almost no discernible effect in non-manic individuals. It is not a sedative, nor a depressant, nor a euphoriant, and these characteristics distinguish lithium from almost every other psychotropic agent known. Lithium blocks both extremes of the manic state: the hyperactivity and expansive emotional phase as well as the depression phase. How can such actions of the drug be explained, which is both a "downer" and an "upper" in bipolar patients, but has no effect in non-manic individuals? Traces of lithium cation occur in animal tissue, but its physiological role (if any) is unknown. Although lithium salts are valuable in long-term treatment of bipolar disorder, they are not useful in controlling acute phases of mania, because the calming effects of the drug are manifested only after several days of dosage. The cause of this extended lag period is unknown.

Exogenous lithium cation demonstrates a multiplicity of effects in the body, but it has been proposed that its psychological effects may be related to one or the other (or both) of two mechanisms. First, lithium cation blocks a step in the *inositol phosphate metabolic pathway* (figure 9.7).

Stimulatory G protein-linked receptors activate the enzyme phospholipase C, which hydrolyzes a membrane phospholipid, phosphatidyl inositol-4, 5-bisphosphate (PIP) to form two new second messengers, a diacyl glycerol and inositol-1,4,5-triphosphate. Lithium disrupts the biochemistry of formation of this second messenger system. Second, lithium cation inhibits other enzymes, resulting in lowered concentrations in the cell of one of the stereoisomers of inositol, *myoinositol*. This reduction of myoinositol content in the cell has been reported to attenuate the brain response to external stimuli, and this has been suggested to be the mechanism of action of lithium cation in affective disorders, since dietary inositol cannot penetrate the blood–brain barrier and the brain cells must synthesize their own. It has been proposed that the mood disorder for which lithium cation is an effective prophylactic agent, may be initiated by an as-yet unidentified group of brain cells that are pathologically overactive, and this hyperactivity is attenuated by lithium's restriction of production of myoinositol. However, presently, the mechanism of action of lithium cation remains uncertain and a direct physiological connection between effects on inositol and/or inositol-derived second messengers and bipolar disorder has not been described.

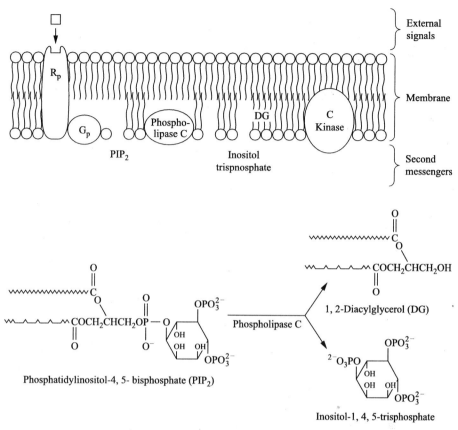

Figure 9.7 The inositol phosphate pathway: a G-protein related receptor system

Lithium cation is readily and almost completely absorbed from the gastrointestinal tract, and the principal excretory route is the urine. Lithium cation penetrates the blood–brain barrier slowly. At steady state the concentration of Li^+ in the cerebrospinal fluid is approximately half that in the blood. Side effects of lithium therapy include nausea, thirst, tremors, and polyuria (elevated urine excretion, a reflection of its high level in the tubular urine). These effects are said to be a reflection of the long plasma half life of lithium cation. Acute lithium intoxication involves severe gastrointestinal upset, tremors, coma, and convulsions. The therapeutic index of lithium cation is remarkably small: 2 or 3.

9.9   Antianxiety Agents

Anxiety is a principal symptom of many psychiatric disorders, and it is an almost inevitable component of many medical and surgical conditions. Anxiety is a universal human emotion. The distinction between a "pathological" and a "normal" state of anxiety is difficult to draw. Evaluation of antianxiety drug candidates in humans is challenging. Several "anxiety scale" tests have been used, based on standard patient questionnaires, and these have frequently proven to be useful in identifying anxiolytic drugs. However, placebo treatment in this type of evaluation has sometimes produced positive responses. Other tests rely on measurement of somatic and autonomic effects associated with anxiety. One such test, the *galvanic skin response*, measures the electrical conductivity of the skin as a measure of sweat production. Former and now obsolete antianxiety therapy utilized a variety of sedative drugs, including barbiturates and "antihistaminics" (H_1 antagonists). Possibly the most useful chemical category of antianxiety agents is the family of benzodiazepines, typified by structures **9.18, 9.19,** and **9.20**. All of the benzodiazepine antianxiety agents are similar in their pharmacological actions, but there are some differences in their pharmacokinetic profiles. The antianxiety effects of the benzodiazepines involve their potentiation of central neural inhibition that is mediated by γ-aminobutyric acid (GABA). Figure 9.8 illustrates a cross section of a gabaergic nerve membrane with its $GABA_A$ receptor and the associated chloride ion channel.

9.18 Chlordiazepoxide   **9.19** Diazepam   **9.20** Oxazepam

Activation of the $GABA_A$ receptor induces a conformational change in the ion channel protein complex, opening the channel and permitting the influx of chloride anions. This influx of large numbers of Cl⁻ leads to inhibition of neurone firing: a so-called hyperpolarization effect. Associated with the $GABA_A$ receptor is a subsite on the protein complex, the so-called *benzodiazepine* (BZD) *receptor*. Interaction of a benzodiazepine

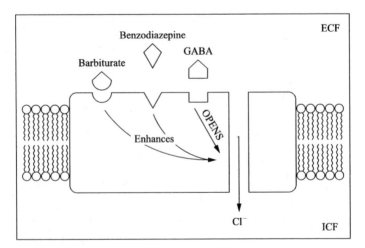

Figure 9.8 The $GABA_A$ receptor. (Reproduced with permission from reference 1. Copyright 1991 Mosby Year-Book Publishers)

drug with this receptor is thought to alter the conformation of the GABA receptor domain such that it is more complementary to the GABA molecule, and GABA binds more tightly to its receptor. There is mutual augmentation of binding between the GABA and the benzodiazepine receptors. Benzodiazepine drugs alone lack the ability to open chloride ion channels, but they act *allosterically* to increase the affinity of GABA for its receptor. They might properly be termed *coagonists* with GABA. Why should there be in the human body a receptor which is specific for a benzodiazepine ring derivative, considering that benzodiazepines are strictly products of the chemist's laboratory, and neither they nor structurally similar congeners have been found in nature? It is probable that the so-called benzodiazepine receptor is a bona fide structural entity which plays a role in the normal physiology of the body. It has been speculated that a 10 kDa peptide isolated from rat brain may represent an endogenous ligand for the benzodiazepine receptor.

Although some biochemical actions of benzodiazepine drugs are at least partially understood, it is less clear how (or whether) these effects on chloride ion transport across membranes are translated into relief of the anxiety syndrome.

As described in more detail in chapter 10 and as shown in figure 9.8, the sedative/hypnotic barbiturates interact allosterically at the $GABA_A$ receptor complex and they potentiate the action of GABA. This action results in a sedative/hypnotic effect. Therefore it is not surprising that the benzodiazepines as a chemical category also have prominent sedative actions and some benzodiazepines are employed exclusively as sedatives, rather than as antianxiety agents. But, it must be emphasized that the benzodiazepines do not relieve anxiety merely by producing drowsiness. Antianxiety effects can frequently be achieved at dose levels that cause minimal drowsiness.

Buspirone **9.21** represents a newer chemical (azaspirodecanediones) and pharmacological category of antianxiety agents. These compounds do not interact with benzodiazepine receptors nor do they reinforce the effects of GABA. Buspirone is a partial agonist at serotonin 5-HT$_{1A}$ receptors, and it interacts with presynaptic dopaminergic and noradrenergic receptors. It does not induce extrapyramidal side effects (vide infra). It does not produce sedation. Anxiolytic effects of buspirone may require days or even weeks to develop.

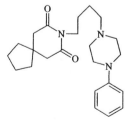

9.21 Buspirone

Antipsychotic Drugs

There is a sizeable number and variety of chemical entities in this category available in the marketplace (structures **9.22–9.26**), but their pharmacological effects and clinical properties are remarkably similar. Pharmacological details and mechanism(s) of action of this structurally diverse category are only partly understood, but some generalizations can be cited:

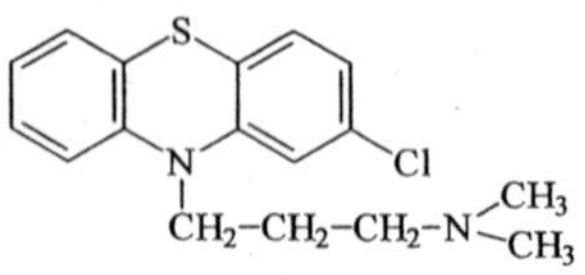

9.22 Chloropromazine

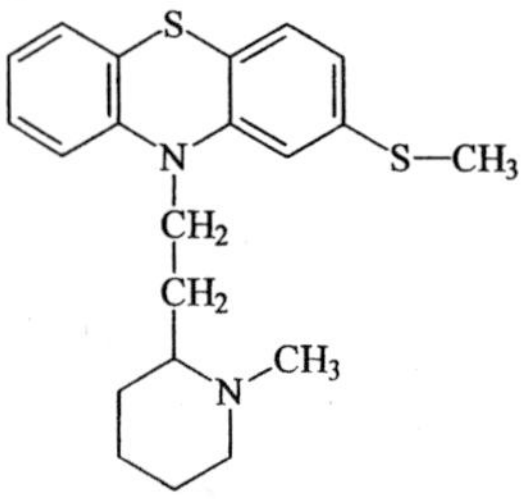

9.23 Thioridazine

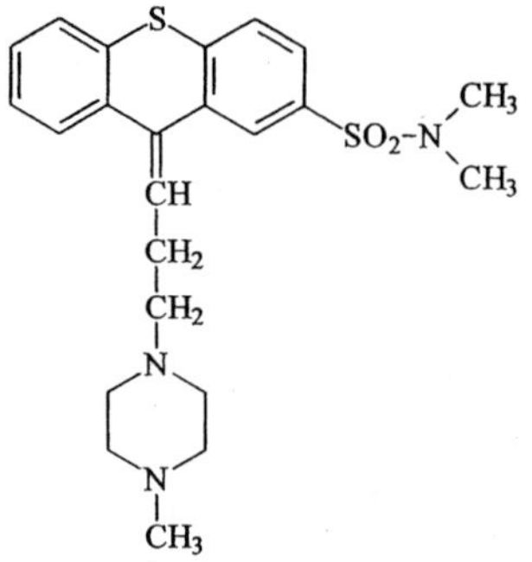

9.24 Thiothixene

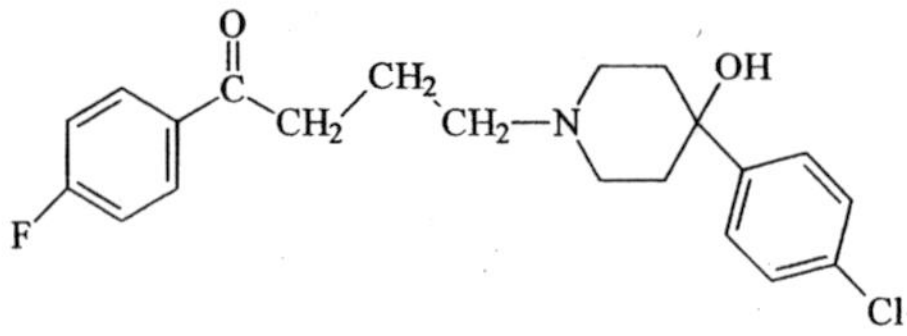

9.25 Haloperidol

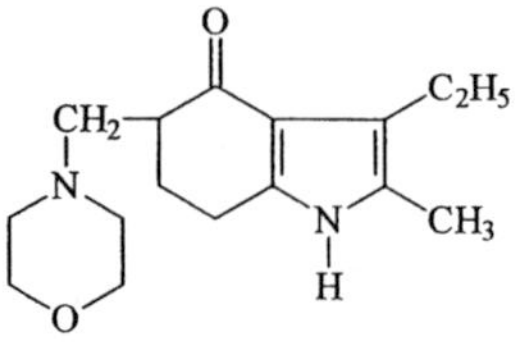

9.26 Molindone

1. the antipsychotic agents have only a limited ability to produce sedation, which is a common, prominent, and persistent side effect of most other categories of psychoactive drugs
2. acute overdosage with the antipsychotic agents has a low potential for lethality due to coma or respiratory depression
3. antipsychotic agents lack euphoriant effects, and they have a low potential for abuse and true addiction, although abrupt withdrawal of drug may result in malaise and difficulty in sleeping
4. routine or early tolerance to the main effects of antipsychotic drugs is not evident, although the individual may become tolerant to some of the side effects, such as hypotension, antimuscarinic effects, and dystonic effects (disordered tone of voluntary muscles).

However, in a relatively small percentage of individuals, tolerance to the antipsychotic activity may develop after several years.

It has long been speculated that a portion of the pathophysiology of psychoses is centered in the limbic system, the area of the brain that in part controls mood, emotion, and memory. Grossly and perhaps somewhat simplistically, the antipsychotic action of the phenothiazine derivatives (**9.22, 9.23**) and probably also the other drugs in this category involves indirect reduction of neuronal activity in the reticular activating system (RAS), and this occurs without causing cerebral cortex depression which results in sedation and sleep. The additional depression of the limbic region contributes to the overall action of the drugs. These regions contain dopaminergic pathways and one school of thought proposes that some psychoses may result from hyperactivity of some of these. The phenothiazine derivatives and congeners (structures **9.22, 9.23, 9.24**) are dopamine pre- and postsynaptic D_2 receptor blockers, as are haloperidol **9.25** and molindone **9.26**. Probably this biochemical action is significant. However, all of these drugs have complex pharmacological profiles in addition to properties related to dopamine, such as blockade of serotonergic 5-HT_2 receptors. Probably the antipsychotic actions reflect a summation of all of these pharmacological properties.

The antipsychotic drug *clozapine* **9.27** is unique in that it binds only weakly to D_2 receptors but it has affinity for D_1 and D_4 receptors. D_4 receptors are up regulated after repeated administration of clozapine. Contemporary speculation implicates G protein-coupled D_4 receptors with the etiology and treatment of schizophrenia. Clozapine also shows high affinity for some subpopulations of serotonin receptors. However, it is not known whether this contributes to its antipyschotic effects.

9.27 Clozapine

9.10.1 Miscellaneous Pharmacological Effects of Antipsychotic Drugs

Most of the antipsychotic agents also are antiemetics because they block D_2 dopamine receptors in the chemoreceptor trigger zone in the medulla oblongata. These agents, especially the phenothiazines, are occasionally used clinically for this purpose. Chlorpromazine **9.22** and several other antipsychotics potentiate the effects of inhalation anesthetics, ethanol, and barbiturates. Uncontrolled, excessive ingestion of this type of drug along with sizeable amounts of an alcoholic beverage is potentially lethal. Chlorpromazine antagonizes the CNS effects of amphetamines because it can block central norepinephrine and dopamine receptors. Most of the antipsychotic drugs are muscarinic receptor blockers; common side effects are dry mouth, flushing of the blush areas of the face and neck, and blurred vision.

It is characteristic of most antipsychotic drugs, irrespective of their chemical nature, that they produce so-called *extrapyramidal motor disturbances* as side effects to their desirable actions. These drugs affect the extrapyramidal tracts, resulting in Parkinsonian symptoms (muscular rigidity, tremors, hypokinesia) and motor restlessness, inability to sit still, restless limbs, and uncontrolled involuntary movements, especially of the face and tongue).

These muscular symptoms are sometimes termed *tardive dyskinesias* (late appearing muscular movement disorders). Extrapyramidal effects are dose-related, and they may be minimized or controlled by dosage adjustment or by switching to other drugs. Clozapine seems less prone to producing tardive dyskinesias. The reason for this is not clear. However, the possibility of extrapyramidal side effects is always a concern to the chemist and the pharmacologist in designing new antipsychotic agents.

9.11   Psychotomimetics

Psychotomimetic drugs (also known as psychedelic or hallucinogenic drugs) affect thought, perception, and mood without causing marked psychomotor stimulation or depression. Thoughts and perceptions tend to become distorted and dream-like and the change in mood is more complex than a simple shift in the direction of euphoria or depression. Categorization of these drugs is very imprecise. The commonly used term *hallucinogenic* is not strictly correct, in that some drugs in this category do not produce frank hallucinations. The perceptual distortions cited above can be produced by very large doses of many drugs. These phenomena may also be seen during toxic withdrawal from such drugs as ethanol. Psychotomimetic effects of some drugs are described else-where (e.g., cannabis, phencyclidine, "ecstasy"). None of the drugs described below are considered to be addicting.

Lysergic acid diethylamide (LSD; from the German *lyserg säure diäthylamid*) **9.28** is the most potent of the psychotomimetics. LSD produces significant psychedelic effects at doses as small as 25–50 µg. It is usually ingested orally and CNS effects are noted after 40–60 min. The peak effect lasts for 2–4 h, then it tapers off over the next 6–8 h. There are perceptual distortions and sometimes hallucinations. Mood changes may include elation, paranoia, depression, and sometimes a feeling of panic (a "bad trip"). Visual effects are prominent; colors seem more intense and shapes appear altered. Visible signs of LSD ingestion include pupillary dilation, increased blood pressure and heart rate, flushing, and salivation.

9.28

LSD is described as an agonist/partial agonist at brain serotonin receptors. Inspection of the chemical structure of LSD reveals the elements of tryptamine with the aminoethyl side chain held in a rigid, specific conformation. Acting at certain 5-HT_{1A} presynaptic receptors in the brain, LSD slows the firing rate of serotonergic neurons. LSD also activates 5-HT_{2A} and 5-HT_{2C} receptors. However, there is much yet to be learned about the pharmacology of LSD.

Claims of the value of LSD in enhancing psychotherapy and for treating addictions and other mental disorders have not been supported by research results. There is no current indication for LSD in therapy, and this statement is probably applicable to other psychedelics also.

Psilocin **9.29** and its phosphate ester *psilocybin* **9.30** are constituents of a mushroom native to Mexico. Both compounds are classed as hallucinogens and their gross pharmacology is described as "LSD-like". Their structural relationship to serotonin is obvious.

Mescaline **9.31** is a constituent of several species of cacti (peyote cactus) indigenous to Mexico.

9.29 R=OH

9.31

9.30 R= —O—P—OH (with =O and OH)

Mescaline has a high affinity for 5-HT$_2$ receptors and this interaction is believed to be an important factor in its CNS effects. However, it also probably interacts with other serotonin receptor subtypes. Large oral doses ($\approx$500 mg) are necessary for hallucinogenic activity. This low potency is a reflection of facile metabolic inactivation by monoamine oxidases. The CNS effects of mescaline are described as being vivid and colorful; perception of the environment is unusually beautiful and users claim increased insight ("mind-expanding experience").

Bibliography

1. Wingard, L. B., *et al. Human Physiology.* Mosby-Year Book Publishers: St. Louis, Mo., 1991; pp. 20–22.
2. Kling, J. *Modern Drug Discovery.* American Chemical Society: Washington, D. C., 1998; p. 34.
3. Tortora, G. J. *Principles of Human Anatomy*, 8th ed. Benjamin/Cummings Science Publishing: Menlo Park, Calif., 1999; pp. 556, 563.
4. Guyton, A. C. *Textbook of Medical Physiology*, 6th ed. Saunders: Philadelphia, Pa., 1981; p. 653.

Recommended Reading

1. Watling, K. J., Ed. *The Sigma–RBI Handbook of Receptor Classification and Signal Transduction*, 5th ed. Sigma–RBI: Natick, Mass., 2006.
2. Dubocovich, M. L. Structure–Activity Relationships. Pharmacology and Function of Melatonin Receptors. In *Trends in Drug Research.* Claassen, V., Ed. Elsevier: Amsterdam, the Netherlands, 1990; Vol. 13, pp. 23–35.
3. Glutamate Receptors. In Rang, H. P., Dale, M. M., Ritter, J. M. and Gardner, P. *Pharmacology*, 4th ed. Churchill Livingstone: Philadelphia, Pa., 2001; pp. 471–478.
4. Birch, N. J. Biomedical Uses of Lithium. In *Uses of Inorganic Chemistry in Medicine.* Farrell, N. P., Ed. Royal Society of Chemistry: Cambridge, UK, 1999; pp. 11–25.

5. Bloom, F. E. Neurotransmission and the Central Nervous System. In *Goodman and Gilman's The Pharmacological Basis of Therapeutics*, 10th ed. Hardman, J. G., Limbird, L. E. and Gilman, A. G., Eds. McGraw-Hill: New York, 2001; pp. 293–320.
6. Baldessarini, R. J. Drugs and the Treatment of Psychiatric Disorders: Depression and Anxiety Disorders. In Recommended Reading ref. 7, pp. 447–484.
7. Baldessarini, R. J. and Tarazi, F. I. Drugs and the Treatment of Psychiatric Disorders: Psychosis and Mania. In Recommended Reading ref. 7, pp. 485–520.
8. Zhang, J. and Snyder, S. H. Nitric Oxide in the Nervous System. *Annu. Rev. Pharmacol. Toxicol.* **1995**, *35*, 213–233.
9. Napoli, C. and Ignarro, L. J. Nitric Oxide-Releasing Drugs. In *Annual Review of Pharmacology and Toxicology*. Cho, A. K., Blaschke, T. F., Insel, P. A. and Loh, H. H., Eds. Annual Reviews: Palo Alto, CAL., 2003; Vol. 43, pp. 97–123.
10. Butler, A. R.; Rhodes, P. S. Nitric Oxide in Physiology and Medicine. In Recommended Reading ref. 6, pp. 58–76.
11. Olah, M. E. and Stiles, G. L. Adenosine Receptor Subtypes: Characterization and Therapeutic Regulation. *Annu. Rev. Pharmacol. Toxicol.* **1995**, *35*, 213–233.
12. Reppert, S. M., Weaver, D. R. and Goodson, C. Melatonin Receptors Step into the Light: Cloning and Classification of Subtypes. *Trends Pharmacol. Sci.* **1996**, *17*, 100–102.
13. Hill, S. J., Ganellin, C. R., Schwartz, J. C., Shankley, N. P., Young, J. M., Schunack, W., *et al.* Classification of Histamine Receptors. *Pharmacol. Rev.* **1997**, *49*(3), 253–278.
14. Anonymous. Nitric Oxide. *Chemistry* December, 1999.
15. Webster, R. A., Ed. *Neurotransmitters, Drugs and Brain Function*. Wiley: Chichester, UK, 2001.
16. Monn, J. A. and Schoepp, D. D. Metabotropic Glutamate Receptor Modulators: Recent Advances and Therapeutic Potential. *Annu. Rep. Med. Chem.* **2000**, *35*, 1–10.
17. Stramford, A. W. and Parker, E. M. Recent Advances in the Development of Neuropeptide Y Receptor Antagonists. *Annu. Rep. Med. Chem.* **1999**, *34*, 31–40.
18. Xiang, J.-N. and Lee, J. C. Pharmacology of Cannabinoid Receptor Agonists and Antagonists. *Annu. Rep. Med. Chem.* **1999**, *34*, 199–208.
19. De Ninno, M. D. Adenosine. *Annu. Rep. Med. Chem.* **1998**, *33*, 111–120.
20. Bernasconi, R., Mathivet, P., Bischoff, S. and Marescaux, C. Gamma-Hydroxybutyric Acid: An Endogenous Neuromodulator with Abuse Potential? *Trends Pharmacol. Sci.* **1999**, *20*, 135–141.
21. Melatonin Receptors. In Recommended Reading ref. 1, pp. 128–129.
22. Langlois, M. and Fischmeister, R. 5-HT$_4$ Receptor Ligands: Application and New Prospects. *J. Med. Chem.* **2003**, *46*, 319–344.
23. Currie, K. S. Antianxiety Agents. In *Burger's Medicinal Chemistry and Drug Discovery*, 6th ed. Abraham, D. J., Ed. Wiley Interscience: Hoboken, N.J., 2003; Vol. 6, pp. 525–597.
24. Altar, C. A., Martin, A. R. and Thurkauf, A. Antipsychotic Agents. In Recommended Reading ref. 25, pp. 599–672.
25. Wattanapitayakul, S. K., Young, A. P. and Bauer, J. A. Genetic Variations in Nitric Oxide Synthase Isoforms. *Pharm. News* **2000**, *7*, 14–20.
26. Tabriizi-Fard, M. A. and Tseng, C.-M. L. Design of Nitric Oxide Synthase Inhibitors. Isoform Selectivity and Pharmacokinetics. *Pharm. News* **2000**, *7*, 22–26.
27. Sigma Receptors. In Recommended Reading ref. 1, pp. 66–67.
28. Augelli-Szafram, C. E. and Schwarz, R. D. Metabotropic Glutamate Receptors: Agonists, Antagonists, and Allosteric Modulators. *Annu. Rep. Med. Chem* **2003**, *38*, 21–30.
29. Vida, J. A. Central Nervous System Depressants: Sedative–Hypnotics. In *Principles of Medicinal Chemistry*, 4th ed. Foye, W. O., Lemke, T. L. and Williams, D. A., Eds. Williams & Wilkins: Baltimore, Md., 1995; pp. 163–179.

10
The Central Nervous System.
II: Sedatives and Hypnotics

10.1 Definitions

A *sedative* drug decreases CNS activity, moderates excitement, and calms the recipient. These effects of a sedative are reversible; the individual can readily be aroused. A *hypnotic* drug produces drowsiness and facilitates the onset of a state of sleep that may resemble natural sleep from which the individual can be aroused. However, it is likely that the subject will immediately fall asleep again. Pain perception is retained. Sedation, hypnosis, and *general anesthesia* may be regarded as increasing depths of a continuum of CNS depression, although the actual mechanisms by which the depression is produced may differ. In general anesthesia, the individual loses the ability to perceive pain and cannot be aroused until the effect of the drug wears off. For some older drugs, the pharmacological effect (sedation–hypnosis–general anesthesia) is a dose-dependent phenomenon.

10.2 Rapid Eye Movement Sleep

Whether hypnotics produce a truly natural sleep is highly debatable. Of all of the body functions, sleep is one of the least well understood. Indeed, authorities differ in their definition of the term *natural sleep*. Physiological sleep is an exquisitely complex process. During physiological sleep, there are cyclical changes in the state of the sleeper, involving periodic episodes of very rapid movement of the eyeballs, "rapid eye movement" or REM. The rapid eye movement phase of sleep produces identifiable changes in the brain wave pattern (electroencephalogram). Approximately 20–25% of the total sleep period is REM sleep, and it occurs in four or five separate periods of 5–30 min duration per night. Figure 10.1 illustrates how sleep increases and decreases in depth throughout the night. Periods of REM sleep become longer as the night progresses. REM sleep is believed to be associated with dreaming and it is an important part of the overall sleep process. The flurry of activity of REM sleep begins in the pons which sends signals that stimulate higher brain centers to cause dreaming. At the same time, the pons sends messages to shut off neurons in the spinal cord to cause muscular paralysis. Thus, in REM sleep the brain is highly active but the body is paralyzed.

183

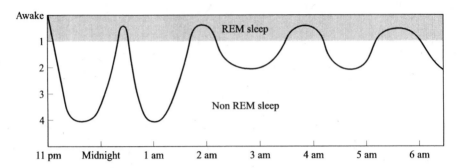

Figure 10.1 REM and non-REM sleep patterns

Cerebral blood flow is greater during REM sleep time than during wakefulness. It has been suggested that REM sleep has a special role in maintenance, restoration, and metabolism of noradrenergic nervous mechanisms in the brain. α Adrenoceptors have been implicated in REM sleep mechanisms. The brain wave pattern during REM sleep is similar to the pattern in the awake individual, even though the individual is deeply asleep. For this reason REM sleep is sometimes termed *paradoxical sleep*. Non-REM sleep is believed to be governed by serotonergic mechanisms. Administration of exogenous serotonin lowers the relative amount of REM sleep.

Hypnotics and most addicting drugs diminish the proportion of REM sleep time, and it is widely accepted that deprivation of REM sleep may have deleterious effects. Subjects deprived of REM sleep for multiple nights experience a rebound effect of an increased relative amount of REM sleep for several nights following the period of deprivation. However, repeated use of most hypnotic drugs leads to tolerance and the REM sleep time usually returns to the normal level after a few weeks. A more realistic term for describing the effect of sedatives and hypnotics is *restful sleep*, to indicate whether the individual wakens from a drug-induced sleep feeling refreshed.

10.3   Ethanol

Despite its reputation as a stimulant, pharmacologically ethanol is primarily a central nervous system depressant whose CNS effects resemble those of the inhalation anesthetics. It is believed that the observed apparent stimulation resulting from ethanol ingestion results from depression of inhibitory control mechanisms in the brain. In severe ethanol intoxication a condition of general anesthesia results. However, there is little margin between a surgical anesthetic dose and a dose that produces a lethal depression of respiration. The mechanism of the central depressant action of ethanol has been the subject of dispute for many years. It was suggested previously (chapter 4) that ethanol dissolves in the lipid phase of neuronal membranes in the central nervous system and thereby disrupts the normal functions of ion channels and of other proteins embedded in the lipid matrix. Another school of thought proposes that ethanol disrupts membrane ion channel function by direct interaction with lipophilic domains of the membrane proteins themselves. Newer studies suggest the involvement of ligand-gated calcium ion channels (GABA$_A$ receptors, glycine receptors, and/or glutamic acid NMDA and kainate receptors) as ethanol targets in the CNS. The long-held belief that ethanol is a structurally nonspecific drug may not be entirely valid. Despite the wide use of ethanolic beverages by the laity for many ailments, legitimate uses in medical practice/therapy are few.

10.4 Non-Barbiturate, Non-Benzodiazepine Sedatives and Hypnotics

Structures **10.1–10.3** are representative members of this category; all three of these have been used clinically for many years, but they have been largely supplanted by other drugs. These agents (**10.1–10.13**) are capable of producing profound hypnosis with little or no analgesia. Characteristic of structurally nonspecific drugs, chloral hydrate **10.1** must be given in large doses (1–2 g). Habitual use may result in true addiction. Chloral addicts take huge doses, up to 10–12 g a day, and sudden withdrawal may result in delerium, seizures, and death. Therapeutic dose levels of chloral hydrate do not shorten REM sleep time, and the drug has the reputation of being less likely than many other sedatives/hypnotics to produce hangover. It produces surgical level general anesthesia only in doses approaching lethality. As cited in chapter 3, chloral hydrate is rapidly metabolically reduced in the liver to trichloroethanol which is chiefly responsible for the sedative/hypnotic activity. A modest percentage of both chloral hydrate and trichloroethanol are oxidized to trichloroacetic acid which is excreted in the urine. The preponderance of the trichloroethanol metabolite is conjugated with glucuronic acid, and this conjugate is excreted in the urine. It is believed that trichloroethanol acts on the reticular activating system to inhibit impulses going upward to the cerebral cortex. The drug exerts barbiturate-like effects at chloride ion channels associated with the GABA$_A$ receptor. It also modifies the function of serotonin 5-HT$_3$ receptors but the significance of this and the details of the mechanism of action of trichloroethanol are not known. Chronic use may cause severe liver damage.

10.1 Chloral hydrate   **10.2** Ethchlorvynol   **10.3** Paraldehyde

Ethchlorvynol **10.2** is representative of a number of tertiary linear or cyclic aliphatic alcohols that have sedative/hypnotic properties. These drugs are used clinically in management of insomnia. Ethchlorvynol produces a mild hangover. It is an inducer of metabolizing enzymes, and it speeds hepatic metabolic inactivation of oral anticoagulants. Acute intoxication with ethchlorvynol resembles that produced by barbiturates (vide infra), except that ethchlorvynol produces more severe respiratory depression.

Paraldehyde **10.3** is a cyclic trimer of acetaldehyde, but its chemical properties are those of a chemically inert aliphatic ether. This drug is a liquid with a pungent, disagreeable odor and taste. It is administered orally or (in solution in a fixed oil) as a retention enema. It is well absorbed across mucous membranes. The majority of a dose of paraldehyde is depolymerized in the liver to reform acetaldehyde, which is rapidly oxidized to acetic acid, thence to carbon dioxide and water. Symptoms of toxic overdosage of paraldehyde include acidosis caused by this excess production of acetic acid. A minor portion of the dose is excreted chemically unchanged from the lungs in the expired air. Paraldehyde is employed in controlling the acute phase of delerium tremens in hospitalized alcoholics. The drug itself is addicting, and some alcoholics prefer it to ethanol, despite its disagreeable odor and taste. Addiction to paraldehyde is said to resemble alcoholism, and abrupt withdrawal can trigger delerium tremens and hallucinations.

10.5   Benzodiazepines

Flurazepam **10.4** and *triazolam* **10.5** typify benzodiazepine derivatives used primarily for their sedative/hypnotic effects rather than for relief of anxiety. Details of the pharmacology of this chemical category were presented in the discussion of their use as anxiolytic agents (chapter 9). All the benzodiazepines exert qualitatively similar sedative effects, but they may differ quantitatively in their pharmacokinetic properties. In modest doses these drugs do not decrease REM sleep time, but in larger (and more commonly used) doses REM sleep time is shortened. As a category the benzodiazepines have very large therapeutic ratios. Frequently, the benzodiazepines may be the sedative/hypnotic agents of choice for relief of insomnia, in preference to barbiturates, based on the following criteria: (1) large therapeutic indices; (2) fewer drug interactions; (3) lower abuse liability, although abrupt withdrawal following long term chronic use may precipitate severe physical withdrawal symptoms similar to those produced by the opioids (see chapter 12). However, in contrast to the opioids, psychological dependence does not seem to be a major problem with the benzodiazepines; and (4) less liklihood of depressing respiration. However, in the presence of other CNS depressants such as ethanol, the benzodiazepines can cause life-threatening respiratory depression. *Flunitrazepam* (rohypnol) **10.6** is considered a drug of abuse, due to its use as a "date rape" drug. The amnesic effect of the drug is probably significant here. Flunitrazepam has a very long half-life in the body, as compared to most of the other benzodoazepines.

10.4   10.5   10.6

Zolpidem **10.7** is typical of a small number of sedative agents which, although they are not benzodiazepine derivatives, are benzodiazepine receptor agonists. These non-benzodiazepine ligands of the benzodiazepine receptor, unlike benzodiazepine drugs themselves, seem to bind exclusively to only one subunit of the GABA$_A$ receptor complex. It has been proposed that the result is a greater specificity of action for the non-benzodiazepine sedatives. The pharmacological effects of zolpidem are very

similar to those of the benzodiazepine sedatives; however, zolpidem's powerful sedative effects mask anxiolytic effects in animal models.

10.7

10.6 Barbiturates

10.6.1 Hangover

Sleep produced by the barbiturates superficially resembles a natural, dreamless sleep.
However, the barbiturates reduce the amount of REM sleep time, and in this respect the
sleep induced is neither natural nor physiological. Frequently, the barbiturates produce a
hangover. The phenomenon of hangover has been difficult to define or to quantify. In the
past, the extent and meaning of the term were based on the subject's description of personal
feelings and impressions upon awakening from a drug-induced sleep. For most individuals, the term hangover has an imprecise, subjective connotation. Psychological studies have
attempted to provide a qualitative and quantitative definition. In a typical study on human
volunteers, there were three groups of test subjects. All members of group 1 took the same
hypnotic-size dose of a barbiturate. All members of group 2 took the same hypnotic-size
dose of a benzodiazepine sedative/hypnotic, and group 3 was composed of self-diagnosed
insomniacs who received no medication of any kind. The two groups were medicated, and
all three groups were sent to bed at the same hour. All three groups were awakened at the
same hour the next morning. Then, all participants were individually subjected to a battery
of psychological performance tests: a *key tapping test* (a test of motor speed: when the
signal is given, tap the telegraph key as rapidly as you can); an *auditory reaction time test*
(when you hear the buzzer, tap the key once; when you hear the bell, tap the key twice.
How long is required for the subject to respond to each signal, and is the response correct?);
and a series of *tests of associative skills* (e.g., digit–symbol substitution: if green is 5 and
brown is 3, how much is green minus brown?). The individuals who had taken a hypnotic
drug the preceding night performed statistically significantly poorer in these tests than did
the control (the self-diagnosed insomniacs). In additional tests, it was demonstrated that a
normal dose of caffeine, such as would be found in two cups of coffee did not significantly
improve performance of individuals in groups 1 and 2 in the tests. These results demonstrate that, although hypnotic drugs lessen the distress of individuals with insomnia,
psychological impairment, compromised motor skills, and lowering of the ability to think
clearly and rapidly are present the next morning and these impairments may last for several
hours after the individual awakens. It is unrealistic to expect any adequate hypnotic drug
to be devoid of pharmacological effect the next morning. There was no obvious difference
between the barbiturate and the benzodiazepine drugs used in the study. This temporary
impairment of psychological and motor functions is the true significance of hangover.

10.6.2 Pharmacological Categories of Barbiturates

Representative barbiturates are shown in figure 10.2. The barbiturates (of which there are many chemical variations available commercially) are grouped according to their duration of action:

1. Long-acting: 6 h or more (barbital **10.8**, phenobarbital **10.9**)
2. Intermediate-acting: 3–6 h (amobarbital **10.10**)
3. Short-acting: under 3 h (pentobarbital **10.11**, secobarbital **10.12**)
4. Ultrashort-acting: (thiopental **10.13**)

The variation in the duration of pharmacological action reflects the speed with which the body disposes of the barbiturate following absorption. Despite the structural similarity of all of the barbiturates, there is no common mode of metabolic disposition. Each chemical entity follows its own biochemical pathway. Barbital **10.8** is excreted unchanged in the urine. Phenobarbital **10.9** is in part hydroxylated on the benzene ring and this pharmacologically inert metabolite is excreted in the urine, but the bulk of a dose of phenobarbital is excreted unchanged. Pentobarbital **10.11** is in part hydroxylated on the 1-methylbutyl side chain, and in part the terminal C-methyl group of this alkyl moiety is oxidized to carboxyl. It is difficult to predict the duration of action of a barbiturate solely by inspecting its chemical structure.

The ultrashort acting barbiturates (e.g., thiopental **10.13**) are sulfur congeners of the oxygen barbiturates. These compounds are not used as sedatives/hypnotics, but rather they are general anesthetics. A discussion of thiobarbiturates is found in chapter 13.

As a class the oxygen barbiturates have small therapeutic indices, making them potentially dangerous drugs. Oxygen barbiturate-mediated loss of pain perception occurs only at dangerously high dose levels. Small doses *increase* the response to painful stimuli, and for this reason they cannot be depended upon to produce sedation or sleep in the presence of pain. As described previously (chapter 9) there is a barbiturate binding site associated with central GABA$_A$ receptors, and activation of this barbiturate receptor facilitates chloride ion migration through the ion channels. Unlike the benzodiazepines, this barbiturate action does not require simultaneous binding of GABA, although barbiturates do indeed potentiate GABA-mediated effects on chloride

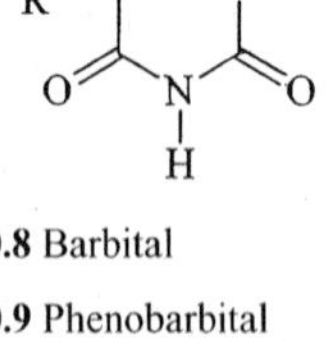

10.8 Barbital	R-R'=C2 H5
10.9 Phenobarbital	R=C$_2$H$_5$, R'=C$_6$H$_5$
10.10 Amobarbital	R=C$_2$H$_5$, R'=1-methylbutyl
10.11 Pentobarbital	R=C$_2$H$_5$, R'=1-methylbutyl
10.12 Secobarbital	R=allyl, R'=1-methylbutyl

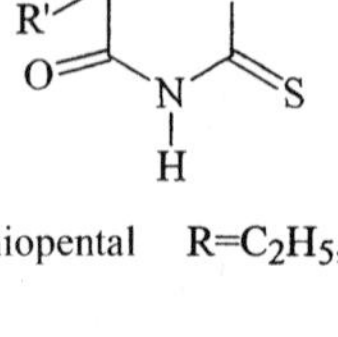

10.13 Thiopental   R=C$_2$H$_5$, R'=1-methylbutyl

Figure 10.2 Representative barbiturates

ion channels. Barbiturates also promote the binding of benzodiazepines. Barbiturates also bind to AMPA and kainic acid subpopulations of glutamate receptors.

10.6.3 Barbiturate Intoxication

Overdosage with barbiturates is common, often by intent. Ethanol and barbiturates make a lethal combination for CNS depression. In severe barbiturate intoxication, circulatory collapse is a major threat: a drastic fall in blood pressure leading to clinical shock, due to a significant extent to depression of blood pressure control centers in the medulla. The barbiturates also depress the respiratory center in the medulla (which normally is stimulated by high levels of carbon dioxide in the blood), so that CNS-controlled maintenance of automatic breathing is lost. The rapidly developing tolerance to the sedative/hypnotic effects of all of the barbiturates is not manifested in the respiratory center. Thus, as tolerance to sedative/hypnotic effects increases, the magnitude of the therapeutic index decreases. As the individual ingests larger and larger doses of barbiturate to attain a desired hypnotic effect, the dose level may approach lethality. This tolerance is an example of *pharmacodynamic* or *tissue tolerance* which differs from the drug disposition tolerance previously described in the discussion of metabolism. Pharmacodynamic tolerance reflects the condition where blood levels of the barbiturate remain constant after chronically ingested doses of the same size, but the drug's ability to produce sleep progressively decreases. The cause of barbiturate pharmacodynamic tolerance is uncertain, but there are two possibilities: (1) as the individual continues to ingest barbiturates day after day, the body calls into play compensatory mechanisms to counteract the sleep-producing effects, and it releases neurotransmitters which promote wakefulness and alertness; (2) the areas of the brain acted upon by the barbiturates become less sensitive to the drug. The receptors become fatigued from the continual bombardment by barbiturate molecules.

10.7   Therapeutic Uses of Barbiturates

The use of barbiturates as sedatives and hypnotics has lost popularity, and most authorities consider that, for sedation, the benzodiazepines are preferable from the standpoints of therapeutic efficacy and safety. The inclusion of a barbiturate in analgesic combinations is described in *Goodman and Gilman's The Pharmacological Basis of Therapeutics* as "counterproductive". Phenobarbital is useful in emergency treatment of convulsions and in relief of the symptoms of epilepsy. The ultrashort acting thiobarbiturates are still employed as general anesthetics (see chapter 13).

Bibliography

1. Carey, J., Ed. *Brain Facts*. Society for Neuroscience: Washington, D.C., 1993; p. 23.

Recommended Readings

1. Vida, J. A. Sedatives–Hypnotics: The Physiology of Sleep. In *Principles of Medicinal Chemistry*, 4th ed. Foye, W. O., Lemke, T. L. and Williams, D. A., Eds. Williams & Wilkins: Media, Pa., 1995; pp. 163–167.

2. Vida, J. A. Sedative–Hypnotics. In *Burger's Medicinal Chemistry and Drug Discovery*, 6th ed. Abraham, D. J., Ed. Wiley-Interscience: Hoboken, N.J., 2003; Vol. 6, pp. 223–234.

3. Charney, D. S., Mihic, S. J. and Harris, R. A. Hypnotics and Sedatives. In *Goodman and Gilman's The Pharmacological Basis of Therapeutics*, 10th ed. Hardman, J. G., Limbird, L. E. and Gilman, A. G., Eds. McGraw-Hill: New York, 2001; pp. 399–427.

4. Peoples, R. W., Li, C. and Weight, F. F. Lipid vs. Protein Theories of Alcohol Action in the Nervous System. *Annu. Rev. Pharmacol. Toxicol.* **1996**, *36*, 185–201.

5. Takaki, K. S. and Epperson, J. R. Pharmacological Interventions in the Sleep Process. *Annu. Rep. Med. Chem.* **1999**, *34*, 41–50.

6. Wilson, E. K. Sleep-Inducing and Sleep-Inhibiting Drug Research. *Chem. Eng. News.* **2003**, *81*(4), 51–55.

11

Analgesics. I: General Considerations and Non-Opioid Analgesics

Definition of Terms

Analgesia is the reduction of pain perception without loss of consciousness. For every individual, the word *pain* has a subjective meaning, but it is extremely difficult to define precisely or to describe verbally. According to one concept, pain is a two-component phenomenon: the initial perception or actual unpleasant sensation, then the psychological, emotional response (reaction component). Pharmacologists studying analgesia invoke two distinguishing terms, *nociception* and *pain*. Nociception is the perception of or the mechanism by which noxious peripheral stimuli are transmitted to the central nervous system. Nociception differs from *pain*, which is the overall subjective experience resulting from the noxious stimulus and includes a strong affective component. The intensity of stimulus at which an individual reacts, the so-called *pain threshold*, varies greatly and may depend upon the individual's personality type. A multitude of adjectives has been employed in attempts to describe varieties of pain: throbbing, gnawing, splitting, stabbing, pinching, crushing, dull-burning. But, do these adjectives have exactly the same meaning to all individuals? One system classifies pain as *acute* (temporary with immediate onset, as in a headache or from some trauma) or *chronic* (continual, as resulting from arthritis or cancer). In contrast to acute pain, most *chronic* pain states (pain that outlasts the precipitating injury) are associated with aberrations in the normal physiological pathways, giving rise to *hyperalgesia*, an increased amount of pain associated with a mild noxious stimulus, *allodynia,* pain evoked by a non-noxious stimulus, or spontaneous spasms of pain having no obvious precipitating stimulus.

Another proposal classes pain according to its point of origin. *Visceral pain* involves the nonskeletal parts of the body: gastric pain, colic, intestinal cramps. The so-called non-narcotic (non-opioid) analgesics are usually ineffective against this type of pain. *Somatic pain* emanates from muscle and bone: toothache, headache, muscle sprains, and arthritic pain. Physiologists now recognize *neuropathic pain*, which is characterized by a spontaneous, burning pain and is caused by injury to a nerve. However, neuropathic pain by definition also occurs in diabetes, cancer, AIDS, and shingles (herpes zoster). As yet, there are no drugs which dependably and adequately control this type of pain.

191

Nociceptors ("pain receptors") are primary afferent neurons that can be activated by harmful or potentially harmful stimuli and then give rise to the sensation of pain. Many of these receptors are ligand gated and they respond to some peptide neurotransmitters (e.g., substance P). Some other nociceptors are proton-activated ion channels.

11.2   Evaluation of Effectiveness of Analgesics in Animals and Humans

A challenge in discovering and developing new analgesic agents concerns the manner in which their effectiveness can be evaluated. This challenge is present at the experimental level with laboratory animals as well as in the clinic with human patients. The uncertainties involve evaluation and quantitation of pain. If it is virtually impossible to construct a valid qualitative verbal definition of pain, how much more difficult it is to quantify the pain phenomenon. Typical laboratory analgesic tests in animals include the hot-plate test, the tail-flick test, and the writhing test.

11.2.1 The Hot Plate Test

This test utilizes a laboratory hot plate set at a predetermined temperature. The test animal, frequently a mouse, is placed on the hot plate and the length of time before the animal displays signs of discomfort is recorded. This procedure is repeated using mice that have been given graduated doses of the candidate analgesic, to observe the lengthening of the time during which the animal can remain on the hot plate before displaying signs of discomfort. Dose–response curves can be constructed, and the data can be subjected to statistical analysis.

11.2.2 The Tail Flick Test

In this test the tails of mice or rats are shaved and are coated with a black, heat absorbing paint. The animal is placed in the beam of an infrared heat lamp, such that the tail receives maximum heat, and the length of time before the animal flicks its tail out of the infrared beam is recorded. The test procedure is repeated using animals that have been dosed with graduated levels of the candidate drug, and the extension of time during which the animal permits its tail to remain in the heat beam is noted.

11.2.3 The Writhing Test

In this test the ability of the candidate analgesic to prevent writhing from intraperitoneal injection into mice of an irritant substance such as acetic acid or formaldehyde solution is measured. Usually, screening of candidate analgesic agents involves performing a series of two or more of these tests, because no one of them alone is dependable nor necessarily confirmatory of analgesic activity. False "positives" and false "negatives" are common, and it is usually considered that only those compounds that demonstrate a positive analgesic effect in two or more of the screening tests merit further study. There is often a poor correlation between the activity of analgesic drugs in animal tests (which measure only antinociceptive activity, but not the psychological, affective component) and their clinical usefulness in humans.

11.2.4 Tests in Humans

Analgesic testing in humans presents an added dilemma. Are results obtained from healthy individuals in whom pain is artificially produced as meaningful as results from patients who are experiencing actual pain? Contemporary thought supports the testing of candidate analgesic drugs in individuals who are in pain. Artificially induced "pain" lacks the affective component of the natural pain syndrome. But now, problems multiply: How can the degree of pain being experienced be quantitatively determined and expressed? How can the degree of relief be expressed quantitatively? What can be used as a biological end point for analgesia, and how can it be determined? The only available measures are patients' verbal communication of their subjective feelings, and these are poorly reproducible experimental end points. Some assay procedure which is more specific and dependable than patients' verbal reports and physicians' observations and impressions is needed, but nothing better is available. The most widely accepted technique for clinical evaluation of analgesics is the *double-blind crossover.* In the double-blind strategy, neither the patient nor the physician knows whether the drug or a *placebo* is being administered. (A placebo is a pharmacologically inert substance, such as lactose, formulated into a dosage form to appear identical to the actual dosage form of the drug.) The crossover strategy further ramdomizes the process so that the participants in the clinical study sometimes receive the candidate drug and sometimes receive the placebo, thereby acting as their own control. Several dose levels of the drug should be used, and in a well-designed study an appropriate reference analgesic drug (e.g., aspirin or morphine) is included for comparison and to verify that the assay is providing an assessment of analgesic effect.

Recognition of and response to pain depends, in part, on the emotional state of the individual. The soldier on the battlefield may not be conscious of the pain of a wound in the heat of the battle. Sexual arousal blocks pain perception. The *placebo effect* is believed to be important in clinical testing of analgesics. Many repeated studies led to the conclusion that 35 patients out of 100 (35%!) will claim partial or complete relief from pain from the placebo dosage, if they were told in advance that they were receiving a very powerful and effective analgesic. A placebo effect has also been invoked for some other pharmacological categories of drugs. In 2001 a report in the medical literature (Recommended Readings, ref. 1, at the end of this chapter) presented a discussion suggesting that the data upon which the concept of the placebo effect is based are suspect, and therefore the placebo effect may be physiologically invalid.

In contrast, an even more recent study led to the conclusion that the placebo effect is a genuine physiological phenomenon. Pain was induced in male volunteers and the subjects were informed that they would be injected with either a placebo or with a drug that would reduce their pain. However, the subjects were unaware that all of them received the placebo. All brains were scanned using positron emission tomography and magnetic resonance imaging to determine activity in brain regions that release endorphins, the body's endogenous peptide analgesics. The scans showed that administration of the placebo resulted in increased endorphin activity, providing for increased pain resistance (see Recommended Reading, ref. 7, at the end of this chapter).

11.3   Physiological and Biochemical Aspects of Pain Production and Recognition

It is beyond the scope of this discussion to present a complete description of the current state of knowledge of this extremely complex topic. However, some appropriate aspects will be cited. Pain receptors in the skin and other tissues are free nerve endings. Three types of stimuli excite pain receptors: mechanical, thermal, and chemical. The mechanisms involved in stimulating nociceptive afferent (sensory) nerve endings are only dimly understood. Among the chemical substances implicated in this initial phase of the pain syndrome are the neurotransmitters serotonin, acetylcholine, and histamine; the kinin peptides ("tachykinins") bradykinin and kallidin; potassium cations; adenosine triphosphate (ATP) and adenosine diphosphate (ADP); lactic acid, whose function seems to be to create a pain stimulus by lowering the pH; and prostaglandins (vide infra) which apparently do not directly create a pain stimulus, but rather, strongly enhance the pain-producing effects of other agents. Bradykinin liberates prostaglandins, and probably other chemical transmitters also, that generate or reinforce the pain response.

As illustrated in figure 11.1, when a tissue is hurt, sensory nerves carry pain signals from the damaged area to the spinal cord, releasing neurotransmitters that activate cells in the dorsal horn of the gray matter. Activated neurons in the white matter carry the pain signal to the brain. Pain impulse transmission in the dorsal horn of the gray matter involves glutamate NMDA receptors and substance P, which functions as a neurotransmitter and for which there are receptors in the spinal cord and elsewhere. Two types of sensory nerve fibers that transmit pain signals from the peripheral regions to the pain processing center in the spinal cord are recognized: Aδ and C fibers. Aδ receptors and fibers transmit acute pain sensations (of the type that would prompt one to remove one's hand from a hot stove). C fibers mediate the diffuse, burning pain that follows later. A third type of sensory fibers (Aβ) responds to touch and other nonharmful stimuli. Much current pain research focuses on ways to interrupt pain signals between the spinal cord and the brain.

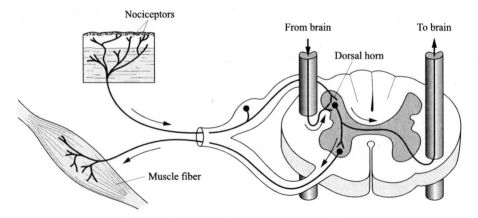

Figure 11.1 Neural transmission of nociceptive impulses. (Reproduced with permission from reference 1. Copyright 1993 Society for Neuroscience)

11.4 Non-Opioid Analgesics and Nonsteroidal Antiinflammatory Analgesics

11.4.1 Salicylates

Many drugs in the inclusive category of nonopioid analgesics are also useful in lowering body temperature in fever. The best known of the antiinflammatory antipyretic analgesics are the salicylates (**11.1–11.5**). Acetylsalicylic acid (aspirin) is the most widely used of the salicylate derivatives. The newest salicylate is *diflunisal* **11.5**. This drug is more potent than aspirin in animal antiinflammatory tests, but it is devoid of antipyretic effects, probably due to its poor penetration of the blood–brain barrier. It is a competitive inhibitor of cyclooxygenase enzymes. Diflunisal does not produce the auditory side effects (tinnitus) characteristric of chronic use of other salicylates and it is said to produce fewer and less intense gastrointestinal and antiplatelet effects than aspirin (vide infra).

11.1   11.2   11.3

11.4   11.5

Minor salicylate drugs

Methyl salicylate **11.4**, a sweet-smelling oily liquid, is also known as oil of wintergreen. This largely obsolete drug persists as an ingredient of some proprietary liniments and ointments recommended for relief of muscular aches and pains. It exerts a topical counterirritant effect. This lipophilic molecule is readily absorbed across intact skin, and once in the blood stream, esterases liberate salicylic acid which is responsible for the pharmacologic effect. Methyl salicylate has no specific pharmacological advantage. Indeed, absorption of it from the gastrointestinal tract is slow and capricious, and it exerts a local irritant effect on the stomach lining. Salicylamide **11.3** survives as a component of a few over-the-counter analgesic products. It is said to be comparable to aspirin as an analgesic, and it is reputed to be less prone than aspirin to causing gastric irritation. However, unlike aspirin, salicylamide has inferior antiinflammatory activity.

Metabolism and excretion

Salicylates (including aspirin) are metabolized chiefly by enzymes of the endoplasmic reticulum of liver cells, to form three principal metabolites: the glycine conjugate **11.6** and two glucuronic acid conjugates **11.7** and **11.8**. A small percentage of a dose of a salicylate is converted into 2,5-dihydroxybenzoic acid (gentisic acid) and/or to 2,3, 5-trihydroxybenzoic acid. All of these metabolites, major and minor, are excreted in the urine, along with a small amount ($\approx 10\%$) of unmetabolized salicylic acid. However, urinary excretion of salicylic acid is highly variable and depends in part on the pH of the tubular urine. In an alkaline environment, as much as 30% of a dose is excreted unchanged (as salicylate anion). In an acidic environment, tubular reabsorption of the neutral, more lipophilic salicylic acid results in urinary excretion of as little as 2% of the dose. The activity of salicylates is said to be enhanced by simultaneous administration of a large dose of p-aminobenzoic acid, and combinations of this compound with various salicylates have been used clinically. This effect of p-aminobenzoic acid has been ascribed to its competitive inhibition of salicylate metabolism.

| 11.6 | 11.7 | 11.8 |

11.4.2 The Inflammatory Syndrome

The inflammatory syndrome includes rheumatoid arthritis which affects the joints, especially those of the fingers, toes, and knees, rendering them inflamed, swollen, and very painful.

Prostaglandins

The underlying causes of rheumatoid arthritis are unknown, but the disease is believed to be an *autoimmune* condition ("autoimmune" = the body's immune system generates antibodies against some of the body's own tissue proteins and the result is severe tissue damage) and *prostaglandins* have been implicated as one chemical factor in the disease (figure 11.2).

Prostaglandins comprise a family of chemically related fatty acid derivatives found in a large number and variety of sites in the body. It is proper to refer to the prostaglandins as a chemical category, analogous to the term "steroid". Prostaglandins affect biochemical processes near their point of release in the body, and they have been termed "local hormones". They are unique in that, unlike hormones and neurotransmitter substances, there seems to be no mechanism for their storage. They are synthesized in response to a stimulus. Once synthesized in the tissue, prostaglandins are released and they function as essential intermediates between the stimulus and its cellular response. In general, prostaglandins are rapidly inactivated, and their action is short-lived. The following physiological actions (from many others) of members of the prostaglandin family are relevant.

Effects on inflammation

Prostaglandins increase capillary permeability, resulting in edema (collection of fluid in the tissues) and painful reddish areas. Why does the body possess a mechanism for

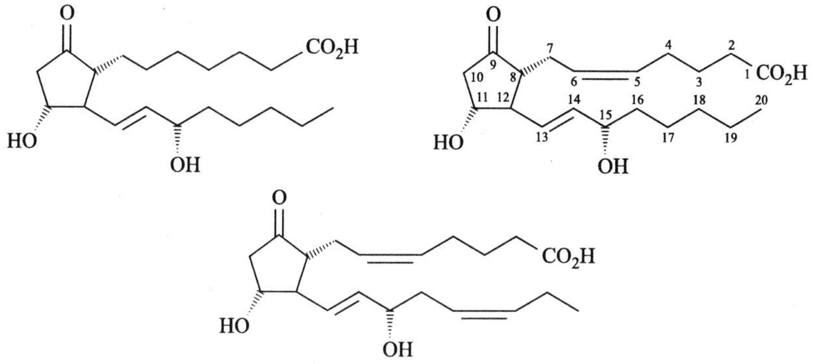

Typical side-chain variations

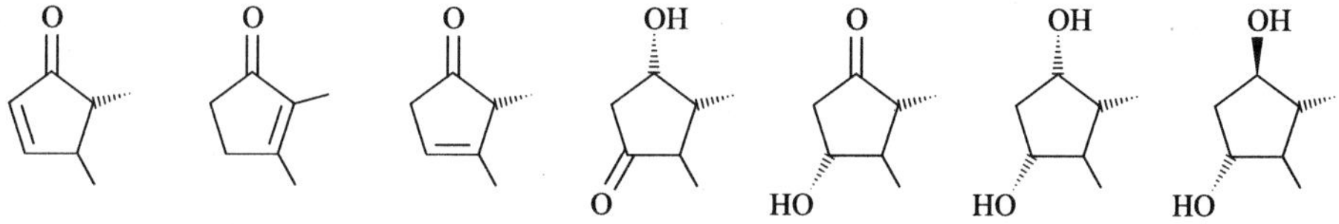

Typical ring variations

Figure 11.2 Representative structures from the family of prostaglandins

producing substances like the prostaglandins which cause such damage? Possibly, the release of small amounts of prostaglandin substances at a joint following an injury, with the production of swelling and pain, may be a defense mechanism to inform the body that the joint has been injured. Inflammatory conditions like rheumatoid arthritis may represent a normal, useful physiological defense mechanism gone out of control. It should be noted that these inflammatory processes differ biochemically/physiologically from allergic responses. They are not mediated by histamine. Despite the well-established relationship of prostaglandin systems to inflammation, it seems likely that hyperrelease of prostaglandin substances is an inadequate and overly simplistic explanation for the etiology of the rheumatoid arthritis syndrome. The nonapeptide bradykinin is believed to be a participant both in the initiation and in the propagation of the inflammatory response, but its role has not been clearly defined. Its effects seem to be a part of a complex cascade of events which involve other chemical mediators also.

Effects on the nervous system

Prostaglandins are essential participants in the central nervous system in physiological control of body temperature. The antipyretic actions originate in the heat regulatory centers of the brain (hypothalamus), resulting in dilatation of surface blood vessels in the skin. More blood flows through these vessels, and heat is dissipated by convection. The effectiveness of antipyretic drugs is believed to be related to their influence on prostaglandin synthesis, utilization, and metabolism. However, these antipyretic drugs do not affect body temperature when it is elevated by such factors as increased ambient temperature or exercise.

Effects on pain

The postulated peripheral role(s) of prostaglandins in nociception have been cited previously. In addition, pain recognition sites in the CNS involve prostaglandin compounds.

Synthesis of prostaglandins

The prostaglandins and closely related *thromboxanes* (*eicosanoids*, "20 carbons") are synthesized in the body from arachidonic acid, a normal body constituent, via a complex series of reactions termed the arachidonic acid cascade, the "cyclooxygenase pathway" (figure 11.3). Prostaglandins and thromboxanes are frequently collectively referred to as *prostanoids*. Arachidonic acid is ubiquitously distributed in esterified form on glycerol carbon-2 of the phospholipid portion of membranes. Arachidonic acid is released by action of a membrane-bound enzyme, phospholipase A_2. Subsequently, the sequence of reactions shown in figure 11.3 occurs. It has been suggested that ethanol facilitates arachidonic acid release and the resulting increase in prostaglandin production is the cause of hangover headache.

Leukotrienes

Fatty acid derivatives, "slow reacting substances" or *leukotrienes*, also arise in the body from arachidonic acid via the "lipoxygenase pathway" (figure 11.4). Structurally, leukotrienes differ from prostaglandins in lacking a carbocyclic or heterocyclic ring. They are also believed to be involved in the inflammatory response.

Mechanism of action of analgesic–antiinflammatory drugs

It seems likely that the nonsteroidal antiinflammatory–analgesic agents owe their analgesic and antipyretic activities and at least a portion of their antiinflammatory activities to their ability to inhibit biosynthesis of prostaglandins by disrupting the arachidonic acid cascade. However, not all of these agents disrupt the sequence of reactions in the same place or in the same manner. Aspirin irreversibly inactivates the enzyme cyclooxygenase by donating its acetyl group for acetylation of a serine hydroxyl moiety at the active catalytic site of the enzyme, which thereby destroys catalytic activity. Obviously, this mechanism is not applicable to the other antiinflammatory drugs. Nevertheless, most of the other antiinflammatory agents shown also inhibit cyclooxygenase. The exception is salicylic acid which, although it is a powerful disrupter of the arachidonic acid cascade, is almost inert in inhibiting cyclooxygenase. These findings relative to salicylic acid and aspirin destroyed hundred-year beliefs concerning salicylate drug design. Salicylic acid, the first of these agents to be used clinically, had the

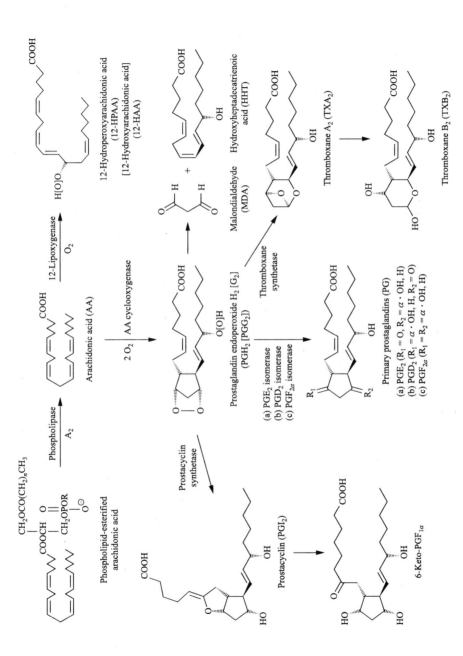

Figure 11.3 Arachidonic acid cascade: cyclooxygenase pathway

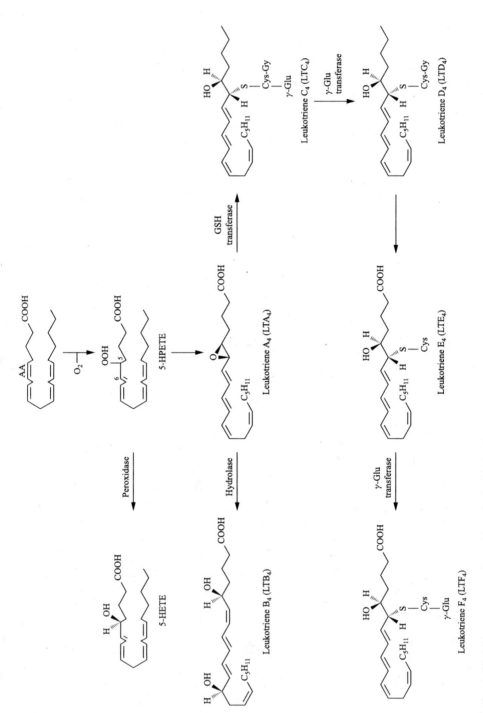

Figure 11.4 Lipoxygenase pathway: biosynthesis of leukotrienes

reputation of causing a high incidence of gastric distress following oral administration. Because phenol itself is a caustic substance, it was speculated that the phenolic hydroxyl group of salicylic acid is responsible for its irritant character. Accordingly, it was rationalized that if the hydroxyl group were masked by conversion to its acetate ester, this derivative should cause much less gastric distress. Indeed, although aspirin (acetylsalicylic acid) retains some tendency to produce gastric discomfort, it produces much less than salicylic acid. It was further hypothesized that once the aspirin molecule is absorbed into the blood, the acetate ester group is rapidly cleaved by plasma esterases to generate salicylic acid which was believed to be the pharmacologically active species. For many years, aspirin was considered to be a latentiated form of salicylic acid. This is not the case, and now aspirin's pharmacological mechanism of action is recognized as being, at least in part, different from that of salicylic acid.

The cyclooxygenase-mediated reaction in the arachidonic acid cascade (figure 11.3) is the rate-limiting step in the process. Two forms of this enzyme have been identified and extensively studied: *COX-1* and *COX-2*. COX-1, a so-called *constitutive* enzyme (i.e., it is always present), is found in most cells and it participates in production of those prostanoids which are involved in regulating vascular responses and in coordinating actions of circulating hormones. COX-1 is also involved in processes leading to production of gastric mucus. Low and medium doses of aspirin selectively inhibit this form of the enzyme. COX-2 is the induced form of the enzyme. There is a link between the inflammatory response and COX-2 induction. Indomethacin, ibuprofen, and high doses of aspirin inhibit both the COX-1 and the COX-2 forms of cyclooxygenase.

Contemporary opinion holds that the antiinflammatory actions of salicylic acid derivatives and perhaps of other nonsteroidal antiinflammatory drugs cannot be explained solely on the basis of inhibition of prostaglandin synthesis. Additional factors, such as membrane-associated processes, probably also contribute to the totality of effects of these drugs. One such process involves aspirin's inhibition of an enzyme involved in phosphorylation of a specific cytoplasmic protein which is intimately involved in triggering of the inflammatory response, once it is phosphorylated.

A relatively new use of aspirin is as a prophylactic agent to prevent heart attacks. This therapeutic use seems to depend upon aspirin's selective blockade of a thromboxane (TXA_2) synthesis by the blood platelets, without preventing production of prostaglandin PGI_2 by endothelial cells (cf. figure 11.3). This inhibition of thromboxane formation in the platelets inhibits platelet aggregation (the tendency of the platelets to become sticky and to clump together), which is an early triggering step in the formation of a clot. A single dose of 650 mg of aspirin doubles the mean clotting time of normal individuals for a period of 4 to 7 days.

11.4.3 Reye's Syndrome

There is a statistical relationship between salicylate ingestion and the occurrence of a comparatively rare but frequently fatal disease in children, *Reye's syndrome*. This condition is caused by influenza and some similar viruses. Salicylates should not be administered to children or adolescents who have chicken pox or influenza. A biochemical/physiological basis for the salicylate–Reye's syndrome relationship has not yet been found.

11.4.4 Nonsalicylate Antiinflammatory Analgesic Agents

In general, the nonsalicylate, nonsteroidal antiinflammatory agents (typified by structures **11.9–11.13**) are markedly hydrophobic aromatic molecules bearing some acidic function. There is a poor correlation between the antiinflammatory and the analgesic properties of most of these drugs. Some are potent analgesics, and some are not. Most of them are extensively bound (>90%) to blood plasma proteins. Most have relatively short serum half-lives and durations of action in the body. In the liver, they are glucuronidated on their carboxyl group, and this metabolite is excreted. In some instances an aromatic ring is metabolically hydroxylated and this metabolite is glucuronidated. *Mefenamic acid* **11.10** is representative of a family of *N*-phenylanthranilic acid derivatives (fenamates) whose chemical structures are reminiscent of the salicylates. Their analgesic–antiinflammatory activity results from their inhibition of cyclooxygenase, as with aspirin. However, the mechanism of this inhibition differs from that of aspirin. Some of the fenamates also act as prostaglandin antagonists. Drugs in this chemical category have no advantage over the salicylates, and they produce a high incidence of potentially serious gastrointestinal side effects.

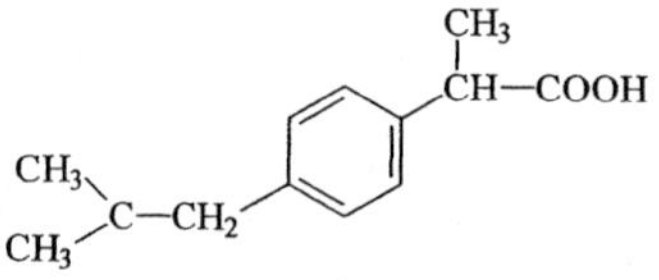

11.9 Ibuprofen

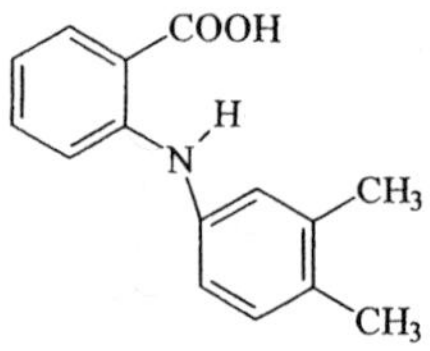

11.10 Mefenamic acid

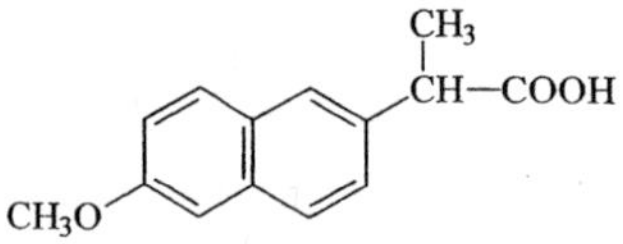

11.11 Naproxen

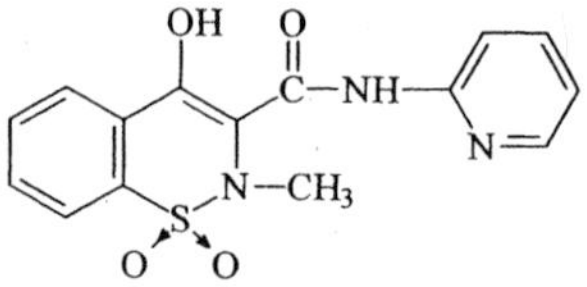

11.12 Piroxicam

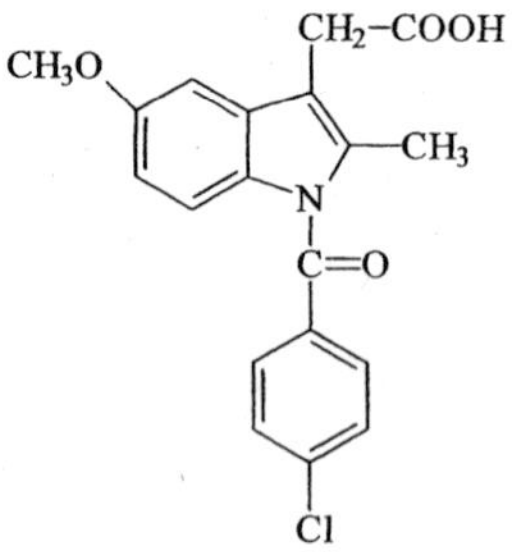

11.13 Indomethacin

Indomethacin **11.13**, one of the most potent inhibitors of cyclooxygenase, has powerful antiinflammatory and analgesic–antipyretic properties. Its overall clinical value is severely limited by its toxicity: 35–50% of patients who receive the drug experience untoward symptoms, and 20% must discontinue its use. Abdominal pain, nausea and vomiting, profound and extensive ulceration of the gastrointestinal tract, dizziness, skin rashes, and acute asthma attacks result from therapeutic dose regimens of indomethacin. Individuals sensitive to aspirin may exhibit cross-reactivity to indomethacin. Because of its severe side effects, indomethacin is not commonly used for its analgesic and antipyretic effects.

Piroxicam **11.12** typifies a newer chemical category, oxicams (trivially called *enolic acids*), which have antiinflammatory, antipyretic, and analgesic properties. This drug inhibits prostaglandin biosynthesis, and it probably has additional, complementary modes of action. It can cause gastric erosions and it prolongs bleeding time. In treatment of rheumatoid arthritis, piroxicam appears to be equivalent to aspirin in efficacy. It is said to be better tolerated than aspirin or indomethacin. Its chief advantage is its long half life which permits once daily dosage.

Ibuprofen **11.9** and *naproxen* **11.11** represent another relatively new chemical category, propionic acid derivatives ("profen" is a generic name for an α-phenylpropionic acid derivative). The profens inhibit cyclooxygenase. They are effective antiinflammatories but, like aspirin, they may produce gastrointestinal erosions and other gastrointestinal side effects. These drugs are used in treating rheumatoid arthritis and as simple analgesics. Smaller dosage size tablets are available as over-the-counter medication under several proprietary names. As a class, the profens bind extensively to blood albumins. However, remarkably, it is reported that this binding does not alter the effects of the anticoagulant drug warfarin whose drug–drug interaction with aspirin was described in chapter 1.

Rofecoxib (Vioxx®) **11.14** and *celecoxib* (Celebrex®) **11.15** are antiinflammatory antipyretic analgesic agents that specifically inhibit COX-2 but do not inhibit COX-1. They have no effect on blood platelet function (see chapter 14). The incidence of gastric ulcer with these drugs is significantly less than with ibuprofen. Rofecoxib was withdrawn from the market in September 2004, after its tendency to produce sometimes fatal cardiovascular side effects was revealed.

11.14   **11.15**

11.4.5 Side Effects of Antiinflammatory Analgesics

It has been mentioned that a commonly encountered side effect of many of the nonsteroidal antiinflammatory–analgesic agents (especially aspirin and other salicylates) is gastric ulceration. Prostaglandins, which are produced by cells of the gastric mucosa, are important gastric mucosal protective agents. These prostaglandins inhibit secretion of gastric hydrochloric acid and they promote secretion of gastric mucus which coats the stomach wall and protects it from damage by hydrochloric acid and other secretions introduced into the stomach lumen as a part of the normal digestive process. Blockade of the arachidonic acid cascade by inhibition of the COX-1 enzyme disrupts the mechanism protecting the stomach wall and results in local damage and irritation by gastric secretions. Probably the direct local irritant effect of the salicylates (the basis for the original design of aspirin) is a less important factor in gastric distress than is the prostaglandin-related systemic effect. *Salicylism* can result from prolonged, repeated ingestion of large doses of salicylates. This condition includes tinnitus (ringing of the ears), vertigo, hearing difficulty, and nausea and vomiting. Asthmatic attacks can result in aspirin-sensitive asthmatics.

11.5   Coal Tar Analgesics

These drugs (figure 11.5) are so designated because they are derivatives of aniline, which was first obtained by destructive distillation of coal tar. Both acetanilide **11.16** and acetophenetidine **11.18** are metabolized to a pharmacologically active metabolite, acetaminophen **11.17**, although probably both acetanilide and acetophenetidine themselves have analgesic activity; in the past they have been widely used for this purpose, but they are no longer available. Acetaminophen was first employed in medical practice before 1900, but it did not become a popular household remedy until the latter part of the twentieth century. It is probably the least toxic of the three drugs shown. The analgesic and antipyretic effects of the coal tar analgesics are superficially like those of aspirin, but their mechanism(s) of analgesic action has/have been considered to be different from that of aspirin and the other non-steroidal antiinflammatory drugs. Their antipyretic and analgesic effects are central in origin. They are only extremely weak inhibitors of peripheral prostaglandin biosynthesis. They have little inhibitory effect on either COX-1 or COX-2 forms of cyclooxygenase.

The coal tar analgesics lack the antiinflammatory properties of the salicylates. They relieve the pain of the inflammatory syndrome, but they do not improve the condition itself, a fact that is not widely recognized by the public. In 2002, a new variant of the cyclooxygenase enzyme family was described: *COX-3*. This enzyme is involved in the synthesis of prostaglandins (presumably only in the CNS) and it plays a role in pain and fever phenomena. COX-3 was reported to have no peripheral role in the inflammatory response. COX-3 does not participate in prostaglandin-mediated secretion of gastric mucus, nor in the thromboxane-mediated clumping of blood platelets in the blood coagulation process. The catalytic activity of COX-3 is inhibited by acetaminophen. This finding rationalizes the observations that acetaminophen does not produce gastric irritation, erosion, or bleeding which are side effects of salicylates. Acetaminophen has

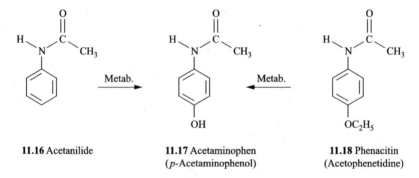

| **11.16** Acetanilide | **11.17** Acetaminophen | **11.18** Phenacitin |
| | (*p*-Acetaminophenol) | (Acetophenetidine) |

Figure 11.5 Interrelationships of coal tar analgesics

no effect on platelets or bleeding time, which are peripheral thromboxane-mediated phenomena.

11.5.1 Chronic Toxicity of Coal Tar Analgesics

Continual, long-term use of any of the coal tar analgesics can cause severe liver and kidney damage. Chronic use of phenacetin (acetophenetidine) may lead to *methemoglobinemia*, a blood dyscrasia in which the normal divalent iron in hemoglobin has been oxidized to the trivalent state, which lacks the ability to bind oxygen. Thus, the oxygen-carrying capacity of the blood is lowered, which is potentially dangerous. Acetaminophen is reputed to be much less likely to cause blood dyscrasias. Acetaminophen has been implicated in liver failure in alcoholics. It is suggested that ethanol induces the liver enzymes that catalyze conversion of acetaminophen into toxic compounds. The "toxic reactive intermediate" formed from acetaminophen in the liver (probably due to cytochrome oxidase enzymes) is a quinone imide **11.19**. This metabolite reacts with the sulfhydryl group in glutathione (cf. chapter 3) in a Michael-type 1,4 addition in a detoxication process (figure 11.6). If the store of glutathione in the liver is depleted by this reaction with excess amounts of the quinone imide metabolite, the quinone imide will then react as a Michael acceptor with sulfhydryl groups of endogenous liver proteins, resulting in hepatic necrosis.

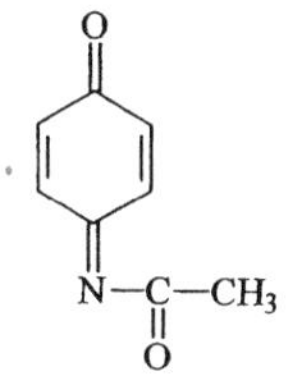

11.19

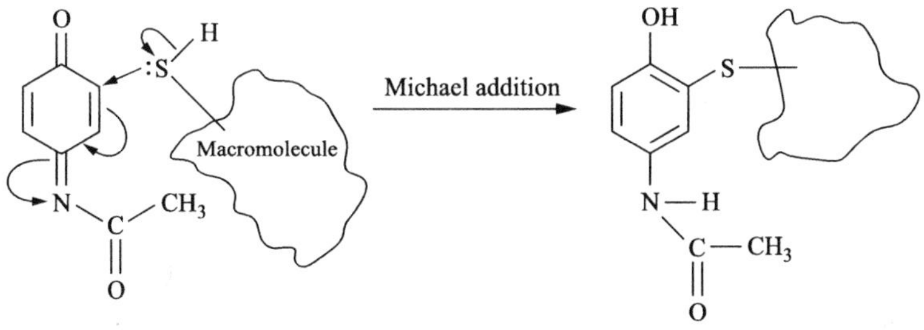

Michael addition

Figure 11.6 Michael addition of glutathions or other S–H containing proteins to quinone imide metabolite of acetaminophen

11.6 Capsaicin

Capsaicin **11.20**, is the component of chili peppers that is responsible for their burning taste. Capsaicin acts on superficial peripheral membrane receptors ("vanilloid receptors") to generate nociceptive and heat sensations. Capsaicin directly opens a cation channel in the membrane of nociceptive nerve fibers, causing an increased flow of Ca^{2+} and Na^+ which are postulated to be related to the pain sensation. With repeated applications of capsaicin to a tissue, the pain-producing effect disappears, and the nociceptive responses to other stimuli also disappear. This is the basis for topical application of capsaicin oint- ments to the skin for the treatment of certain kinds of neuropathic pain. However, it is still arguable whether occupation, e.g., of only capsaicin ("vanilloid") receptors, is suffi- cient to relieve complex pain states. Among its several documented pharmacological actions is capsaicin's ability to trigger release of substance P (the pain-generating peptide neurotransmitter) from peripheral sites and also from afferent neurons in the spinal cord. Thus, in the course of time, these neurons become depleted of substance P, and some time (days or even weeks) is required for recovery after cessation of application of capsaicin. This has been proposed as the pharmacological basis for the topical applica- tion of capsaicin ointment for relieving the pain of neuralgia and rheumatoid arthritis. The ninth edition (1996) of *Goodman and Gilman's The Pharmacological Basis of Therapeutics* assessed the efficacy of capsaicin in relieving pain as "debatable". The tenth edition (2001) of this book omitted all mention of capsaicin.

11.20

11.7   Classification of Pain

In sum, it has been demonstrated that certain types of pain sensations are mediated by biochemical processes involving prostaglandins and/or thromboxanes. It is noteworthy that morphine does not inhibit prostaglandin or thromboxane biosynthesis, and appar- ently it has no direct effect on prostaglandin-based pain mechanisms, centrally or peripherally. It can be concluded that biochemically/physiologically, morphine produces analgesia which is different from that produced by the salicylates, the other antiinflammatory analgesics, and the coal tar analgesics. There is a genuine physiolog- ical difference between *somatic pain* (prostanoid-related) and *visceral* pain (not prostanoid-related). Thus, while dental pain is relieved by aspirin, this drug is ineffec- tual in relieving pain of a gallbladder attack.

Bibliography

1. *Brain Facts*, 2nd ed. Society for Neuroscience: Washington, D. C. 1993; Fall, p. 17.

Recommended Reading

1. Hrobjartsson, A. and Gotzsche, P. C. A Challenge to the Validity of Placebo Effect. *N. Engl. J. Med.* **2001**, *344*, 1594–1602.

2. Vogelson, C. T. Searching for the Placebo Effect. *Modern Drug Discovery.* **2002**, July, 27–28.

3. Williams, M., Kowaluk, E. A. and Arneric, S. P. Emerging Molecular Approaches to Pain Therapy. *J. Med. Chem.* **1999**, *42*, 1481–1500.

4. Rang, H. P., Dale, M. M., Ritter, J. M. and Gardner, P. Analgesic Drugs. In *Pharmacology.* Churchill–Livingstone: New York, **2001**; pp. 579–603.

5. Jenkins, D. W., Humphrey, P. P. A. and Coleman, R. A. Agents Acting on Prostanoid Receptors. In *Burger's Medicinal Chemistry and Drug Discovery*, 6th ed. Abraham, D. J., Ed. Wiley: New York, **2003**; Vol. 4, pp. 265–315.

6. Roberts, L. J. III and Morrow, J. D. Analgesic–Antipyretic and Antiinflammatory Agents and Drugs Employed in Treatment of Gout. In *Goodman and Gilman's The Pharmacological Basis of Therapeutics*, 10th ed. Hardman, J. G., Limbird, L. E. and Gilman, A. G., Eds. McGraw-Hill: New York, **2001**; pp. 687–731.

7. Zubieta, J.-K., Bueller, J. A., Jackson, L. R., Scott, D. J., Xu, Y., Koeppe, R. A., *et al.* Placebo Effects Mediated by Endogenous Opioid Activity on μ-Opioid Receptors. *J. Neurosci.* **2005**, *25*, 7754–7762.

12
Analgesics. II: Opioid Analgesics

Terminology

The following semantic discussion is paraphrased from *Goodman and Gilman's The Pharmacological Basis of Therapeutics*, 10th edition, p. 569 (Recommended Reading ref. 4). *Opiates* are drugs derived from opium, and the term encompasses a wide variety of semisynthetic congeners of the opium alkaloids as well as the naturally occurring compounds. The term *opioid* is even more inclusive and is applied to all drugs, natural and synthetic, that have morphine-like pharmacological properties. For many years the term *narcotic* has been used to designate opiate analgesics. It is now employed in a legal context to refer to a wide variety of substances with abuse or addictive potential, and the word is no longer useful in a pharmacological context. The generic term *endorphin* is applied to the families of endogenous opioid peptides in toto, but it also refers to a specific endogenous opioid, β-endorphin.

Morphine-Like Analgesics

Morphine-like drugs produce analgesia by acting mainly on the central nervous system. The prototype opioid analgesic is morphine **12.1**, the principal alkaloidal constituent of the pod latex of the opium poppy, *Papaver somniferum*. Morphine is not maximally effective when it is given orally. It undergoes significant first-pass metabolic inactivation in the liver. The bioavailability of oral preparations of morphine is estimated at 25%. The principal hepatic metabolite is the position 3-glucuronic acid conjugate **12.4** which is pharmacologically inert and is excreted chiefly through the kidney by glomerular filtration. This morphine 3-glucuronide also undergoes a degree of enterohepatic circulation, and a small amount is excreted in the feces. Another morphine metabolite, the 6-glucuronide **12.5**, is formed in significant amounts in the body. This compound, although quite polar, can cross the blood–brain barrier to exert significant clinical effects. The 6-glucuronide has pharmacological actions undistinguishable from those of morphine; additionally, it is twice as potent as morphine in humans. With chronic administration of morphine, the 6-glucuronide accounts for a significant portion of the analgesic effect. Morphine is effective against both visceral and somatic pain, a property not possessed by the analgesics discussed in the previous chapter. Morphine and the other opioid analgesics are less dependable against neuropathic pain. Morphine depresses the cough center in the medulla oblongata, and it is an excellent antitussive (anticough) agent.

12.1 Morphine R=R'=H
12.2 Codeine   R=CH₃, R'=H

12.3 Heroin   R=R'=C—CH₃

12.4 Morphine-3-glucuronide R'=H R=

12.5 Morphine-6-glucuronide R=H R'=

Morphine produces euphoria which is usually followed by CNS depression and in overdose, a potentially fatal depression of the respiratory center in the medulla oblongata results. The pharmacological mechanism for respiratory depression involves inhibition of the responsiveness of the respiratory center to high blood levels of carbon dioxide, which is a physiological stimulus to the respiratory center. Euphoria is an important and frequently desirable component of morphine's pharmacological profile, since the agitation and anxiety associated with painful illness or trauma are reduced. Morphine produces constipation, which was once thought to result from indirect stimulation of β_1 adrenoceptors in the intestinal wall, but is now recognized to be caused chiefly by morphine's interaction with specific receptors (probably the μ and δ types) in the wall of the intestine and also in the spine, causing hypomotility of the colon. In many parts of the world, tincture of opium is a standard treatment for simple diarrhea. *Diphenoxylate* **12.6** is typical of synthetic opioids that are 4-phenylpiperidine derivatives (vide infra) and are employed exclusively as antidiarrheal agents. Diphenoxylate is approximately one order of magnitude more potent an antidiarrheal agent than morphine. After absorption, the ester linkage is cleaved enzymatically and the free carboxylic acid has actions of a typical opioid agent. The antidiarrheal action results from stimulus of μ receptors. "Diarrhea-level" doses of diphenoxylate show little of the other morphine-like actions. However, in doses approximately 25 times the antidiarrheal dose, diphenoxylate is a typical opioid with all that this connotes. A popular proprietary product contains diphenoxylate in combination with a subtherapeutic dose of atropine, which is included to discourage inappropriate use of the product. In order to ingest sufficient diphenoxylate to support an opioid habit, the individual will also be ingesting sufficient atropine to produce unpleasant symptoms of antimuscarinic overdosage (hyperthermia, tachycardia, urinary retention, flushing, dryness of the skin and mucous membranes), all of which make the overall diphenoxylate experience much less pleasurable.

12.6

In analgesic doses, morphine stimulates the chemoreceptor trigger zone in the medulla oblongata to produce nausea and vomiting. These responses are transient and they usually disappear with repeated administration of morphine. As mentioned in chapter 1, the emetic center in the medulla oblongata is physiologically outside the blood–brain barrier, hence the rapid emetic response to morphine. However, once a sufficient amount of morphine has penetrated the blood–brain barrier, it exerts an anti-emetic effect there.

Chronic administration of therapeutic doses of morphine results in tolerance to its analgesic effects and requires ever larger doses to attain the same level of analgesia. Most disturbing is morphine's ability to cause physical and psychic dependence: addiction. Psychic dependence (craving for the drug) is probably more important than the physical withdrawal syndrome as a factor in dependence in humans. This craving can persist for months or even years following cessation of opioid use.

Because of the large number of problems associated with the use of morphine as an analgesic, massive attempts have been made to modify the molecule to retain analgesic effects but to diminish or abolish the unwanted side effects. The chemical/pharmacological strategy has concentrated largely on the separation of analgesic effects from addictive properties. The fundamental uncertainty is whether the two effects are indeed separable or whether analgesia and addiction are inextricably biochemically linked. A product of an early molecular modification of morphine was its diacetate ester **12.3**, heroin. As an analgesic, this compound is somewhat more potent than morphine. Taken intravenously, heroin produces a rapidly developing sensation of euphoria combined with a peculiar, pleasurable warm glow over the entire abdominal region which users liken to the sensation of orgasm. The summation of these simultaneous pleasurable sensations is the "joy pop" that addicts initially experience from intravenous heroin injection. Morphine is somewhat hydrophilic and it penetrates the blood–brain barrier relatively slowly. Masking its two hydrophilic hydroxyl groups by acetylation increases lipophilicity considerably, and heroin rapidly and efficiently penetrates the blood–brain barrier by passive diffusion. Once in the brain, esterases cleave the ester linkages and a mixture of 6-acetylmorphine (which retains opioid analgesic activity) and morphine itself is generated. The pharmacological effects of heroin are due to these metabolites. The CNS effects of heroin can be noted in less than a minute following intravenous injection. Heroin is a classic example of drug latentiation; it possesses no special, unique pharmacological properties compared to morphine, other than greater ease of penetrating the blood–brain barrier. Addiction to heroin or to other short-acting opioids usually becomes incompatible with a productive life. Physical symptoms of opioid withdrawal include drug craving (psychological effect), dilation of the pupils of the eye, sweating, vomiting and diarrhea, intestinal cramps, rapid heart rate, piloerection (goose pimples), and elevation of blood pressure.

The 3-methyl ether of morphine, codeine **12.2**, has less tendency to produce addiction than morphine, but it is decidedly less potent as an analgesic. Codeine, like morphine, has a prominent antitussive effect, and it has been widely used clinically for this purpose. Because the 3-hydroxyl group is masked as its methyl ether, codeine does not undergo the kind of first-pass metabolic inactivation (3-glucuronidation) that inactivates morphine, and codeine is effective orally. It is approximately 60% as effective orally as parenterally. Codeine itself is metabolized to a variety of inactive products, but approximately 10% of the dose is O-demethylated to produce morphine itself. Codeine has very poor binding affinity for analgesic receptors, and pharmacologists consider that the analgesic property of codeine results from the modest amount of morphine that is produced by metabolism. Codeine can be viewed as being a latentiated form of morphine. However, the antitussive effects of codeine are believed to be produced by codeine itself, involving receptors that bind codeine specifically, and that differ from the antitussive receptors that bind morphine. Ethnic Chinese produce less morphine from codeine than do caucasians, and they are less sensitive to morphine's effects than are caucasians. This reduced sensitivity to morphine may be due to decreased production of morphine-6-glucuronide.

Structures **12.7**, **12.8**, and **12.9** illustrate representative further modifications of the morphine molecule. Although these are clinically useful analgesics, the prime goal, namely the separation of potent, high-level analgesia from addiction liability, has not been attained with any of them. In recent years oxycodone (oxycontin) **12.7** has become notorious as a drug of abuse, a substitute for heroin. This drug has a high abuse potential. It is well absorbed from the gastrointestinal tract and it is metabolized in part to oxymorphone (by O-demethylation), which is the pharmacologically active agent.

Opioid analgesic effects can be obtained with molecules simpler than morphine, and some molecular modification strategies involve synthesis of fragments of the morphine molecule, typified by levorphanol **12.10**. This is a morphinan derivative that lacks the furan ring and the alcoholic hydroxyl group of morphine, and it is a potent morphine-like analgesic, acting primarily at μ receptors. If the enantiomer of levorphanol is converted into its methyl ether, the product, *dextromethorphan*, is devoid of analgesic effect. It is not addicting nor does it support a heroin habit but, like morphine and codeine, it is a potent antitussive agent. However, dextromethorphan's antitussive action is mechanistically different from those of codeine and morphine. Whereas the cough suppressant effect of morphine and codeine is blocked by administering a morphine antagonist drug such as naloxone (vide infra), opioid antagonist drugs do not block or reverse the antitussive effect of dextromethorphan. Dextromethorphan is an NMDA (glutamate) receptor antagonist, but pharmacologists do not accept that this is necessarily the basis for its antitussive action. The mechanism of the antitussive effect of dextromethorphan remains unclear. This drug is the key ingredient of many proprietary nonprescription cough mixtures. It produces few side effects and its toxicity is low.

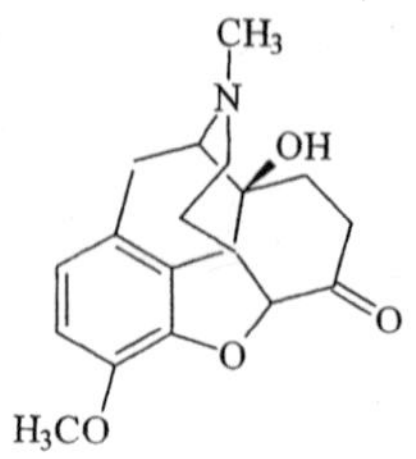

12.7

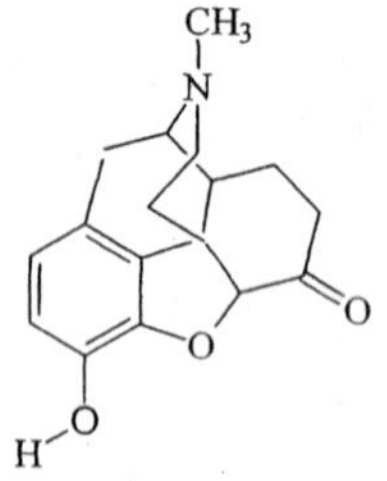

12.8 Dihydromorphinone

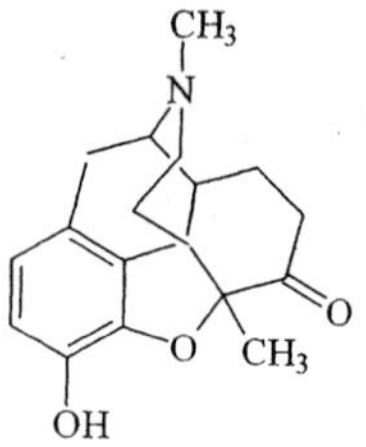

12.9 Metopon

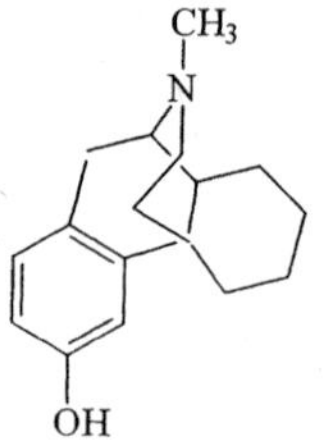

12.10

Compounds **12.11** to **12.14** possess true opioid analgesic effects, similar to those of morphine. The gross dissimilarity of chemical structures of synthetic and semi-synthetic morphine-like analgesics is striking: simpler fragments of the morphine molecule (**12.10**), 4-phenylpiperidine derivatives (**12.11**, **12.12**), and non-heterocyclic systems (**12.13**). Regardless of the heterogeneity of chemical structures, all of these are potent analgesics, qualitatively similar to morphine, and all are addicting. All support a heroin habit. Fentanyl **12.14** is approximately 100 times more potent, on a molar basis, than morphine as an analgesic. It is a μ agonist and in addition to its clinical use as an analgesic, it is used intravenously as an adjuvant with general anesthetics (see chapter 13).

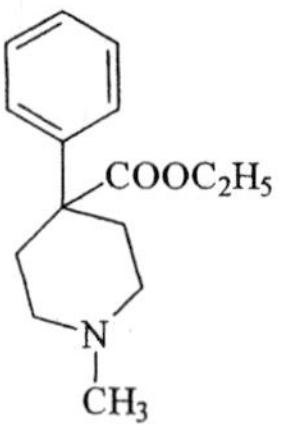

12.11

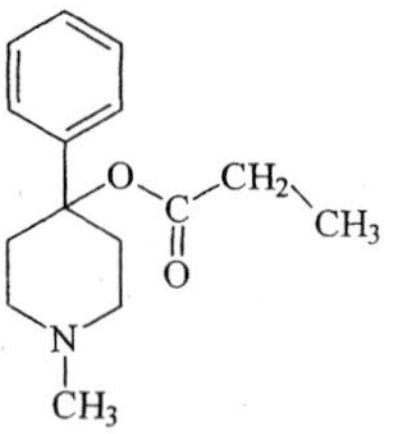

12.12

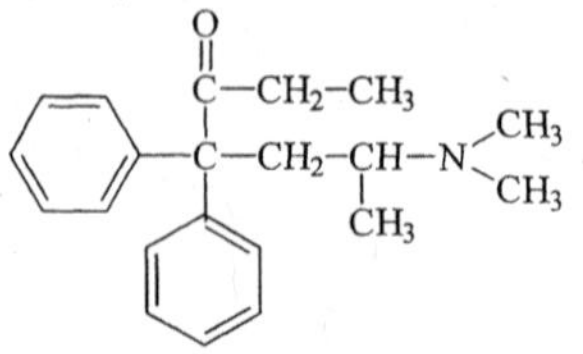

12.13

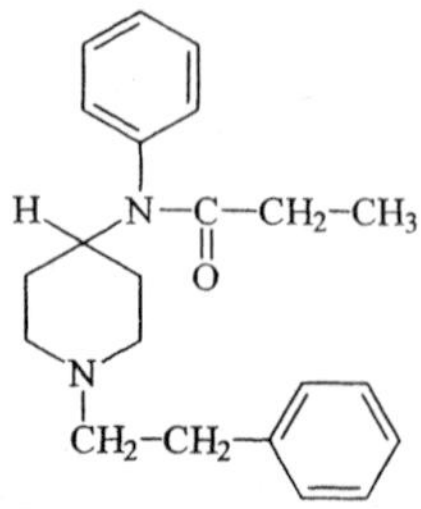

12.14

Noteworthy pharmacological properties of methadone **12.13** are its analgesic efficacy, its high oral potency, its extended duration of action in suppressing withdrawal symptoms in physically dependent individuals, and its low tendency to produce drug tolerance. This drug is the basis of the so-called *methadone treatment* for addicts. Typically, addicts report periodically to a methadone clinic for an oral dosage form of methadone. Oral administration of methadone prevents the withdrawal syndrome and the addict need not inject heroin to avoid becoming ill. However, orally administered methadone does not produce the "joy pop" sensation, and injecting heroin into a methadone-dosed individual does not produce the expected and customary pleasurable experience. In essence, the methadone-treated individual has exchanged a heroin habit for a methadone habit. However, regular use of methadone is less debilitating than heroin and it is possible that the methadone-dosed individual can attend school productively and/or demonstrate adequate job performance. Differences in responses to heroin and to methadone have been illustrated graphically (figure 12.1). As shown, the individual who injects heroin frequently alternates between being sick and being "high". In contrast, the methadone-treated individual remains in the "normal" physical and mental region (indicated in gray) after dosing once daily. Note that this figure describes mental and physical state, and not blood levels of the drugs. A deterrent to the greater success of the methadone treatment strategy is the low level of participation and compliance by addicts who crave the "joy pop" sensations of heroin injection.

The compound *MPTP* **12.15** first gained attention when it was discovered as a contaminant of some street drugs, 4-phenylpiperidine-derived "designer drugs" similar to meperidine and alphaprodine, produced as heroin substitutes by illicit chemical laboratories. MPTP is probably a product of an elimination side reaction occurring in the piperidine ring in the course of synthesis of the designer drug, and its presence results from inadequate purification of intermediates and/or the final product. Intravenous administration of MPTP causes permanent Parkinsonian symptoms. Victims of the MPTP syndrome were monitored for several years following induction of their symptoms; the Parkinsonian symptoms persisted unabated. The biochemical cause of this condition involves metabolism of MPTP, mediated by monoamine oxidase (MAO-B), to produce two products, **12.16** and **12.17**, which inhibit oxidation processes in the mitochondria in the

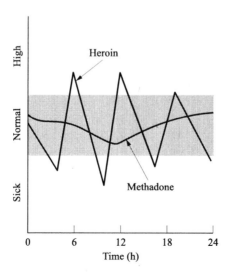

Figure 12.1 Differences in responses to heroin and to methadone

dopaminergic neurons in the nigrostriatal pathway, resulting in their degeneration, grossly quite similar to the condition noted in idiopathic Parkinson's disease. MPTP seems to be selective in destroying nigrostriatal neurons, and it does not affect dopaminergic neurons elsewhere. Parkinsonian symptoms resulting from MPTP do not appear until approximately 80% of the dopaminergic neurons in the nigrostriatal pathway have been destroyed. Inhibitors of MAO-B block the toxicity of MPTP. The drug produces Parkinsonian symptoms in mice and primates, but not in rats; this is a reflection of the inability of the rat body to metabolize MPTP to the toxic products and it further illustrates species differences in metabolism. Sufferers from the MPTP syndrome are treated with standard anti-Parkinsonian drugs and many experience relief from symptoms.

12.15 "MPTP"   12.16   12.17

12.3   Opioid Drug Antagonists

N-allylnormorphine (nalline, nalorphine) **12.18**, a semisynthetic compound in which the *N*-methyl group of morphine is replaced by allyl, has morphine-like analgesic effects. This compound is also a direct pharmacological *antagonist* to morphine and to other opioid analgesics, both natural products and synthetics. When given following a dose of morphine, *N*-allylnormorphine reverses morphine's analgesic effect, terminates euphoria, and counteracts any respiratory depression caused by the morphine. So specific are these actions of *N*-allylnormorphine that if it is administered to an individual who presumably is suffering from an overdose of an opioid analgesic and if the individual's condition does not rapidly improve, it may be concluded that the subject's condition is not caused by opioid overdosage. *N*-Allylnormorphine precipitates the symptoms of the withdrawal syndrome in addicts. But, as mentioned, the drug itself is a moderately potent opioid analgesic and it can itself produce physical dependence. It is classed as a partial agonist/partial antagonist. *N*-allylnormorphine is not employed clinically as an analgesic because it produces a high incidence of nausea and vomiting and it causes dysphoria and/or hallucinations.

CH_2—CH=CH_2

N-allylnormorphine

12.18

Another synthetic compound, *levallorphan* **12.19**, the *N*-allyl homolog of levor-phanol, has a pharmacologic profile quite similar to that of *N*-allylnormorphine. *Naloxone* **12.20** and *naltrexone* **12.21** differ in that they are "pure" opioid antagonists with almost no analgesic effects and then only at very large doses. These latter two drugs do not produce euphoria, nor are they addicting. They are used clinically as a part of a comprehensive strategy for treatment of opioid dependence. They are also widely used to reverse the respiratory depression caused by opioid overdosage. Naltrexone is also used clinically as an adjunct in treating alcoholism. It blocks some of the reinforc-ing properties of ethanol, and it seems to reduce the craving for alcohol. There is some indication that naltrexone blocks ethanol's activation of dopaminergic pathways in the brain which are thought to be involved in the craving syndrome.

N-allylnormorphine
12.18

12.19

12.20 Naloxone R= —CH$_2$-CH=CH$_2$

12.21 Naltrexone R= —CH$_2$—

12.4 Analgesic Receptors

Morphine and the other opioid drugs described herein, both natural products and synthetic compounds, produce analgesic effects because they act as agonists at specific receptors in the central nervous system (and perhaps elsewhere). The population of so-called analgesic receptors is heterogeneous. Classical pharmacological studies and cloning studies have identified μ receptors and κ receptors. Analgesia is associated with both μ and κ receptors, but most of the clinically useful opioid analgesics (including morphine) are selective for the μ receptor. However, at high doses the opioid agents lose their selectivity and stimulate both μ and κ types. Both of these opioid receptors are heterogeneous, but there seems to be no agreement regarding exact classification of recep-tor subtypes. Two subtypes of μ receptors have been proposed: μ$_1$ which are located in the brain and are the principal mediators of the analgesic effect of morphine and μ$_2$ which are located in the spinal cord and which participate in mediation of the respiratory depression and constipation side effects of morphine. It has been speculated that eupho-ria is mediated through μ receptors and that dysphoria (malaise, depression) is mediated through κ receptors. Thus, opioid drugs may vary in the degree of euphoria that they elicit, depending on the relative magnitudes of their effects on μ and κ receptors. The literature is complex and confusing about κ receptor subtypes; some pharmacologists categorize κ$_1$, κ$_2$, and κ$_3$, but this system is not universally accepted. All of the opioid receptors are of the G protein-linked type.

Earlier literature implicated another population of receptors, σ receptors, in analgesia. It has been shown subsequently that these are not involved in analgesia, even though some opioid drugs interact with them; σ receptors in the CNS appear to be a site of action of certain psychotomimetic agents.

Mixed agonist–antagonist drugs such as *N*-allylnormorphine and levallorphan compete with morphine at μ receptors but do not themselves have agonist activity at these receptors. Instead, their analgesic agonist effects are expressed through κ receptors at which morphine has only weak activity. The "pure" antagonists naloxone and naltrexone act at both μ and κ receptors and they do not display agonist effects at any opioid receptor sites.

12.4.1 Pentazocine, a κ Agonist

Pentazocine **12.22** is a moderately potent/active opioid analgesic. Initially, it was considered to have almost no dependence liability, and it was introduced into the marketplace subject to no special controls. Subsequently, it was found that pentazocine is capable of producing physical and psychic dependence, although the risk of dependence development is lower than for μ-receptor agonists (morphine-like drugs).

12.22 Pentazocine

Pentazocine is a κ receptor agonist, which accounts for its analgesic effects. It is a weak antagonist or partial agonist at μ receptors. Consistent with these properties, pentazocine does not prevent nor improve the morphine withdrawal syndrome. When it is administered to an individual who is addicted to morphine or to other μ agonists, pentazocine can precipitate the withdrawal syndrome. Long term administration of pentazocine itself results in dependence, and naloxone induces withdrawal symptoms in pentazocine-addicted individuals. Addicts administer pentazocine intravenously (extemporaneously extracting the drug from tablets, the most readily available dosage form). A common practice is the intravenous injection of pentazocine combined with *tripelennamine* **17.6**, an antiallergic histamine H_1 receptor blocker. Intravenous tripelennamine is reputed to produce euphoria which adds to the effects produced by pentazocine. A prominent side effect of tripelennamine is inhibition of central re-uptake (uptake-1) of norepinephrine. This may account in part for the euphoria. To discourage the illicit use of pentazocine tablets as a source of injectible drug, tablets for oral use also contain naloxone hydrochloride equivalent to 0.5 mg of naloxone base. Following oral administration, the naloxone is destroyed by first-pass metabolism in the liver, and it has no systemic pharmacological effect. However, if an extract of the tablet is administered intravenously, the naloxone is not rapidly destroyed; it antagonizes the systemic effect of the pentazocine and no euphoria results.

12.5 Endogenous Analgesic Receptor Agonists

The morphine-like analgesics exert their pain-abating effects because they react with specific receptors in the brain and spinal cord and thereby trigger a series of physiological responses. A philosophical question that troubled chemists and pharmacologists for well over a century was, "Why do our bodies contain receptors that produce dramatic responses in relief of pain, and why are these receptors so exquisitely structurally and stereochemically selective for activation by an obscure chemical constituent of poppy juice?" Many years ago it was speculated that the mammalian body may produce morphine as a normal physiological component. Convincing evidence for this intriguing concept was lacking, and the theory was rejected by most chemists and pharmacologists. However, more recent and more convincing studies have led to claims that both morphine and codeine have been detected as endogenous components in some animal species and in humans, and reasonable biosynthetic pathways beginning with dietary constituents and endogenous substances have been proposed. These more recent claims and hypotheses are also not universally accepted.

The so-called opioid receptors in animals and humans, found both centrally and peripherally, are normal physiological sites of action for several endogenous agonist peptides, which are neurotransmitters. Two pentapeptides, *met-enkephalin* **12.23** and *leu-enkephalin* **12.24**, and families of several larger peptides, *dynorphins* and *endorphins* (e.g., **12.25** and **12.26**) have been isolated from brain tissue. There is a common five-amino acid-sequence in these molecules: tyrosine–glycine–glycine–phenylalanine–leucine (or methionine). Endorphins interact with opioid receptor(s) and produce patterns of analgesic effects similar to (but not always identical with) those seen with the opioid analgesic drugs. β-Endorphin is the most potent of the endogenous opioid peptides discovered thus far. Tolerance to and physical dependence upon exogenously administered enkephalins and endorphins have been demonstrated, as has a cross-tolerance between these peptides and exogenous morphine. The analgesic and other effects of at least some of the endorphins are reversed by naloxone.

H-tyr-gly-gly-phe-met-OH   H-tyr-gly-gly-phe-leu-OH

12.23 Met-enkephalin   **12.24** Leu-enkephalin

H-tyr-gly-gly-phe-leu-arg-arg-ile-arg-pro-lys-leu-lys-trp-asp-asn-gln-OH

12.25 A typical dynorphin

H-lys-arg-tyr-gly-gly-phe-met-thr-ser-glu-lys-ser-glu-thr-pro-leu-val-thr-leu-phe-

lys-asn-ala-ile-ile-lys-asn-ala-tyr-lys-lys-gly-glu-OH

12.26 β-Endorphin

Physiologically, the effects of the endorphins are terminated by enzyme-mediated inactivation reactions involving two peptidase enzymes, commonly referred to as enkephalinases. Experimental synthetic nonpeptide compounds have been identified which inhibit these enzymes, thus protecting the endorphins from destruction. One such compound (*thiorphan* **12.27**) exhibits antinociceptive activity in mice, which is blocked by naloxone. The bioavailability of thiorphan is improved by protecting the thiol and carboxyl groups by esterification, to form *acetorphan* **12.28**. This derivative is rapidly hydrolyzed in vivo by esterases to regenerate thiorphan which increases local levels of enkephalins at both μ and δ opioid receptors to produce, like morphine, an antidiarrheal effect.

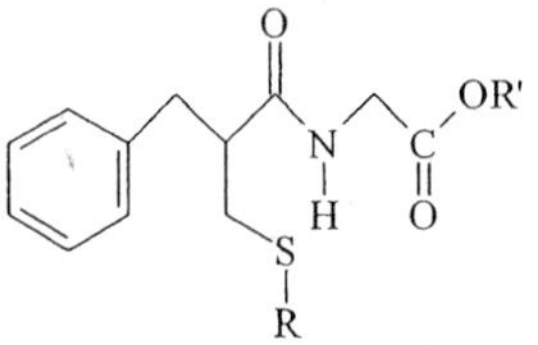

12.27 Thiorphan   R=R'=H

12.28 Acetorphan   R= CH₃—C—   R'= —CH₂—

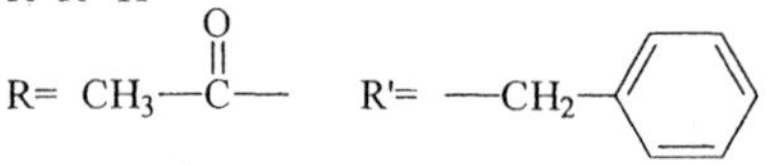

Studies of the endorphins resulted in identifying yet another opioid receptor, the δ, for which the endogenous peptides are ligands. Subtypes of the δ receptor have been defined. Morphine and almost all of the natural product and the synthetic and semisynthetic nonpeptide opioids have little effect at δ receptors. Met-enkephalin, leu-enkephalin, and β-endorphin exert a potent ligand effect at δ receptors and also at μ receptors but they show little or no activity at the population of κ receptors. In contrast, some of the dynorphins exert their greatest ligand activity at some κ receptors and they are relatively weak binders at μ and δ receptors. In addition to their relationship to analgesia, δ receptor agonists have a *stimulatory* effect on respiration, the opposite to the effect of morphine and other μ receptor agonists. The "pure" μ and κ receptor antagonists naloxone and naltrexone also display antagonist effects at δ receptors. For some years pharmacological research was impeded by the chemists' and pharmacologists' inability to identify specific or highly selective nonpeptide δ agonists. Selective nonpeptide δ agonists are now known; one of the first, **12.29**, exhibited 500- to 2600-fold selectivity for δ over μ receptors in binding assays and 200-fold selectivity in smooth muscle assays. Nonpeptide δ-selective antagonists have also been found.

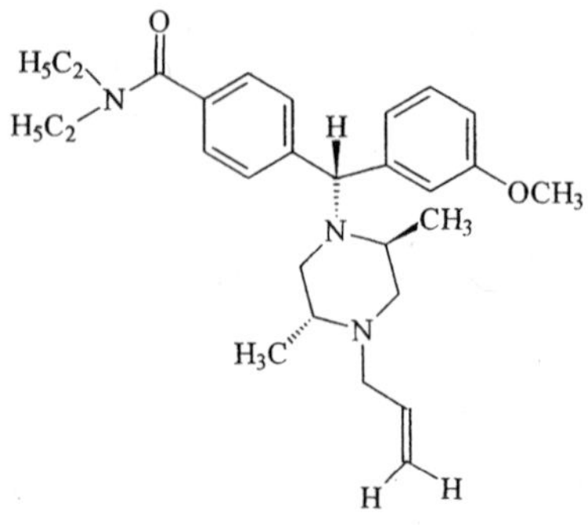

12.29

A new family of analgesic peptides (*endomorphins*) was originally isolated from bovine brains. These are tetrapeptides, two of which have the amino acid sequences tyrosine–proline–tryptophan–phenylalanine and tyrosine–proline–phenylalanine–phenylalanine. Note that these sequences differ from the sequence (tyrosine–glycine–glycine–phenylalanine) described previously as the putative pharmacophore of endogenous opioid peptides. These endomorphins have high affinity and specificity for the μ receptor, and in animal tests they were comparable to morphine as analgesics.

Yet another "analgesic receptor" has been revealed by cloning techniques: the so-called *opioid receptor-like* binding site (the "ORL-1" receptor, also named the *NOP receptor*). This is a G protein-linked receptor. None of the native opioid peptides discussed previously (endorphins, enkephalins, etc.) bind to this receptor, nor do most of the natural product or synthetic agonists which are selective for μ, κ, or δ receptors. An endogenous 17-amino acid peptide, *nociceptin* (also called *orphanin FQ*) has been found in mammalian bodies, and this is a ligand for the NOP receptor. This receptor is involved in modulating pain mechanisms at the level of the spinal cord. It also appears to be involved in the rewarding and reinforcing properties of drugs of abuse. It modifies the responses to both ethanol and morphine. In rats, nociceptin reduces the activity evoked by noxious stimuli, and these effects are not antagonized by naloxone. Nociceptin does not show affinity for μ, κ or δ receptors.

12.6   Analgesic Mechanisms of Endorphins, Opioids, and Some Other Drugs

The endorphins are considered to be neurotransmitters, but their multifaceted physiology is incompletely understood. One of their principal roles involves diminution of pain perception, possibly by both central and peripheral mechanisms, but it seems likely that there are additional physiological roles for these substances. Production of analgesia by opioids and endorphins is biochemically extremely complex. The G protein-linked μ, κ, and δ receptors are negatively coupled to inhibit adenylate cyclase activity, and the receptors are also involved with passage of K^+ and Ca^{2+} across membranes. The endorphins and the opioid analgesics possibly modulate dopaminergic and cholinergic sites in the cerebral cortex and the limbic region of the brain (where mood and emotion are controlled). It is intriguing that epibatidine **8.15** (cf. chapter 8), an acetylcholine nicotinic receptor agonist, demonstrates antinociceptive effects which are not blocked by naloxone but are antagonized by nicotinic receptor antagonists. Some synthetic congeners of epibatidine are also nicotinic agonists and they too display potent analgesic effects. Certain $GABA_A$ and $GABA_B$ receptor agonists have antinociceptive effects in rodents. Opioid receptors on the terminals of afferent (sensory) nerves mediate inhibition of the release of substance P, whose role as a neurotransmitter in pain perception was cited in chapter 11. Morphine also antagonizes the effect of substance P by exerting postsynaptic inhibitory actions on interneurons in the spinal cord. Clonidine **7.13**, an α_2-adrenoceptor agonist which inhibits cyclic AMP formation, also affects cellular mechanisms that mimic opioid effects and alleviates many of the symptoms of opioid withdrawal. Clonidine itself is said to possess analgesic activity and a cross-tolerance occurs between it and opioid analgesics. It is difficult to integrate these extensive pharmacological data and observations into a cohesive mechanistic explanation for opioid analgesia.

12.7   Opioid and Analgesic Peptide Tolerance and Dependence

Tolerance refers to the increase in dose required to produce a given level of pharmacological response. It develops rapidly in the case of the opioid analgesics. *Dependence* is a different phenomenon, involving physical dependence associated with a well-defined physical withdrawal syndrome and psychological dependence, manifested as a craving for the drug involved. It has been stated that psychological dependence is the more important. In the case of tolerance, causative factors such as accelerated metabolic inactivation of the drug, reduced affinity of the drug for its receptors, diminished release of endorphins, and down-regulation of receptors, have been rejected. Cross-tolerance occurs between drugs acting at the same receptor subtype, but not between drugs acting at different receptors. It has been suggested that changes in gene expression, leading to changes in levels of G-proteins and adenylate cyclase are involved. Nitric oxide production has been implicated in tolerance to morphine. Noradrenergic pathways in some brain areas may be involved in the abstinence syndrome. CNS dopaminergic and perhaps cholinergic pathways have also been proposed as participants in the dependency syndrome.

The role of endorphins in the physiology of addictive states is still largely unknown and poorly understood, despite a voluminous literature on the subject. The putative addictive properties of the endorphins generate an intriguing question: "Why are humans not addicted to their own endogenous analgesic peptides?" It may be that the rapid and efficient normal physiological enzymatic inactivation of the endorphins is a protective mechanism against dependence.

12.8   Physiological Implications of the Existence of Endogenous Analgesic Substances

There are profound implications to the concept that animal (including human) bodies synthesize and utilize analgesic substances. A physiological mechanism exists for terminating the pain sensation. As has been described repeatedly for other biological control phenomena such as the peripheral autonomic nervous system's innervation of organs and glands, there is a set of checks and balances for pain. The body synthesizes and utilizes chemical substances that create pain sensations, and the effects of these are antagonized by endogenous analgesic substances. The purported stoicism of some individuals in the presence of pain-producing stimuli probably reflects a genuine physiological phenomenon and is not merely a measure of courage. These individuals may be experiencing a conditioned reflex to liberate analgesic peptides, and they truly feel diminished pain. A similar conditioned reflex may explain the placebo effect in analgesic testing in humans, as well as the frequently cited use of acupuncture as a substitute for drug-induced anesthesia/analgesia. Does release of opioid peptides or other similar neurotransmitter substances explain the "runners' high" syndrome?

12.9   Mechanistically Novel Approaches to Analgesic Drug Therapy

In addition to the opioid analgesics, the cyclooxygenase-related agents, nicotinic agonists such as epibatidine (cf. chapter 8), and cannabinoids, compounds affecting the following physiological systems have been reported to demonstrate analgesic effects:

- Antagonists of glutamate NMDA receptors. Numerous studies have suggested a role for excitatory amino acids in the development and maintenance of chronic neuropathic pain. Reversible antagonists of glutamate NMDA receptors and/or NMDA ion channel blockers may have some value in relief of neuropathic pain.
- Blockers of certain voltage-gated calcium channels.
- Inhibitors of adenosine kinase, the enzyme involved in inactivation of extracellular endogenously released adenosine. Adenosine has been implicated in antinociceptive effects in animal models of inflammatory and neuropathic pain.
- Antagonists of certain subpopulations of adenosine triphosphate receptors. ATP has been implicated as a participant in the induction and transmission of pain signals.

The potential for future therapeutic use of these systems remains to be demonstrated.

Bibliography

1. O'Brien, C. P. In *Goodman and Gilman's The Pharmacological Basis of Therapeutics*, 10th ed. Hardman, J. G., Limbird, L. E. and Gilman A.G. Eds. McGraw Hill: New York, 2001; p. 632.

Recommended Reading

1. Aldrich, J. V. Physiology and Pharmacology of Narcotic Analgesics. In *Burger's Medicinal Chemistry and Drug Discovery*, 6th ed. Abraham, D. J., Ed. Wiley Interscience: Hoboken, N.J., 2003; Vol. 6, pp. 341–358.
2. Watling, K. J. *The Sigma–RBI Handbook of Receptor Classification and Signal Transduction*, 5th ed. Sigma–RBI: Natick, Mass., 2006; pp. 204–205.
3. Tseng, L. F., Ed. *The Pharmacology of Opioid Peptides*. Harwood: Chur, Switzerland, 1995.
4. Gutstein, H. B. and Akil, H. Opioid Analgesics. In *Goodman and Gilman's The Pharmacological Basis of Therapeutics*, 10th ed. Hardman, J. G., Limbird, L. E. and Gilman, A. G., Eds. McGraw Hill: New York, 2001; pp. 569–619.
5. O'Brien, C. P. Drug Addiction and Drug Abuse. In Recommended Reading ref. 4, pp. 621–642.
6. Kowaluk, E. A., Lynch, K. J. and Jarvis, M. F. Recent Advances in Development of Novel Analgesics. *Annu. Rep. Med. Chem.* **2000**, *35*, 21–30.
7. Aslanian, R., Hey, J. A. and Shih, N.-Y. Recent Developments in Antitussive Therapy. *Annu. Rep. Med. Chem.* **2001**, *36*, 31–40.
8. Buschmann, H., Christoph, F., Friderichs, E., Maul, C. and Sundermann, B. *Analgesics. From Chemistry and Pharmacology to Clinical Application*. Wiley-VCH: Hoboken, N.J., 2002.
9. Rang, H. P., Dale, M. M., Ritter, J. M. and Gardner, P. *Pharmacology*, 4th ed. Churchill-Livingstone: New York, 2001; pp. 579–603.
10. Butera, J. A. and Brandt, M. R. Current and Emerging Opportunities for the Treatment of Neuropathic Pain. *Annu. Rep. Med. Chem.* **2003**, *38*, 1–10.

13

General and Local Anesthetics

Introduction

General anesthetics render the individual unaware of and unresponsive to painful stimuli. They are given systemically (by inhalation or by intravenous injection), and their principal effects are manifested on the central nervous system. *Local anesthetics* reversibly block the action potentials responsible for nerve impulse conduction. These effects can be almost universally manifested in the body, and a local anesthetic in contact with a nerve trunk blocks both sensory and motor nerve impulses. Most of the drugs categorized as local anesthetics are intended for parenteral administration, usually subcutaneously or directly into a tissue or an organ. Some local anesthetic agents (e.g., cocaine **9.14**) are too toxic to be given by injection. They are employed as *topical anesthetics*, for application to a surface for an anesthetic effect on a mucous membrane: nose, mouth, throat, esophagus, trachea, bronchi of the lungs, or genitourinary tract.

General Anesthetics

13.2.1 Inhalation Anesthetics

Figure 13.1 illustrates the variety of substances that demonstrate general anesthetic activity following inhalation of their vapors. These are gases or low boiling liquids, and they have been considered to be the quintessential structurally nonspecific drugs. Their chemical composition, molecular shape, electronic configuration, and stereochemistry have seemed relatively unimportant, but the nature of certain of their physical chemical properties is critical. They all are decidedly lipophilic and hydrophobic. With the exception of nitrous oxide, these agents are considered to be obsolete drugs.

The blood–brain barrier is freely permeable to the inhalation anesthetics. A considerable knowledge of respiration physiology and an appreciation of the physical chemical parameters involved in gas exchanges between inspired air, expired air, blood, and tissues is required in order to use inhalation anesthetics appropriately. The main factors that determine the speed of induction and of recovery from general anesthesia are: (1) properties of the anesthetic: the blood:gas partition coefficient (i.e., solubility in blood)

222

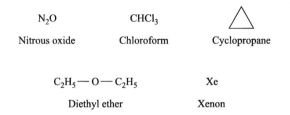

Figure 13.1 Agents that have been used as general inhalation anesthetics

and oil:gas partition coefficient (i.e., solubility in fat); and (2) physiological factors such as alveolar ventilation rate (rate of air flow into and out of the lungs) and cardiac output. The measure of potency of inhalation anesthetics is the *minimum alveolar concentration* (MAC), the minimum concentration of the anesthetic in the alveoli of the lungs at one atmosphere pressure that produces immobility in 50% of patients or animals exposed to a noxious stimulus such as a surgical incision. The MAC is inversely proportional to potency. This definition is based on the premise that at or near equilibrium the partial pressures of the anesthetic in the lung and in the brain are almost equal and that there is a rapid equilibration between alveoli, blood, and brain.

There is no universally accepted, satisfactory explanation for the general anesthetic effects of these drugs; for most of the twentieth century it was believed that all general anesthetics have a common mechanism of action. This is likely not true. Theories include the following.

Lipid solubility–viscosity hypothesis

Based on the observation that inhalation anesthetics are lipophilic, it was proposed that these compounds dissolve in the lipid matrix of nerve membranes and lower the viscosity of the matrix. (Recall that the *fluid mosaic membrane model* postulates that membranes have properties of viscous fluids.) It was proposed that diminished viscosity of the membrane disrupts normal nerve impulse transmission, resulting in anesthesia. Many pharmacologists no longer accept the validity of this lipid solubility–viscosity hypothesis.

Binding to hydrophobic domains

There is increasing evidence that inhalation anesthetics act by binding to hydrophobic domains of membrane protein molecules, although the question of which specific proteins are involved is unsettled. The marked differences in anesthetic potencies between the enantiomers of isoflurane (**13.3**, figure 13.2) have been interpreted to be compelling evidence for the hypothesis that proteins rather than lipids are the prime sites of anesthetic action. There is general agreement that inhalation anesthetics operate by influencing synaptic transmission rather than by affecting axonal conduction. It is further maintained that their mechanism of action involves promoting neurotransmitter release at inhibitory synapses and/or by inhibiting excitatory synapses.

Modification of serotonin receptor function

Recent studies reveal that certain inhalation anesthetics (halothane **13.1** and isoflurane **13.3**, figure 13.2) modify the function of 5-HT$_3$ subtypes of serotonin receptors. However, the significance of this is unclear. Other recent work suggests that inhalation anesthetics activate GABA$_A$ and glycine receptors on neurons. On the basis of these studies, the designation of inhalation anesthetics as structurally nonspecific may not be valid.

Stages of anesthesia

When a relatively slowly acting inhalation anesthetic, such as diethyl ether, is administered, well-defined *stages of anesthesia* can be observed as its concentration in the blood increases:

1. *Stage I: Analgesia.* A mild depression of higher centers in the cerebral cortex. The individual is conscious but drowsy. Response to painful stimuli is reduced.
2. *Stage II: Excitement.* Depression of motor centers in the cerebral cortex. The individual is unconscious and does not respond in a reflex manner to nonpainful stimuli but does respond to a painful stimulus. The individual may move, talk incoherently, hold the breath, or vomit. Stage II is potentially dangerous, and modern anesthetic techniques and strategies are designed to shorten or eliminate it.
3. *Stage III: Surgical anesthesia.* This is the useful stage for surgery. Spontaneous movement ceases, and respiration becomes regular. This stage is divided into four planes, representing increasing depth of anesthesia.
4. *Stage IV: Medullary paralysis.* Vasomotor control ceases. Depression of the CNS reaches the respiratory control apparatus in the medulla oblongata. Efferent impulses to the respiratory muscles (diaphragm and intercostals) stop. Respiration ceases and death ensues rapidly.

Current inhalation anesthetics

Figure 13.2 illustrates some representative, currently employed inhalation anesthetics. These agents have replaced older drugs, such as diethyl ether and cyclopropane, which in addition to other possible disadvantages, form explosive mixtures with air and thus present a considerable hazard in the operating room which is equipped with electrical/electronic apparatus and instrumentation.

Halothane **13.1** is not flammable; it produces relatively rapid induction of stage III anesthesia, and rapid awakening occurs when its administration is stopped. Up to 80% of absorbed halothane is excreted unchanged in the expired air. Of that portion not exhaled, approximately 50% is metabolized by oxidases of the endoplasmic reticulum of liver cells. The most significant metabolite is trifluoroacetic acid which is excreted in the urine. With repeated administration of halothane, resulting elevated amounts of trifluoroacetic acid can react with liver proteins to produce trifluoroacetyl derivatives that can trigger an immune response. This is thought to be the cause of liver damage reported in some individuals following repeated administration of halothane.

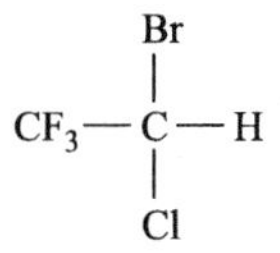

$$CF_3-\overset{\overset{\displaystyle Br}{|}}{\underset{\underset{\displaystyle Cl}{|}}{C}}-H$$

13.1 Halothane

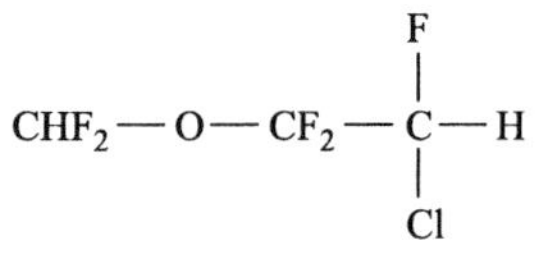

$$CHF_2-O-CF_2-\overset{\overset{\displaystyle F}{|}}{\underset{\underset{\displaystyle Cl}{|}}{C}}-H$$

13.2 Enflurane

$$CHF_2-O-CHCl-CF_3$$

13.3 Isoflurane

$$N_2O$$

13.4 Nitrous oxide

$$CHF_2-O-CHF-CF_3$$

13.5 Desflurane

Figure 13.2 Representative currently used inhalation anesthetics

Additionally, repeated administration of halothane can induce microsomal enzymes, resulting in even greater production of trifluoroacetic acid. A rare but serious side effect of halothane (and of some other inhalation anesthetics) is *malignant hyperthermia*, a dramatic rise in body temperature associated with muscle contractures and acidosis, and caused by excessive release of Ca^{2+} within the muscle fibers. The condition can be fatal; it is treated with a drug (dantrolene) that blocks calcium channels.

The polyfluorinated ethers enflurane **13.2**, desflurane **13.5**, and isoflurane **13.3** are metabolized to a lesser extent than halothane, and most of a dose of these drugs is removed in the expired air. This is especially true for isoflurane and desflurane, for which less than 1% metabolic degradation is claimed. Isoflurane is the most widely used inhalation anesthetic in the United States.

Nitrous oxide **13.4** is one of the oldest inhalation anesthetics, and it is still employed. It is odorless, tasteless, and not flammable although, like oxygen, it supports combustion. No metabolites of it have been reported, and presumably it is removed from the body 100% unchanged. It demonstrates a rapid onset of action, but it has low potency. Even a mixture of 80% nitrous oxide and 20% oxygen (the maximum possible concentration without lowering oxygen intake) does not produce stage III surgical anesthesia. Only analgesia results. Because of its rapid production of CNS depression, nitrous oxide is frequently employed preliminary to administering a more potent/active inhalation anesthetic whose onset of action is slow. Given for brief periods, nitrous oxide is devoid of any serious side effects. However, prolonged exposure to very low concentrations of nitrous oxide may affect protein and DNA synthesis, due at least in part to its inhibition of methionine synthase.

13.2.2 Intravenous Anesthetics

Even so-called rapid onset inhalation anesthetics, such as nitrous oxide, require some minutes to act, and the individual may experience some degree of stage II anesthesia (excitement). The intravenous anesthetics produce unconsciousness in approximately 20 s. These drugs are frequently used to induce anesthesia in conjunction with more active

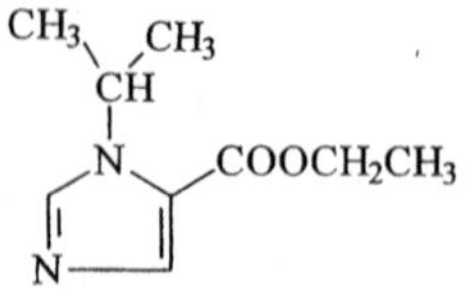

13.6 Etomidate 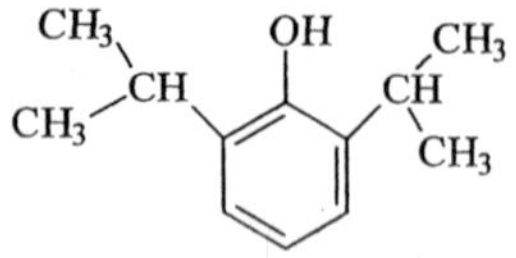   **13.8** Ketamine   **13.7** Propofol

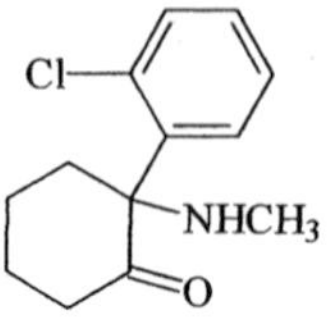

Figure 13.3 Representative intravenous anesthesia-inducing agents

inhalation anesthetics, as was described for nitrous oxide. This use of thiobarbiturates, such as thiopental, was discussed in chapter 10.

Figure 13.3 shows some intravenous anesthesia-producing agents. *Ketamine* **13.8** is structurally related to the drug of abuse phencyclidine **9.5**. Both drugs block the glutamic acid NMDA receptor, which results in diminishing the influx of Ca^{2+} and prevents excitatory synaptic transmission (cf. figure 9.2). Ketamine's relationship to phencyclidine is further demonstrated in its psychotropic activity; hallucinations, irrational behavior, and/or delerium are common during recovery from ketamine-induced anesthesia.

Etomidate **13.6** has been used to replace thiobarbiturates because it exhibits a greater margin between an anesthetic dose and the dose which produces respiratory and cardiovascular depression. An intravenous dose of etomidate produces sleep lasting approximately 5 min.

Propofol **13.7** is structurally unique, compared to other intravenous anesthesia-inducing agents. This compound is a liquid with poor water solubility and with no appropriate water-solubilizing group. It is administered intravenously as an aqueous emulsion. It has been suggested that propofol's pharmacologic action results from its ability to enhance GABA-ergic effects in the brain at the site of $GABA_A$ receptors. Although it apparently does not bind to benzodiazepine receptors, its actions have been described as being similar to those of the benzodiazepine drugs. Propofol is rapidly metabolized and hence it permits a rapid recovery without producing hangover. Propofol is often used in maintenance of anesthesia as well as for induction.

Fentanyl **12.14** is representative of a number of opioids which are used in combination with inhalation anesthetics. Morphine has been used for this purpose, but fentanyl is 50–100 times more potent than morphine and it lacks several of the undesirable cardiovascular-related side effects of morphine. It also produces a shorter duration of respiratory depression than morphine.

Diazepam **9.19** is representative of several benzodiazepine drugs used in conjunction with some anesthetics. The mechanism of anesthetic activity of the benzodiazepines is believed to be related to the drug's interaction with the benzodiazepine receptor adjacent to the GABA-A receptor site.

13.3   Local and Topical Anesthetics

The primary site of action of local anesthetics is the cell membrane, where they prevent the generation and conduction of a sensory (afferent) nerve impulse. It is widely accepted that local anesthetics interact with one or more specific binding sites within voltage-gated sodium channels, to block them by physically plugging the transmembrane pore. Evidence suggests that some local anesthetics have an additional secondary, nonspecific component of activity involving membrane proteins, similar to the interaction described for inhalation anesthetics. Most of the local anesthetics are esters or amides that contain a highly lipophilic moiety and a basic amino group which is often at the terminus of a side chain or, less commonly, part of a saturated heterocyclic ring (figure 13.4).

Local anesthetics (such as procaine **13.9**, tetracaine **13.10**, and lidocaine **13.11**) must be formulated as their water-soluble salts to be administered parenterally. Local anesthetic activity is pH-dependent and it increases in an alkaline environment where the relative amount of uncharged amino groups is large. To penetrate the axon membrane and reach the inner end of the sodium channel (where the local anesthetic binding sites are located), the molecule must exist in the uncharged, maximally lipophilic state. Some local anesthetics appear to arrive at their sodium channel binding sites by actual movement through the gating portion of the open state of the channel. Evidence indicates that the local anesthetic molecule's basic amino moiety takes on a proton from its environment and regains its cationic nature before binding to its ion channel receptor. *Benzocaine* (**13.12**, figure 13.4) is an exception. This is a poorly soluble compound whose extremely weak basic amino group is not protonated under physiologic conditions. Benzocaine is used as a topical anesthetic ingredient in a variety of proprietary nonprescription products. In medical practice, it is applied directly to wounds and ulcerated surfaces. Because of its low water solubility it is only slowly dissipated from the site of application, which provides a long-lasting anesthetic effect coupled with low toxicity.

The duration of action of a local anesthetic is proportional to the time of its contact with the affected nerve. Cocaine **9.14**, by virtue of its inhibition of the uptake-1 mechanism for norepinephrine, produces vasoconstriction in areas where it is applied, and it impedes

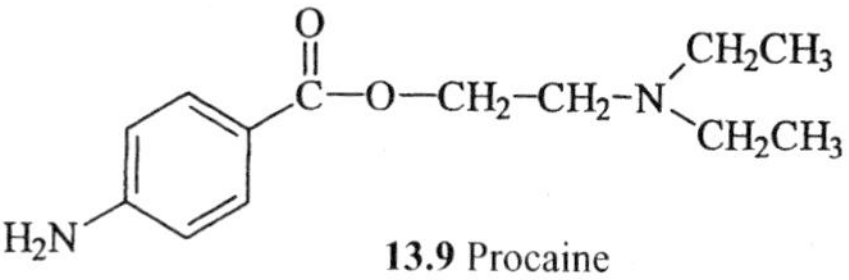

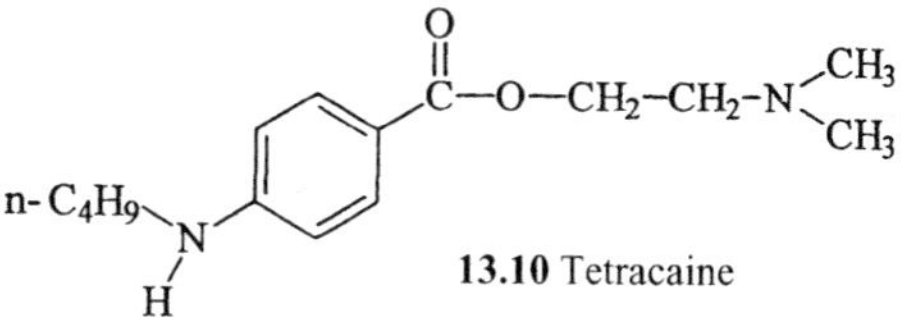

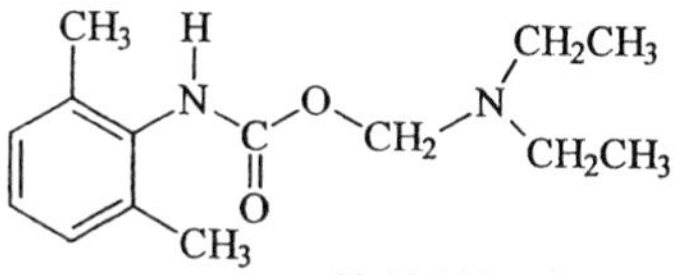

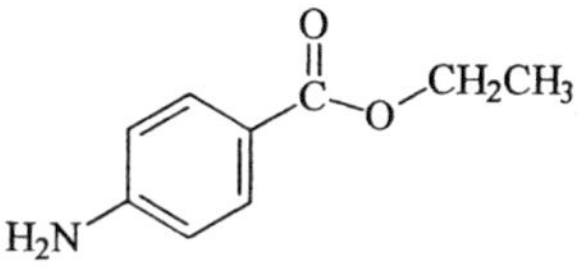

Figure 13.4 Representative local and topical anesthetics

its own dissipation from this site via the circulatory system. Presumably, the local anesthetic effect of cocaine is similar to that of the agents in figure 13.4. Its limited clinical value has been described earlier in this chapter. Other local anesthetics lack cocaine's vasoconstrictor activity, and for these (specifically the ones to be administered parenterally) it is customary to prolong the local anesthetic effect by adding a small amount of a vasoconstructor drug (epinephrine) to the injected solution of the local anesthetic.

Following absorption, local anesthetics can cause stimulation of the central nervous system, resulting in restlessness and tremors which can develop into convulsions. This stimulation is followed by profound depression, and death may result from respiratory failure. Inspection of the structures of the typical local anesthetics in figure 13.4 leads to the conclusion that they all possess a degree of lipophilicity, and penetration of the blood–brain barrier by passive diffusion should be expected. The biphasic CNS response (stimulation followed by depression) to local anesthetics is believed to be caused solely by depression of neuronal activity. The initial CNS stimulant response results from a selective depression of inhibitory neurons. Additionally, cocaine has powerful effects on mood and behavior because of its ability to elevate levels of dopamine, norepinephrine, and serotonin in synaptic clefts in the CNS.

Recommended Reading

1. Beattie, C. History and Principles of Anesthesiology. In *Goodman and Gilman's The Pharmacological Basis of Therapeutics*, 10th ed. Hardman, J. G. and Limbird, L. E.; Gilman, A. G., Eds. McGraw-Hill: New York, 2001; pp. 321–335.
2. Evers, A. S. and Crowder, C. M. General Anesthetics. In Recommended Reading ref. 1, pp. 337–365.
3. Catterall, W. and Mackie, K. Local Anesthetics. In Recommended Reading ref. 1, pp. 367–384.
4. Rang, H. P., Dale, M. M., Ritter, J. M. and Gardner, P. Local Anesthetics and Other Drugs that Affect Ion Channels. In *Pharmacology*, 4th ed. Churchill Livingstone: Philadelphia, Pa., 2001; pp. 635–646.
5. Fee, J. P. Howard and Bovill, J. G., Eds. *Pharmacology for Anesthesiologists*. Taylor & Francis: Boca Raton, Fla., 2005.

III

Pharmacology of Some Peripheral Organ Systems

14

The Cardiovascular System. I: Anatomy and Physiology

Aspects of Functional Anatomy of the Heart

The human heart is a four-chambered organ, consisting of the right and left atria (auricles) and the right and left ventricles (figure 14.1). The right auricle receives blood from all parts of the body via the venous system (superior vena cava). Once filled with blood the atrial muscle contracts and forces blood through a one-way valve (the tricuspid) into the lower right chamber, the right ventricle. Now, the right ventricle contracts and forces blood through another one-way valve (the pulmonary) out into the pulmonary artery, thence to the right and left lungs. After appropriate gas exchange in the lung vasculature (oxygen entering the blood and carbon dioxide leaving the blood and being blown out in the expired air), the newly oxygenated blood returns to the left atrium of the heart. The filled left atrium now contracts and forces its contents through another one-way valve (the mitral) into the left ventricle. The filled left ventricle contracts, expelling its contents into the principal artery leaving the heart (the aorta), whence oxygenated blood is delivered to all parts of the body. The left ventricle is thicker-walled than the right, because it must create a hydrostatic pressure four times that produced in the right ventricle, to force blood out into the systemic circulation against a higher resistance. The output of blood from the left ventricle averages 4.2–5.6 L min^{-1}, but it rises to as much as 12 L min^{-1} during strenuous activity. The arteries carry oxygenated blood, and the veins transport deoxygenated blood. The one exception to this is the pulmonary vasculature which carries blood between the heart and the lungs. The pulmonary vein carries oxygenated blood from the lungs to the heart and the pulmonary artery carries low-oxygen tension blood from the heart to the lungs. The heart muscle itself receives blood from the coronary arteries, which branch from the aorta.

Cardiac muscle is distinct, anatomically and physiologically, from skeletal and smooth muscle. Found only in the heart, this muscle tissue is striated like skeletal muscle, but its contraction is not under conscious control. It has the same biochemical mechanism of contraction as skeletal and smooth muscle. In cross section, the heart muscle fibers present a different appearance from peripheral skeletal muscle fibers.

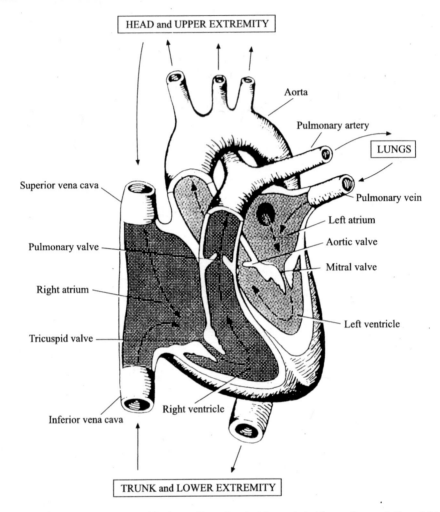

Figure 14.1 Functional anatomy of the heart. (Reproduced with permission from reference 1. Copyright 1981 Saunders)

The heart muscle fibers not only participate in the physiological process of muscle contraction, but they also carry nerve impulses on their membranes.

A small mass of unique cardiac tissue, the *sinoatrial node* ("SA node") is located in the wall of the right atrium. This anatomic entity generates nerve impulses which pass from cell to cell, travelling along the muscle cell membranes and inducing the heart to contract. Anatomically, the individual heart muscle cells are very close together, and this proximity permits the electrical impulses to pass readily from one cell to the next. Thus, only a small area needs to be stimulated initially for the entire mass of the heart muscle to be excited. The SA node is the pacemaker of the heart. The heart is self-stimulating. Although a variety of physiological systems (including nerve impulses from the central nervous system) participate in regulating the *rate* of heart beat, the actual stimulus for beating originates within the heart itself. The heart can be separated entirely from the body, devoid of any nervous connections, and it will continue to beat for some time.

Another small mass of specialized tissue, the *atrioventricular node* ("AV node") is located in the wall of the heart between the right atrium and the right ventricle. A nerve impulse traveling from the atria to the ventricles passes through this node and the impulse is delayed slightly before passing down to the ventricles to stimulate them to contract. This phenomenon permits the atria to constrict prior to ventricular contraction. Thus, heart function consists of the two atria contracting first, then after a slight delay, the two ventricles contract. This provides for pumping efficiency. It is obvious that if the two atria and the two ventricles all contracted simultaneously, the heart would be completely ineffective as a pump.

14.2 Hypertension

14.2.1 Physiological Regulation of Blood Pressure

As described previously, the heart actively pumps oxygenated blood out to the tissues via the arteries, and oxygen-depleted blood returns (rather passively) to the heart via the veins. The arterial blood pressure, whether normal or abnormal, is ultimately determined by two parameters:

1. *Cardiac output*: the volume of blood expelled by the heart into the arteries per unit of time. Factors determining cardiac output include the rate of heart beat and the force of the individual contractions.
2. *Peripheral resistance*: resistance to the passage of blood through the vessels. Factors determining peripheral resistance include the volume of circulating blood, the degree of dilation or contraction of blood vessel walls, and the viscosity of the blood. A more viscous blood requires higher pressure to circulate it. The degree of elasticity of the large arteries determines in part the rise in *systolic pressure* (when the heart is contracting) and the fall of *diastolic pressure* (between beats when the ventricles are relaxed. A stiffened artery results in higher systolic and lower diastolic pressure.

The blood pressure of most individuals remains amazingly constant and this is especially remarkable, considering the continual, minute-to-minute large variations in heart rate, thus in cardiac output. Quantitative maintenance of blood pressure is accomplished in part by a negative feedback system, a principal component of which is the *baroreceptor reflex*. Baroreceptors (pressure receptors) are stretch receptors embedded in the walls of certain arteries (carotid artery and the aortic arch). These receptors monitor hydrostatic pressure through specialized nerve endings that are sensitive to mechanical deformation of the arterial wall. If the volume of blood expelled into the artery from the heart is so great that it causes the arterial wall to stretch, this generates in the baroreceptor an afferent nerve impulse which travels to the brain where the impulse is received and is translated into a series of efferent nerve impulses which move down appropriate nerve fibers from the spinal cord to the arterial walls, signaling the smooth muscle fibers there. The *vasomotor tone* is reduced; relaxation of the muscles in the arterial walls causes the arteries to dilate, lowering peripheral resistance and lowering blood pressure. Other efferent signals are sent to the heart itself, causing it to slow its rate of beating. Thus, cardiac output is lowered, which contributes to the overall lowering of blood pressure. In contrast, if the initial stimulant event were a lower than desirable blood pressure, the baroreceptors would recognize a diminution of the stretching of the arterial wall, and an opposite set of responses would occur. Peripheral resistance and cardiac output would be induced to rise and arterial blood pressure would increase.

The blood vessels and the heart itself are innervated by autonomic motor neurons. Cholinergic fibers from the vagus nerve lead to the heart, and stimulation of these nerves causes slowing of the heart rate. Significant autonomic innervation of the arterial smooth muscle involves α_1 adrenoceptors, whose stimulation produces vasoconstriction. Stimulation of β adrenoceptors in arterial smooth muscle produces vasodilation. In the heart muscle, α adrenoceptors are of less significance. However, stimulation of the β adrenoceptors in the heart (chiefly β_1) augments both *chronotropic* (heart rate) and *inotropic* (contractile force) effects. Broadly based stimulation of the peripheral noradrenergic system elevates blood pressure. The resulting excessive force with which the heart expels blood and a less stretchable arterial system (caused by stimulation of α adrenoceptors) exert additive effects to raise both systolic and diastolic pressures. Cholinergic nerves are not important in controlling vascular smooth muscle.

However, the body's methods for overall control of blood pressure are much more complex than has been described thus far. The baroreceptor reflex and the noradrenergic and cholinergic systems act in concert with yet other blood pressure-regulating mechanisms: the rennin–angiotensin system (vide infra), action of the steroidal hormone aldosterone, pathways in the central nervous system mediated by epinephrine (separate and distinct from norepinephrine-mediated pathways), and levels of sodium ion in the extracellular fluid. Increased sodium concentrations intensify the effects of norepinephrine on the heart muscle and on the blood vessel walls, especially in borderline hypertensive patients. It will be recalled that β-adrenoceptor activation involves the production of a second messenger, cyclic AMP, and its subsequent influence on sodium ion migration across the membranes of muscle cells. This rationalizes the value of low salt diets for hypertensive patients.

A series of 21-amino acid peptides, the *endothelins*, is produced in a variety of cell types in the body (including the endothelium, hence the name). One of these peptides is said to be the most potent vasoconstrictor substance known, producing pressor effects at nanogram-level doses. Stimulus for its release from endothelial cells in blood vessels is crushing of the tissue (as in trauma). The resulting vasoconstriction prevents extensive bleeding. The physiology and pharmacology of the endothelins are incompletely understood, but there is some evidence that they and their heterogeneous population of G-protein-linked receptors, among other physiological functions, participate in blood pressure regulation. This belief is not universally accepted by pharmacologists and physiologists. Some natural product and synthetic endothelin receptor antagonists have been identified and their possible value in treating essential hypertension and other cardiovascular conditions has been suggested. One of the endothelins has been implicated in eclampsia, a disease of the last trimester of pregnancy, characterized by hypertension, edema, and seizures. Eclampsia is a leading cause of maternal and newborn fatality in industrialized countries.

14.2.2 Clinical Categories of Hypertension

It has been estimated that 15–25% of the adult population of most countries has elevated blood pressure. Two major types and one rare type of hypertension are recognized.

Primary (essential) hypertension

This includes approximately 90% of all clinical cases. The cause(s) is/are unknown. Cerebral blood flow, cardiac output, kidney function, and functions of the hormonal systems of the body are normal, but peripheral resistance is elevated. Essential hypertension is incurable, but it is usually highly controllable with proper drug therapy, when that therapy is combined with programs of exercise, changes in dietary habits, and weight loss.

Secondary hypertension

Here the etiology is understood, at least to some extent. Secondary hypertension includes elevated blood pressure due to hormonal imbalance, psychogenic causes, metabolic disorders, and brain tumors. The most common form of secondary hypertension is *renal hypertension*, which is related to the action of a proteolytic enzyme, *renin*, which is liberated into the blood stream from the kidney (figure 14.2).

The mechanism of renin release is complex and is only partly understood. Activation of β_1 adrenoceptors is one portion of the overall process. Once liberated into the blood, renin acts on a specific substrate, a globulin protein, *angiotensinogen*, which is a normal component of the blood, to release a decapeptide fragment, *angiotensin I*, which has no appreciable effect on blood pressure. Another blood proteolytic enzyme, the so-called *converting enzyme*, cleaves two amino acids from angiotensin I to produce *angiotensin II* which is very active pharmacologically. This substance increases the force of the heart beat (positive inotropic effect) and it constricts the small arteries, and these combined actions increase blood pressure. Angiotensin II exerts these effects by interaction with specific cell surface receptors of the G-protein-linked type. These receptors are heterogeneous, and two major subtypes have been characterized: AT_1 and AT_2. Two additional subtypes, AT_3 and AT_4, have been claimed but they are as yet incompletely described. The biological actions of angiotensin II involve interaction with the AT_1 receptor subtype which in itself is not homogeneous. The physiological role(s) of AT_2 receptors is/are uncertain and confusing.

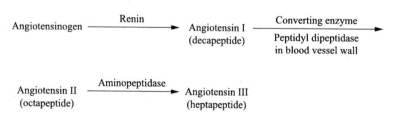

Figure 14.2 The renin–angiotensin system

The onset of pressor effects of angiotensin II is rapid, less than 10 s after the release of renin from the kidney. The pressor effect also rapidly declines because enzymatic degradation of angiotensin II occurs within minutes. Also, angiotensin II causes feed-back inhibition of renin release. Angiotensin II is a fantastically potent pressor agent. In one assay it was some 200 times more potent than epinephrine on a molar basis. A blood aminopeptidase utilizes angiotensin II as a substrate, removing one more amino acid to form a heptapeptide *angiotensin III*, which is also pharmacologically active but is considerably less potent than angiotensin II in elevating blood pressure. Angiotensin III stimulates release of aldosterone. Further enzymatic cleavage of the angiotensin III peptide chain occurs in the blood, giving rise to a hexapeptide angiotensin IV, which is apparently not directly involved in blood pressure regulation, but has other physiologi-cal functions which are not well understood. The rennin–angiotensin system acts syner-gistically with the noradrenergic nervous system in blood pressure regulation and it represents a principal stimulus to the release of the steroidal hormone aldosterone from the adrenal cortex. This hormone decreases the excretion of sodium ions and water by the kidney, and this effect further elevates blood pressure by maintaining a high volume of circulating blood and maintaining a high sodium ion concentration in the body fluids. Angiotensin II and angiotensin III are equally potent in their effect on aldosterone release. The active angiotensins also enhance peripheral noradrenergic activity by inhibiting the uptake-1 active transport of norepinephrine into the nerve terminal. Angiotensin II may also act in the central nervous system to elevate blood pressure. These secondary sites of action also contribute to the hypertensive effect of the angiotensins.

Pulmonary hypertension

This is a rare disease of unknown cause. The condition is defined as inappropriately high pulmonary arterial pressure for a given level of blood flow through the lungs. Normally, the pulmonary circulation is a low pressure, high flow system. Pulmonary hypertension seems to be most common in young adults, and it can lead to right heart failure which is frequently fatal. The disease is progressive and it has no cure. Therapeutic strategies for pulmonary hypertension have not been perfected, but it appears that long term therapy with prostaglandin PGI_2 is effective. Some current studies suggest that sildenafil (Viagra®) may be useful in treatment of pulmonary hypertension (see chapter 9).

14.2.3 Therapeutic Challenges and Strategies in Hypertension

As a matter of therapeutic strategy, of the two possible ways of lowering blood pressure (lowering cardiac output or lowering peripheral resistance), the more appealing is lowering peripheral resistance. Lowering cardiac output can diminish the amount of oxygenated blood that is transported to the tissues, and this may not be beneficial. In general, it is most desirable to utilize a drug that lowers peripheral resistance (by dilating the arteries) but also maintains the cardiac output at a relatively high level to meet the oxygen needs of the tissues.

Essential hypertension has many possible biochemical/physiological causes. In contemplating the large number and complexity of mechanisms that participate in the human body to regulate blood pressure (e.g., the peripheral and central noradrenergic nervous systems, central epinephrine-mediated nerve pathways, the cholinergic system, the baroreceptor reflex, the rennin–angiotensin system, aldosterone activity, sodium ion concentration in the extracellular fluid, possibly the endothelins and, doubtless, others), it seems obvious that if any of these individual regulatory mechanisms/systems fails or malfunctions, the result may be persistent hypertension. Thus, for any hypertensive individual, the physician probably can never determine exactly what physiological mechanisms are malfunctioning, and the physician usually does not attempt to do so. Therefore, "essential hypertension" is a widely inclusive generic term reflecting many possible biochemical/physiological causes. And, it is unreasonable to expect that *any* single antihypertensive drug will have a beneficial effect in all patients. It is characteristic of essential hypertension therapy that a considerable degree of trial-and-error in drug selection and in dosage regimens may be required to establish the best drug therapy for each individual patient. Combinations of two or even three drugs (having different mechanisms of hypotensive effect) are frequently employed.

14.2.4 Antihypertensive Drugs

Diuretics

The most commonly used drugs in this category are *hydrochlorothiazide* **14.1** (a "thiazide" diuretic) and *chlorthalidone* **14.2**. The exact mechanism for lowering of arterial pressure by these diuretic agents is uncertain. For hypertension therapy, they are customarily administered in low doses, and initially they decrease cardiac output and lower blood volume. With continual use, cardiac output returns to predrug level. It has been noted that these relatively long-acting diuretics produce a slowly developing decrease in extracellular sodium levels due to enhanced urinary excretion ("saluresis"), and this is proposed to be the cause of the hypotensive effect. The blood pressure lowering of hydrochlorothiazide and chlorthalidone can be reversed by administering large amounts of sodium chloride. The fact that these diuretics must be administered for several days before a hypotensive result is noted is consistent with this proposed mode of their action. The more active "high ceiling" diuretics (e.g., ethacrynic acid **16.6** and furosemide **16.7**) are occasionally employed as antihypertensive agents, but generally they are not as satisfactory as the thiazides and chlorthalidone by once-daily dosage for routine control of hypertension; their short duration of action does not readily provide for a consistent, prolonged net loss of sodium. Used as the sole hypotensive drug, hydrochlorothiazide or chlorthalidone may disrupt the normal electrolyte balance in the body by causing excess loss of potassium. As described subsequently, diuretics are often used in combination with other hypotensive agents.

14.1 Hydrochlorothiazide   **14.2** Chlorthalidone

Sympatholytic agents

These are drugs that oppose the effects of noradrenergic stimulation. Under this broad category there are several specific mechanisms of hypotensive action. α–*Methyldopa* **14.3** is one of the most widely used antihypertensive agents. This synthetic drug is absorbed across the intestinal wall by active transport; it exerts its hypotensive effect in the central nervous system. It penetrates the blood–brain barrier by an active transport mechanism, presumably the DOPA transporter. In the central nervous system, α-methyldopa is a substrate for L-aromatic amino acid decarboxylase ("DOPA decarboxylase") and it is converted into α-methyldopamine **14.4**, thence by a second enzyme-mediated step into α-methylnorepinephrine **14.5**. This latter metabolite is the pharmacologically active species and it is stored in the synaptic storage vesicles of CNS adrenergic neurons, substituting for norepinephrine itself. When the neuron releases its neurotransmitter in response to a nerve impulse, α-methylnorepinephrine is released and it produces its hypotensive effect by acting as an $α_2$-adrenoceptor agonist. This interaction with presynaptic α receptors diminishes the activity of noradrenergic pathways which are responsible for vasoconstrictor actions on the arterial walls. The net effect is vasodilation and a fall in blood pressure. Peripherally, α-methylnorepinephrine diminishes the release of renin from the kidney, but this is believed to be a relatively minor component of its hypotensive action. α-Methyldopa is frequently used in combination with a thiazide diuretic. Given alone for a prolonged period of time, α-methyldopa causes retention of sodium and this elevated sodium level counteracts the hypotensive effect of the drug (this is an example of so-called "pseudotolerance"). Concurrent administration of a saluretic diuretic obviates this problem. When it is deemed advisable to administer α-methyldopa by injection, the hydrochloride salt of its ethyl ester is used. α-Methyldopa hydrochloride is poorly soluble in water, but the ethyl ester salt is quite soluble. Once in the blood stream, the ethyl ester moiety is enzymatically cleaved to generate the free acid.

1.43 α-Methyldopa

14.4 α-Methydopamine

hydroxylase

14.5 α-Methylnorepinephrine

Clonidine **7.13** is an $α_2$-adrenoceptor agonist whose hypotensive effects are produced in part at the same central nervous system locations and in the same manner as those described for the α-methylnorepinephrine metabolite of α-methyldopa. However, most authorities concede that the antihypertensive mechanism(s) of clonidine is/are incompletely understood. Intravenous administration of clonidine produces an acute rise in blood pressure, which has been attributed to the drug's interaction with post synaptic α_2 adrenoceptors in peripheral blood vessel walls to produce vasoconstriction. However, this peripherally produced vasoconstricion is transient and is followed by the more prolonged centrally mediated hypotensive response. The initial hypertension is generally not seen when the drug is administered orally.

Congeners of clonidine currently in use are *guanfacine* **14.6** and *guanabenz* **14.7**. These drugs have mechanisms of action and uses similar to those of clonidine.

14.6 Guanfacine

14.7 Guanabenz

β -Adrenoceptor blocking agents are widely used to control hypertension. Typical examples of "β blockers" are *propranolol* **7.18**, *labetalol* **7.20**, *nadolol* **14.8**, and *pindolol* **14.9**. These drugs do not produce a fall in blood pressure in those individuals

14.8 Nadolol

14.9 Pindolol

whose pressure is normal. However, they have a hypotensive effect in hypertensive individuals, in whom they lower peripheral resistance and reduce cardiac output. No single satisfactory explanation for the reduction of peripheral resistance by these β blockers has been forthcoming. Simple, broadly based blockade of β adrenoceptors seems to be an inadequate explanation. Release of renin from the kidney is under noradrenergic (β receptor) control, and propranolol inhibits release of renin, with resultant fall in angiotensin II levels and inhibition of its subsequent multiple effects on circulatory control as well as inhibition of aldosterone release. The possibility of a contributory central nervous system mechanism for antihypertensive β blockers has been suggested. Propranolol, nadolol, and pindolol block β_1 and β_2 adrenoceptors equally well. Thus, these drugs are contraindicated in asthmatic patients in whom the β_2-adrenoceptor blocking effects constrict the bronchi and thus precipitate an asthmatic attack (see chapter 7, Table 7.1). Some β-blocker drugs, typified by *metoprolol* **14.10**, are selective (although not specific) for blocking β_1 receptors. Thus, these are less hazardous for asthmatic patients.

14.10 Metoprolol

Drugs that selectively block α_1 adrenoceptors without affecting α_2 adrenoceptors represent yet another approach to hypertension therapy. Typical drugs in this category are *prazosin* **14.11** and *terazosin* **14.12**. Initially the α_1 antagonists reduce peripheral resistance, which causes a reflex increase in heart rate and in renin activity, which latter two effects promote a *rise* in blood pressure. However, during long term therapy, heart rate effects and elevated renin activity return to normal while vasodilation persists, resulting in a fall in blood pressure. The most prominent side effect of these α_1 blockers is a sometimes severe *first-dose phenomenon*. This is a dose-dependent excessive postural hypotension with resultant loss of consciousness (*syncope*). This side effect can be minimized by giving the initial (low) dose at bedtime.

14.11 Prazosin   **14.12** Terazosin

Vasodilators

Minoxidil **14.13** is a prodrug. A relatively minor hepatic metabolite, the N-O-sulfate ester **14.14**, is the hypotensively active entity. The major hepatic metabolite of minoxidil is the glucuronic acid conjugate of the *N*-oxide moiety, and this has greatly lowered hypotensive activity. Minoxidil sulfate activates ATP-modulated potassium channels in vascular smooth muscle. The resulting potassium efflux causes hyperpolarization, relaxation of the vascular smooth muscle, and a hypotensive effect. Minoxidil causes water and sodium retention, an indirect result of reflex stimulation of renal tubular

14.13 Minoxidil   **14.14** Minoxidil *N-O* sulfate metabolite

α adrenoceptors. This fluid retention can be counteracted by concurrent administration of a diuretic. However, a thiazide diuretic (hydrochlorothiazide) is not sufficiently efficacious, and a high ceiling "loop" diuretic is necessary (see chapter 16). Minoxidil is reserved for treating severely hypertensive individuals who do not respond to other drugs. Although the hypotensive effect of minoxidil is impressive, adverse reactions to it (a variety of effects on heart activity) require its withdrawal in some patients. The most common and most obvious result of chronic use of the drug is *hypertrichosis*: abnormal growth of hair at the temples, forehead, eyebrows, projecting parts of the ears, forearms, legs, and other portions of the body surface. Hypertrichosis occurs in almost all individuals who take the drug for more than 4 weeks. Many patients reject the drug for aesthetic reasons. The excess hair growth is not a reflection of altered endocrine activity, but rather it is thought to result from enhanced cutaneous blood flow to the hair follicles, a result of potassium and calcium channel activation. A dilute solution of minoxidil in aqueous ethanol and propylene glycol (*rogaine®*) is marketed as a non-prescription product for topical application to the scalp for reversing male pattern baldness. Sulfation of minoxidil occurs in the hair follicles and in the keratinocytes. The drug is not well absorbed across intact skin, and topical application produces no hypotensive effect in most individuals. The hair growth produced by minoxidil is reversible, and if dosage or topical application of the drug is stopped, the newly grown hair is lost.

14.15 Hydralazine

Hydralazine **14.15** causes direct relaxation of the smooth muscles of the arteries. The molecular mechanism of this effect is not known. In the bowel and/or the liver, hydralazine is metabolically acetylated at the terminal nitrogen of the hydrazine moiety and this metabolite is pharmacologically inert. The rate of N-acetylation is genetically determined: fast acetylators require larger doses than do slow acetylators. Hydralazine is not used as the sole drug in long term treatment of hypertension, due to development of tachyphylaxis. It also frequently causes a variety of potentially serious side effects involving immunological reactions, including drug-induced lupus of unknown cause.

Sodium nitroprusside, $Na_2Fe(NO)(CN)_5$, dilates small arteries and small veins. It is an unusually powerful hypotensive agent. It is metabolized in vascular smooth muscle to produce nitric oxide, the pharmacologically active species. Nitric oxide's action in stimulating guanylate cyclase, which has been described previously (chapter 9), ultimately produces the vasodilation. The hypotensive effect of sodium nitroprusside is of very short duration (approximately 3 min), and its use is limited to hypertensive emergencies. It is administered by intravenous drip, which permits minute-to-minute control of blood pressure. The in vivo destruction of nitroprusside anion generates, in addition to nitric oxide, five equivalents of cyanide anion which is metabolically converted into the less toxic thiocyanate (CNS^-) anion and this is eliminated in the urine by glomerular filtration. However, renal excretion of thiocyanate is slow (mean elimination half time of 3 days in individuals with normal kidney function), and prolonged (more than 24 h) intravenous infusion of sodium nitroprusside can lead to thiocyanate intoxication. Accumulation of toxic levels of cyanide anion can be prevented by concomitant administration of sodium thiosulfate. Sodium nitroprusside is ineffective orally. It is rapidly destroyed by the acidity of the stomach, generating hydrogen cyanide, cyanogen, and other products, some of which are also toxic.

Angiotensin converting enzyme (ACE) inhibitors

Captopril **14.16** competitively inhibits the proteolytic enzyme ("converting enzyme") that catalyzes the conversion of angiotensin I into angiotensin II (figure 14.2). Captopril and the other ACE inhibitors have no effect on the pressor actions of angiotensin II; their mode of action strictly involves inhibition of formation of angiotensin II. The level of angiotensin II falls and blood pressure falls. Captopril was the first generation of this category of antihypertensives. Its clinical value is limited by some of its side effects (skin rash and temporary, reversible loss of taste sensation). Typical newer ACE inhibitors are *enalapril* **14.17** and *lisinopril* **14.18**. Enalapril, a monomethyl ester, is a latentiated drug; it (but not the free dicarboxylic acid form) is a substrate for an active transport system to carry it across the intestinal wall. After absorption, blood esterases cleave the methyl ester moiety and liberate the active form of the drug, the dicarboxylic acid *enalaprilate*.

14.16 Captopril   **14.17** Enalapril   **14.18** Lisinopril

It might be assumed that these converting enzyme inhibitors would be useful only in renal hypertension, and initially this was believed to be the case. However, these drugs are effective in lowering blood pressure in the majority of essential hypertensive patients. The blood pressure lowering mechanism(s) in these individuals is/are not known. The converting enzyme has in vivo substrates in addition to angiotensin I, and inhibition of the enzyme's catalytic ability may produce effects unrelated to the renin-angiotensin system. For example, the converting enzyme accepts the nonapeptide *bradykinin* as a substrate and destroys its physiological activity. Bradykinin is a powerful vasodilator, and it also stimulates prostaglandin biosynthesis. It has been proposed that some prostaglandin-related mechanisms are involved in lowering blood pressure. Protection of bradykinin from metabolic destruction may be a direct or an indirect contributor to the hypotensive effect of the ACE inhibitor drugs. One of the side effects of the ACE inhibitors is the production of a dry cough, which some pharmacologists ascribe to the accumulation of bradykinin in the lungs.

A newer therapeutic strategy for hypertension involves the use of angiotensin II receptor ("AT$_1$") blockers, typified by *losartan* **14.19**. Losartan does not inhibit the converting enzyme, nor is it a partial agonist at AT_1 receptors; it is a competitive blocker at this receptor. Losartan itself has a rather short half-life in the body, but its hypotensive effect is prolonged such that it can be given only once daily. This seeming inconsistency is explained by the fact that the drug is metabolized to the 5-carboxylic acid **14.20**, which is also an AT_1 receptor blocker. Because losartan is also pharmacologically active, it is not categorized as a true prodrug. Losartan does not produce the annoying cough that the ACE inhibitors do. AT_1 receptor blockers, like the ACE inhibitors, are useful in essential hypertensive patients.

14.19 R=CH₂OH
14.20 R=COOH

The angiotensin converting enzyme is not the only one in the body that is capable of forming angiotensin II. A *chymase* (which is not inhibited by the ACE inhibitors) provides an alternate route from angiotensin I to angiotensin II. The clinical significance of this alternate pathway is not known. However, if this alternate route is significant, AT_1 receptor antagonists could prove to be more useful clinically than the ACE inhibitors.

14.3 Hyperlipidemia/Atherosclerosis

14.3.1 Pathology

Almost all of the lipids in the plasma are carried as special macromolecule complexes called lipoproteins. *Hyperlipoproteinemias* encompass those metabolic disorders that involve elevation of plasma lipoprotein concentration. The term *hyperlipemia* designates elevated plasma triglyceride concentration, and the term *hyperlipidemia* is applied to both conditions. *Atheroma* is the term given to the characteristic change that takes place in a blood vessel wall that begins with a fatty streak or deposition of lipid material in the smooth muscle wall, and develops into a fibrous plaque that becomes laden with lipids (cholesterol and triglycerides) and cell debris, and eventually with calcium. The plaque narrows the diameter of the artery and also acts as a site upon which a thrombus (blood clot) can develop.

Development of atherosclerosis ("hardening of the arteries"), a leading cause of death in the United States, involves certain plasma lipoproteins. Those lipoproteins that contain *apolipoprotein B-100* are the vehicles in which cholesterol is transported into the artery wall. These so-called atherogenic lipoproteins are the *low-density* (LDL) *intermediate-density* (IDL), and very low-density (VLDL) lipoproteins. Cholesteryl esters, which are found in the foam cells in the atheroma, also appear in the extracellular matrix and they induce collagen production by the fibroblasts. Macrophages and smooth muscle cells also play a key role in atherogenesis. Oxidation of lipoproteins leads to their uptake by special receptors on the smooth muscle cells, forming foam cells in which cholesteryl esters accumulate. Other contributory factors in atherosclerosis include smoking, diabetes, hypertension, and low levels of *high-density lipoprotein* (HDL) which is involved in removing cholesterol from the atheroma. Arterial thrombosis rarely occurs in healthy vessels, but it is associated with atherosclerosis. The development of atherosclerotic plaques is not usually associated with clinical signs of thrombosis. However, disruption of an atherosclerotic plaque plays a fundamental role in development of myocardial infarcts, angina, and stroke. Atherosclerosis in both coronary and other peripheral arteries is a dynamic process. Net regression of coronary artery lesions is associated with lipid-lowering therapy. The formation of new lesions is also decreased. Atherosclerotic lesions are induced to form in two ways: by enriching the diet in cholesterol or by repeated injury to an arterial wall. In either case the earliest lesions are focal, and they occur at vessel orifices and branches, points where normal arterial shear forces are disrupted by turbulent flow conditions.

14.3.2 Chemistry and Physiology of Lipoproteins

As illustrated in figure 14.3, most of the plasma lipoproteins are spherical and have a hydrophobic core region containing triglycerides and cholesteryl esters. Surrounding this core is a monomolecular layer of phospholipid and unesterified cholesterol. The *apolipoproteins* (proteins associated with lipids) are located on the surface. Some lipoproteins (B-proteins) contain very high molecular weight apolipoproteins that,

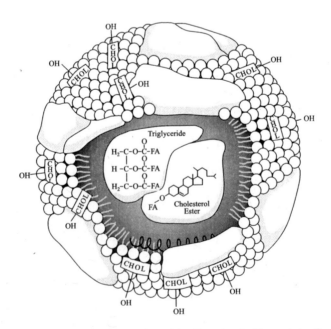

Figure 14.3 Hypothetical model of a lipoprotein particle. (Reproduced with permission from reference 2. Copyright 1998 Lippincott-Raven)

unlike some lower molecular weight apolipoproteins, do not migrate from one particle to another. There are two important forms of apolipoprotein-B:

1. *Apo B-48* is formed in the intestine and is found in the chylomicrons.
2. *Apo B-100* is formed in the liver and is found in VLDL, IDL, and LDL.

The chylomicrons are the largest of the lipoproteins. They are formed in the intestine, and they carry dietary triglycerides, together with some esterified cholesterol. Phospholipids, free cholesterol, and newly synthesized apolipoprotein B-48 form the surface layer of the chylomicron. The chylomicron is absorbed from the intestine into the lymphatic system and thence into the blood stream. Triglycerides are removed from the chylomicrons by a hydrolytic mechanism involving the lipoprotein lipase system. As the core triglycerides are removed, the size of the chylomicron diminishes and the surface lipids are transferred to HDL. The remnant of the chylomicron is taken up into the liver cells; the cholesteryl esters are hydrolyzed; and the free cholesterol is in part excreted in the bile and in part is metabolized. Chylomicrons are not normally present in the serum of individuals who have fasted. Figure 14.4 illustrates this role of HDL in mobilizing and excreting cholesterol. HDL is composed of phospholipids, unesterified cholesterol, and several different proteins. In its nascent form (top diagram) HDL exists as discs, and the lipid bilayer forms an oily, hydrophobic interior. While moving through the bloodstream and absorbing cholesterol, HDL becomes rounder, larger, and more spherelike (bottom diagram). The liver recognizes and takes up this spherelike form of HDL by an as-yet unknown mechanism.

The VLDL are secreted by the liver, and they carry triglycerides synthesized there. After leaving the liver, the triglycerides are hydrolyzed by lipoprotein lipase, and the freed fatty acids are oxidized in the tissues or are stored in the adipose tissue. Now the depleted VLDL are termed intermediate-density lipoproteins (IDL). Some of these are taken up the liver and the rest lose more of their triglycerides to form LDL. The proteins of HDL are secreted by the intestine and the liver. Much of the lipid content of HDL derives from the surface monolayers of the chylomicrons and from lipolysis of the VLDL. The high-density lipoproteins also acquire cholesterol from the peripheral tissues.

The risk of atherosclerotic heart disease increases with the LDL cholesterol level. Although there is only a moderate statistical relationship between hypertriglyceridemia and coronary heart disease, atherosclerosis is strongly linked to hypertriglyceridemia in some families. Some types of hypercholesterolemia are genetically transmitted as a dominant trait. Often the serum triglyceride levels are in the normal range in hypercholesterolemic individuals. Defects in the ligand domain of Apo-B (the region that binds to the membrane-associated LDL receptor) inhibit the transport of LDL from the blood into the cell, resulting in hypercholesterolemia. In some human genetic lines, the serum LDL level is elevated, but the cholesterol level approaches a normal range. Rare genetic disorders are associated with low levels of HDL in serum.

14.3.3 Therapy of Hyperlipidemia

Decisions concerning drug therapy should be based on the specific metabolic defect. Diet is a necessary adjunct for all types of drug therapy, and in some instances diet may be sufficient to correct the hyperlipidemic condition. Daily multigram doses of nicotinic

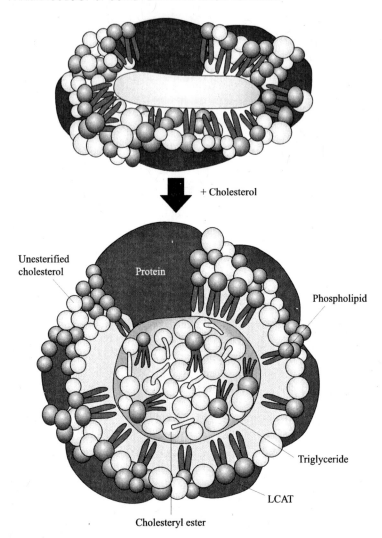

Figure 14.4 Structural effects of absorbing cholesterol onto high-density lipoprotein. (Reproduced with permission from reference 3. Copyright 1999)

acid **14.21** inhibit VLDL secretion, which in turn diminishes production of LDL. Increased clearance of VLDL via lipoprotein lipase activity contributes to nicotinic acid's triglyceride-lowering effect. There is no effect on bile acid production. Decreased biosynthesis of cholesterol in the liver also occurs, which results in increased uptake of LDL from the blood into the liver cells. When nicotinic acid therapy is initiated or when

14.21

the dosage is increased, each dose of the drug produces cutaneous vasodilation and a sensation of warmth. This is a prostaglandin-mediated response and a small preliminary dose of aspirin prevents it. Tachyphylaxis to these adverse side effects usually occurs within a few days. Nicotinic acid also produces pruritis. Chronic intake of therapeutic level doses of nicotinic acid for hyperlipidemia can cause severe hepatotoxicity. Sustained release dosage forms of nicotinic acid are said to be somewhat less prone to producing severe liver toxicity, however this opinion has been challenged in the literature. Recent studies report that dose levels of nicotinic acid as used in humans have been associated with birth defects in experimental animals, and some authorities suggest that pregnant women should not take this drug.

14.22

An alternate approach to hyperlipidemia is the use of an orally administered bile acid-binding resin such as *cholestyramine* **14.22**, a high molecular weight copolymer of polystyrene and divinylbenzene containing approximately 4 meq of quaternary ammonium groups per gram of resin. These agents (large molecule, water-insoluble anion exchange resins) bind endogenous bile acids in the lumen of the intestine and prevent their being reabsorbed. Fecal excretion of bile acids is increased up to 10-fold by using a resin. Thus, more of the body's cholesterol is required for replacement synthesis of bile acids by the liver. This process is normally controlled by the bile acids by negative feedback. Because they are nonselective ion exchange resins, these drugs may bind any negatively charged molecules and also some neutral compounds that they may encounter. They decrease absorption of digitalis glycosides, some anticoagulants, propranolol, furosemide, tetracyclines, and also some other hypolipidemic agents such as gemfibrizil. Patients should take other drugs 1 h before or 4 h after taking a resin. For oral administration these bile acid sequestrant resins are slurried with water or fruit juice. The inconvenience of administration is compounded by the tendency of these resins to produce bloating and dyspepsia.

Gemfibrozil **14.23** is representative of a number of *fibric acid* derivatives. It was once thought that the mechanism of action of fibric acid derivatives involves

14.23

increased lipolysis of lipoprotein triglyceride by lipoprotein lipase. However, current thought concludes that their actions are more complex and their mechanism of action is unclear. Gemfibrozil is useful in hypertriglyceridemias in which VLDL predominates. The drug is efficiently absorbed from the intestine when it is given with a meal, but it is less efficiently absorbed when it is given on an empty stomach. The drug is extensively and tightly blood protein-bound. It undergoes enterohepatic circulation and it readily penetrates the placenta. A small percentage of a gemfibrozil dose is metabolically hydroxylated on the ring methyl groups, but most is excreted unchanged in the urine. The plasma half-life is short (approximately 1.5 h).

14.24 Lovastatin   14.25 Atorvastatin

Lovastatin **14.24** and atorvastatin **14.25** are representative of the *statins*, a series of competitive inhibitors of HMG-CoA reductase, the rate-limiting enzyme in the biosynthesis of cholesterol. The lactone ring of lovastatin is hydrolytically cleaved in the body to form the pharmacologically active β,δ-dihydroxy acid **14.26**. This portion of the molecule is structurally similar to 3-hydroxy-3-methylglutaryl Coenzyme A (HMG-CoA) **14.27**, the precursor to mevalonate **14.28** in the biosynthetic route to steroid systems (including cholesterol).The active form **14.26** of lovastatin competes with HMG-CoA for the catalytic site of the enzyme, thus disrupting the reaction and inhibiting the biosynthesis of cholesterol precursors.

14.26 Lovastatin active form   14.27 HMG-CoA   14.28 Mevalonate

Lovastatin induces an increase in high-affinity LDL receptors, increasing clearance of LDL, which lowers the plasma pool of LDL-cholesterol. The drug is effective orally, and it undergoes marked first-pass extraction by the liver. For this reason, the major site of its pharmacological action is the liver. The liver is also the major route for excreting lovastatin and its metabolites (enterohepatic circulation). Only a very small percentage of them is found in the urine. Atorvastatin is transformed in the liver to *o*- and *p*-hydroxylated derivatives which account for the majority of the drug's inhibitory action on HMG Co-A reductase. Atorvastatin has a much longer half life (≈ 20 h) than the other statins ($\approx$ 1–4 h), and atorvastatin's half-life of enzyme inhibitory effect is even longer (up to 30 h). The statins are not recommended for use in individuals with liver disease.

14.4 Myocardial Infarction

An *infarct* is broadly defined as an obstruction of flow in a blood vessel. Within the context of this discussion, the term is limited to formation of a clot in a coronary blood vessel (thrombosis) which greatly restricts the flow of blood to some region of the heart muscle. Death of an area of the myocardium occurs. This is a heart attack. Another danger in myocardial infarction is that the clot or a portion of it may detach from the arterial wall and float freely in the cardiac vessel. This is an *embolism*, and the condition is usually fatal due to eventual blockage of a region of the vessel which is too small to permit further passage of the embolus. Drug therapy of myocardial infarction involves prophylaxis (prevention of clot formation) or acceleration of resorption of a clot already formed.

14.4.1 Physiology of Blood Coagulation

Hemostasis is the spontaneous arrest of bleeding from a damaged blood vessel. The immediate hemostatic response of a damaged vessel is vasospasm. The biochemistry of formation of a blood clot is exquisitely complicated and not all of the individual steps in the process are known or understood. The *blood platelets*, small round or oval discs found in large numbers in the blood, are formed in the bone marrow. They have no nuclei, and they cannot reproduce. Within seconds of the vasospasm, the blood platelets in the vicinity of the damaged area of the vessel become sticky and they adhere to the injured surface of the blood vessel, and to each other. This change in physical state of the platelets is mediated by the release of thromboxanes. Next, the platelets lose their individual membranes and form a gelatinous mass. This plug quickly arrests bleeding, but it must be reinforced by *fibrin* for long-term effectiveness. Fibrin reinforcement results from local stimuli to coagulation. A complex of substances called *prothrombin activator* is formed in response to rupture of the blood vessel. Prothrombin is a plasma protein of the α-globulin type, and its continuous formation in the liver involves the participation of vitamin K. The prothrombin activator acts on prothrombin, cleaving it roughly in half, and one of these cleavage products is *thrombin* which is itself a proteolytic enzyme. *Fibrinogen* is a high molecular weight protein formed in the liver and transported to the blood, where it is a normal constituent. Thrombin, when liberated, utilizes fibrinogen as a substrate to remove four low molecular weight fragments from each fibrinogen molecule, thus forming a molecule of *fibrin monomer* which interacts with other fibrin monomer molecules to form a polymer. This polymerization occurs within seconds, forming long threads of fibrin which comprise the framework for the clot. The blood platelets entrapped within the clot release a *fibrin stabilizing factor*. However, before this factor can exert an effect on fibrin, it must itself be activated by the thrombin which has already been released. The activated fibrin stabilizing factor operates as an enzyme to induce formation of covalent bonds and multiple cross-linkages between the interacting fibrin strands. Thus, the blood clot is composed of a

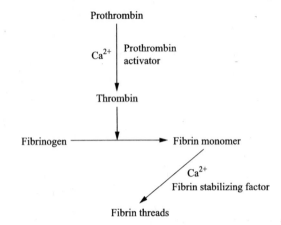

Figure 14.5 Principal steps in blood coagulation

three-dimensional network of fibrin threads in which blood cells, platelets, and plasma are entrapped. The fibrin threads adhere to damaged surfaces of blood vessels, preventing blood loss. Figure 14.5 summarizes principal steps in the formation of a clot. Within a few minutes after the clot is formed it begins to retract, squeezing out the entrapped fluid, which is called *serum*, because all or most of the fibrinogen and other clotting factors have been removed. Serum differs from blood *plasma* in that serum cannot clot. Once a clot has started to develop, it initiates a cycle to promote more clotting. One of the causes of this phenomenon is thrombin which, once formed, has a direct proteolytic effect on prothrombin itself, thus liberating more thrombin to extend the clotting process. There are endogenous plasma protease inhibitors which rapidly inactivate thrombin. There is also another mechanism for inhibiting clot formation: a system of enzymes (e.g., urokinase) which digest fibrin (fibrinolysis).

As a part of the process of conversion of soluble fibrinogen into insoluble fibrin, several circulating blood proteins interact in a cascading series of proteolytic reactions. For example, a clotting factor zymogen (*factor VII*) undergoes proteolysis and becomes an active protease (*factor VIIa*). This protease in turn activates the next clotting factor (*factor IX*) until eventually a solid fibrin clot is formed.

14.4.2 Anticoagulant Drugs

Heparin **14.29**, a heterogeneous mixture of sulfated mucopolysaccharides containing 2000–4500 monosaccharide units, is a powerful anticoagulant whose mechanism of action is related to the action of an endogenous plasma protease inhibitor, *antithrombin III* (heparin cofactor). Heparin forms an equimolar stable complex with antithrombin. This complex induces a conformational change which makes antithrombin's active site more complementary to the topography of thrombin. Once an antithrombin–thrombin complex is formed (which inactivates the catalytic activity of the thrombin), heparin is released and is available for binding to more antithrombin III. High doses of heparin interfere with blood platelet aggregation and thereby prolong bleeding time. However, the extent to which heparin's blood platelet effect contributes to its pharmacological effect is unclear.

14.29 Heparin pentasaccharide sequence

Heparin is usually administered by intravenous infusion, but for prophylaxis of thrombosis or embolism, it can be administered subcutaneously. Its value rests in part on its rapid onset of action when it is administered intravenously. The drug is ineffective orally. Heparin does not penetrate the placental barrier; therefore the drug is useful as an anticoagulant during pregnancy. An excessive anticoagulant effect of heparin is best treated by discontinuing the drug treatment. However, in severe overdosage where bleeding occurs, a specific antidote, *protamine*, is indicated. Protamine is a peptide that forms a stable complex with heparin, which complex is devoid of anticoagulant activity. Protamine itself has *anticoagulant* activity due to its effect on blood platelets, fibrinogen, and other plasma proteins. Therefore, it is essential to administer the minimum amount of protamine required to neutralize the plasma heparin. Heparin fragments, prepared by treatment of native heparin with enzymes, nitrous acid, or hydrogen peroxide, are known as *low molecular weight heparins*. This material contains up to 25 molecular fragments having molecular weights between 4000 and 9000. This material is said to be clinically superior to native heparin and is slowly replacing it in clinical practice.

Warfarin (coumadin) **14.30** is typical of "oral anticoagulants". These agents act only in vivo and they have no effect on clotting if they are added to blood in vitro. The oral anticoagulants are antagonists of vitamin K. Several of the coagulation factors (II, VII, IX, X) which are synthesized in the liver are biologically inactive until they are carboxylated on several of their terminal glutamic acid residues. The resulting dicarboxylic acid derivatives bind Ca^{2+}, which is an essential intermediate step in conferring catalytic activity on the molecule. As shown in figure 14.6, this reaction in the endoplasmic reticulum requires carbon dioxide (the source of the carboxyl group), elemental oxygen, and the reduced (hydroquinone) form of vitamin K.

14.30

Carboxylation is directly related to the oxidation of vitamin K hydroquinone to the epoxide. The hydroquinone form must be regenerated from the epoxide to maintain the peptide carboxylation reaction. This process proceeds through the quinone. The reductase enzyme(s) which effect this reaction are targets for warfarin, which blocks their action. The body contains other reductases which catalyze reduction of vitamin K

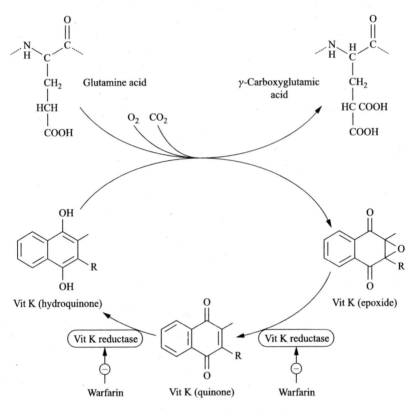

Figure 14.6 Mechanism of action of vitamin K and the site of action of oral anticoagulants

oxide, but these require higher concentrations of the substrate. These alternate reductases are less sensitive to warfarin, which may be the reason that the anticoagulant effects of large doses of warfarin are reversed with similarly large doses of vitamin K. Warfarin administered orally as its water-soluble sodium salt is 100% bioavailable. Approximately 99% of a dose is bound to plasma proteins, which accounts for its small volume of distribution and its long half-life (36 h). There is a massive roster of drugs that affect the pharmacodynamics of oral anticoagulants. The role of salicylates was described in chapter 1. Acute intake of ethanol increases the risk of spontaneous hemorrhage in individuals who are on warfarin therapy. Chronic intake of barbiturates leads to enzyme induction phenomena which can lead to accelerated metabolic inactivation of warfarin and a diminished pharmacological effect. Oral anticoagulants can pass the placental barrier; they are not given in the first months of pregnancy because they are teratogenic. They appear in milk during lactation. This may be significant because the newborn infant does not yet adequately synthesize vitamin K in the intestine. However, it is customary to administer vitamin K to infants, so that warfarin treatment of the mother does not generally represent a threat for the breast-fed infant.

The action of oral anticoagulant drugs should be monitored by their effect on the *prothrombin time*, which is the time required for clotting of citrated plasma (citrate sequesters Ca^{2+}) after adding specified standard amounts of Ca^{2+} and reference standard thromboplastin. Thromboplastin is a saline extract of brain tissue that contains tissue clotting factors and phospholipids. The prothrombin time ratio is expressed as the ratio of the prothrombin time of the patient to the prothrombin time of a pool of plasma from healthy subjects on no medication.

14.4.3 Thrombolytic Drugs

Plasmin, the body's enzyme that dissolves fibrin, is activated by hydrolytic removal of one amino acid from its inactive precursor, *plasminogen*, catalyzed by *tissue plasminogen activator.* The fibrinolytic system is regulated so that unwanted fibrin thrombi are removed, but fibrin in wounds persists to maintain hemostasis. Therapy with thrombolytic drugs dissolves the fibrin deposits at sites of vascular injury and pathological thrombi. Hence, these drugs can produce serious hemorrhage.

Streptokinase is a protein produced by β -hemolytic streptococci. It has no enzymatic activity, but it forms a stable noncovalent 1:1 complex with plasminogen which induces a conformational change on the plasminogen molecule, facilitating its conversion into active plasmin. Streptokinase is administered intravenously. Because it is a foreign protein in the body, its side effects include allergic reactions and (rarely) anaphylaxis. Tissue plasminogen activator is occasionally employed therapeutically to dissolve thrombi.

14.4.4 Antiplatelet Drugs

Inhibition of the arachidonic acid cascade is the basis for the use of aspirin as a prophylactic agent to prevent heart attack. The blood platelets, which are involved in an early phase of the clotting phenomenon, contain the components of the arachidonic acid cascade. Aspirin's inhibition of thromboxane A_2 production by inhibition of the COX-1 enzyme isoform (cf. chapter 11) retards platelet aggregation, thus disrupting the coagulation process. Since platelets do not synthesize new proteins, the action of aspirin on platelet cyclooxygenase is permanent, lasting the lifetime of the platelet (7–10 days). Thus, repeated doses of aspirin produce a cumulative effect on platelet function. Aspirin is maximally effective as an antithrombotic agent at doses much lower than those required for other actions of the drug. There are several other drugs marketed as antiplatelet agents, but it has not been shown that any of these is superior to aspirin.

Bibliography

1. Guyton, A. C. *Textbook of Medical Physiology,* 6th ed. Saunders: Philadelphia, Pa., 1981; p. 153.
2. Cocolas, G. H. In *Wilson and Gisvold's Textbook of Organic Medicinal and Pharmaceutical Chemistry,* 10th ed. Delgado, J. N. and Remers, W. A., Eds. Lippincott-Raven: Philadelphia, Pa., 1998; p. 615.
3. Garber, K. *Modern Drug Discovery* **1999,** *September/October* 68.

Recommended Reading

1. Oates, J. A. and Brown, N. J. Antihypertensive Agents and the Drug Therapy of Hypertension. In *Goodman and Gilman's The Pharmacological Basis of Therapeutics,* 10th ed. Hardman, J. G. Limbird, L. E. and Gilman, A. G., Eds. McGraw-Hill: New York, 2001; pp. 871–900.
2. Masaki, T. Possible Role of Endothelin in Endothelial Regulation of Vascular Tone. *Annu. Rev. Pharmacol. Toxicol.* **1995,** *35,* 235–255.
3. Griendling, K. K., Lassèque, B. and Alexander, R. W. Angiotensin Receptors and Their Therapeutic Implications. *Annu. Rev. Pharmacol. Toxicol.* **1996,** *36,* 281–306.
4. Timmermans, P. B. M. W. M. and Smith, R. O. Antihypertensive Agents. In *Burger's Medicinal Chemistry and Drug Discovery,* 5th ed. Wolff, M. E., Ed. Wiley–Interscience: New York, 1996; Vol. 2, p. 265.
5. Majerus, P. W. and Tollefsen, D. M. Anticoagulant, Thrombolytic, and Antiplatelet Drugs. In Recommended Reading ref. 1, pp. 1519–1538.
6. Jackson, E. K. Renin and Angiotensin. In recommended Reading ref. 1, pp. 809–842.
7. Mahley, R. W. and Bersot, T. P. Drug Therapy for Hypercholesterolemia and Dyslipidemia. In Recommended Reading ref. 1, pp. 971–1002.
8. Guyton, A. C. and Hale, J. E. Hemostasis and Blood Coagulation. In *Textbook of Medical Physiology,* 9th ed. W. B. Saunders: Philadelphia, Pa., 1996; pp. 390–399.
9. Highsmith, R. F., Ed. *Endothelin. Molecular Biology, Physiology, and Pathology.* Humana Press: Totowa, N.J. 1997.
10. Rang, H. P., Dale, M., Ritter, J. M. and Gardner, P. Hemostasis and Thrombosis. In *Pharmacology,* 4th ed. Churchill-Livingstone: Philadelphia, Pa., 2001; pp. 310–327.
11. Bialecki, R. A. Recent Advances in Pulmonary Hypertension Therapy. *Annu. Rep. Med. Chem.* **2002,** *37,* 41–52.
12. Billman, G. E. and Altschuld, R. A. Myocardial Infarction Agents. In *Burger's Medicinal Chemistry and Drug Discovery,* 6th ed. Abraham, D. J., Ed. Wiley: Hoboken, N.J., 2003; Vol. 3, pp. 155–192.
13. Watling, K. J., Ed. Endothelin Receptors. In *The Sigma–RBI Handbook of Receptor Classification and Signal Transduction,* 6th ed. Sigma–RBI: Natick, Mass., 2001; pp. 98–99.
14. Bisacchi, G. S. Anticoagulants, Antithrombotics, and Hemostatics. In Recommended Readings ref. 12, pp. 283–338.
15. Sierra, M. L. Antihyperlipidemic Agents. In recommended Readings ref. 12, pp. 339–383.

15

The Cardiovascular System. II: Arrhythmias and Antiarrythmic Drugs

15.1 Arrhythmias

15.1.1 Pathology

The arrhythmias, disorders of heart rate and beating rhythm, represent extremely serious cardiac conditions. They include:

- *Premature contractions* (extra-systoles), which originate either in the atria or the ventricles.
- *Atrial fibrillation*, which consists of irregular, convulsive movements of the atria. The number of impulses is very great and the individual heart muscle fibers act independently, not in concert.
- *Ventricular fibrillation*, which is twitching of the ventricular muscle. The impulses traverse the ventricles so rapidly that coordinated contractions cannot occur.
- *Atrial flutter*, which denotes atrial contractions that are extremely rapid (180–400 min^{-1}) but are rhythmic and uniform in amplitude. The ventricles are unable to respond to each atrial impulse, so that the entire coordinated rhythmicity of the heart is disrupted.

Grossly, the cause(s) of arrhythmias may be: (1) altered formation of impulses in the SA node (the pacemaker of the heart); (2) altered conduction of impulses from the SA node through the heart muscle; or (3) a combination of the two. As described in chapter 14, a nerve impulse originating in the sinoatrial node moves from cell to cell across the entire heart, triggering contraction. Subsequent to this contraction, the muscle must relax so that the chambers of the heart can dilate and again fill with blood. Because the sinus node-generated impulse travels from cell to cell, the possibility exists that the impulse could circle back and reactivate cells that were activated earlier ("re-entry"). Normally, re-entry does not occur because the cells, once stimulated, become refractory for a period of time long enough for the original signal to die and the heart muscle cells do not contract until they are stimulated by a new signal from the SA node. However, in some instances re-entry occurs. A causative factor is a refractory period for the heart cells that is shorter than the time for conduction of the nerve impulse. Thus, any situation that shortens the refractory period or speeds the rate of impulse conduction favors the re-entry phenomenon. It is believed that almost all fibrillations and flutters are caused by re-entry. However, the ultimate causes and dysfunctions of cardiac rhythmicity are poorly understood.

15.1.2 Drug Therapy of Arrhythmias

The following is a classification of antiarrhythmic drugs according to their mode of action:

- *Class I*: drugs that block voltage-regulated sodium channels, thus reducing the excitability of the heart.
- *Class II*: β-adrenoceptor blockers.
- *Class III*: drugs that prolong the action potential on the heart cell membrane, and thus indirectly prolong the refractory period of the heart muscle. Excess impulses are blocked.
- *Class IV*: drugs that block voltage-regulated calcium channels and thus impair impulse propagation in the SA and AV nodes and also in some myocardial regions.

This system of classification has been criticized for being pharmacologically simplistic. However, it remains useful for didactic purposes.

Class I antiarrhythmic drugs

Quinidine **15.1** is a diastereomer of the antimalarial drug quinine. Quinidine increases the refractory period of the heart, that period of time between contractions when the heart muscle does not respond to stimuli. This effect discourages re-entry and tends to cause the arrhythmia to cease. Sodium channels exist in three distinct functional states: resting, open, and refractory. As described in chapter 1, channels switch rapidly from resting to open in response to a stimulus. This is the process of activation. Class I drugs bind to sodium channels most strongly when they are either in the open or the refractory state. Quinidine is classed as an open-state sodium channel blocker. It slows the passage of sodium and potassium cations across the myocardial cell membranes, presumably similar to the action of the local anesthetics. Quinidine's potassium channel blocking properties are responsible for its prolonging action potentials in most heart cells. The sodium channel blockade and the prolongation of duration of the

15.1 Quinidine

action potential are believed to be the causes of the prolongation of refractoriness in myocardial tissue. Because these effects on ion channels inhibit the propagation of action potentials on the membrane, quinidine has been described as a "membrane-stabilizing agent." Contemporary thought rejects this term as simplistic and misleading. Quinidine blocks α adrenoceptors, and the result is hypotension. The drug also blocks that branch of the vagus nerve innervating the heart, and this is an undesirable property of the drug because the normal result of vagal stimulation is a slowing of the heart rate. This vagal effect of quinidine is termed *anticholinergic* because the vagal fibers involved are cholinergic. Thus, the direct effect of quinidine on the heart itself is to slow heart rate, but its indirect effect through the vagus nerve is to speed heart rate. Improper doses of quinidine can *cause* arrhythmias. In these toxic responses, it is believed that the action of quinidine on the vagus nerve predominates over its effects on the myocardium.

Quinidine is well absorbed from the gastrointestinal tract, and it is extensively bound to plasma proteins. Up to 50% of a dose of quinidine is metabolized to a variety of products: quinoline or quinuclidine ring-hydroxylated compounds, an O-demethylated compound, and an oxidation product of the vinyl moiety. Some of these metabolites display greatly lowered activity, and some are completely inactive. It is unclear how much the metabolites of quinidine contribute to its overall therapeutic effect. Metabolic action on quinidine is inducible by drugs, such as phenobarbital and the anticonvulsant drug phenytoin. This must be taken into account when establishing dosage regimens for the drug. Quinidine itself is a potent *inhibitor* of one form of cytochrome oxidase. This is significant if some other drug, which is normally metabolically inactivated by this enzyme, is administered concurrently with quinidine. Quinidine has a small therapeutic ratio and it has the reputation of being potentially dangerous. It is rarely used in some parts of the world. However, in the United States, the drug still enjoys some popularity for treating certain types of arrhythmias.

The local anesthetic procaine **13.9** has pronounced antiarrhythmic activity when given intravenously. However, it is an excellent substrate for blood and other tissue esterases, and its duration of effect is too brief for therapeutic utility. Procaine's amide congener, *procaineamide* **15.2**, is chemically more stable and it is not a substrate for blood esterases. It is effective orally. Like quinidine, procaineamide is an open-state sodium channel blocker and it blocks outward currents of potassium. Its pharmacological effects on the heart are similar to those of quinidine. However, procaineamide lacks quinidine's α-adrenoceptor blocking ability and it has a greatly diminished ability to block vagal transmission.

15.2 Procaine amide R=H
15.3 N-Acetylprocaine amide R=$\overset{\text{O}}{\overset{\|}{\text{C}}}-\text{CH}_3$

The major metabolite of procaineamide, N-acetylprocaineamide **15.3**, which is formed in the liver in an N-acetyltransferase enzyme-mediated reaction, lacks the sodium channel blocking action. However, this metabolite is equipotent to procaineamide in prolonging action potentials. Procaineamide is rapidly eliminated, partly by renal excretion of unchanged drug and partly by the liver metabolism described previously. Both procaineamide and N-acetylprocaineamide can accumulate to toxic levels in individuals with kidney hypofunction. Individuals who are genetically slow acetylators (see chapter 3) may accumulate high levels of unmetabolized procaineamide which can produce toxic side effects.

Remarkably, another local anesthetic, *lidocaine* **15.4**, is also an effective antiarrhythmic agent. Like quinidine and procaineamide, it is classed as a sodium channel blocker. Lidocaine, blocks both open and refractory channels. It has no effect on the vagus nerve. The utility of lidocaine is limited by its short duration of action. When given by mouth, it is efficiently absorbed across the wall of the gut, but it undergoes extensive first-pass metabolism to form an N-de-ethylated product **15.5** which is subsequently hydrolytically cleaved by hepatic amidases to N-ethylglycine **15.6** and 2,6-xylidine **15.7**.

15.4 Lidocaine   **15.5** N-dealkylated metabolite

15.8 Mexiletine   **15.6** Ethylglycine   **15.7** 2,6-Xylidine

The N-de-ethylated metabolite **15.5** has some antiarrhythmic activity, but it is not clinically useful because of its extremely rapid metabolic inactivation by hydrolysis of the amide group. The primary amine congener of lidocaine is devoid of antiarrhythmic activity. Even when given by intravenous drip, lidocaine's duration of action is short because of hepatic metabolic inactivation.

Several analogs of lidocaine, for example *mexiletene* **15.8**, were designed to provide enhanced resistance to first-pass metabolic inactivation, to make chronic oral therapy effective. This drug is effective orally; it is used to control ventricular arrhythmias. Mexiletene eventually undergoes metabolism by liver enzymes which are inducible by several other drugs, including the anticonvulsant drug phenytoin. The mexiletine metabolites (ring- and ring methyl group hydroxylated derivatives) have no antiarrhythmic activity.

Class II antiarrhythmic drugs

The β-adrenoceptor blockers have been discussed previously (chapter 7). Drugs such as propranolol **7.18** are useful in preventing mortality in patients recovering from myocardial infarction, and it is claimed that the protective action of these drugs reflects their ability to prevent ventricular arrhythmias. These β blockers also correct those atrial arrhythmias that are triggered by increased noradrenergic activity. In contrast to the enantiospecificity shown by the enantiomers of propranolol with respect to β-adrenoceptor blockade, the two enantiomers are equally active as antiarrhythmic agents.

Class III antiarrhythmic drugs

The prolongation of the refractory period of the heart produced by drugs in this category is believed to be an effect on ion channels: inhibition of repolarizing K^+ currents and/or activation of slow inward Na^+ current. These drugs have been described as sodium and/or potassium channel blockers. However, mechanisms in this category are complex and are not well understood. Typical drugs are *amiodarone* **15.9** and *bretylium* **15.10**.

15.9 Amiodarone   15.10 Bretylium

Amiodarone exhibits a multiplicity of pharmacological effects, none of which is clearly linked to its antiarrythmic action. The drug could be viewed as an analog of the thyroid hormone, and some of its actions on the heart may be a reflection of its interaction with thyroid hormone receptors. Amiodarone is extensively bound in the tissues. It has a long elimination half-life (10–100 days) and it accumulates in the body during dosing. It normally requires days or weeks for its action to develop.

Bretylium prolongs action potentials in Purkinje fibers by an unknown mechanism, although blockade of K^+ channels has been suggested. Bretylium initially induces norepinephrine release and inhibition of its reuptake, and the resulting hypertensive effect may be dangerous in some patients. Virtually 100% of a dose of bretylium is excreted unchanged in the urine. Bretylium's clinical utility is limited to treatment of certain types of ventricular tachycardia or fibrillation. It is almost always administered by injection. Oral dosage forms are said to be unsatisfactory in a significant percentage of patients, due to poor absorption of this hydrophilic quaternary ammonium molecule.

Class IV antiarrhythmic drugs

Voltage gated calcium channels in the cell membrane transport calcium cations into the cytoplasm and this in itself triggers the release of additional Ca^{2+} into the cytoplasm. A drug that blocks these calcium channels, thus preventing calcium ion passage, can reduce the rate of conductance in the AV and SA nodes. This encourages resumption of normal rhythmicity. Three commonly employed antiarrhythmic calcium channel blockers are *verapamil* **15.11**, *nifedipine* **15.12** and *diltiazem* **15.13**. Diltiazem's pharmacology is similar to that of verapamil, but diltiazem has relatively more smooth muscle relaxing effect and it produces bradycardia.

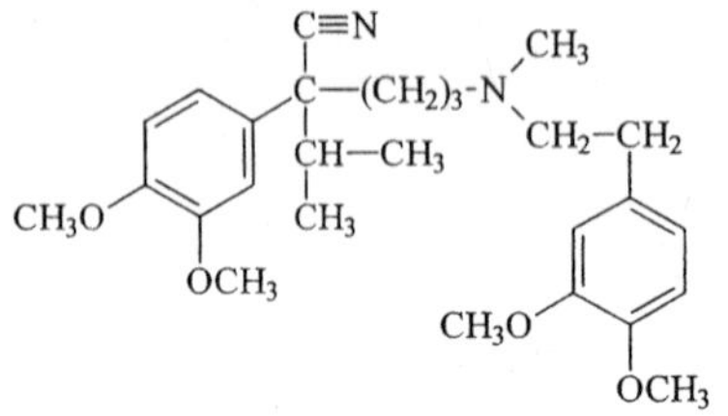

15.11

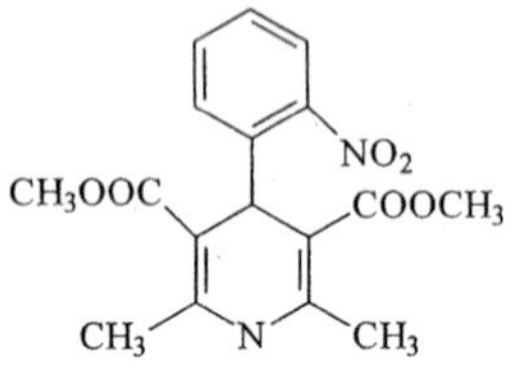

15.12

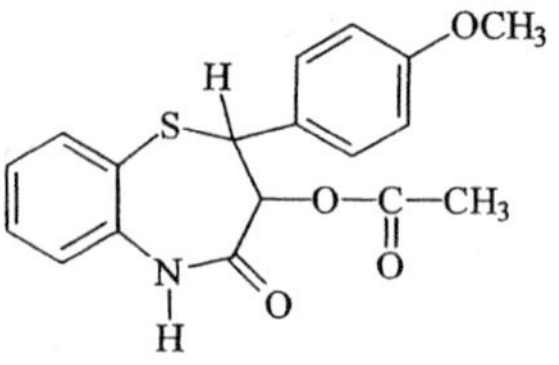

15.13

15.2   Myocardial Ischemia

15.2.1 Aspects of Pathology and Etiology

Angina pectoris is an older term applied to a substernal (beneath the sternum or breastbone) intense, severe, strangling pain. The condition is more properly called *myocardial ischemia* ("ischemia" is a deficiency in blood supply to some body part caused by some obstruction or constriction of blood vessels). Anginal pain is often associated with coronary artery disease. Two principal types of angina are recognized: exercise-induced (the most common) and nonexercise-induced (Prinzmetal's or variant angina). Ischemic myocardial pain results from an imbalance between oxygen supply and oxygen demand by the heart muscle. The cause of Prinzmetal's angina is ascribed to reduction of blood flow and not to increased oxygen demand. The physiological mechanism by which ischemia evokes pain is not well understood. Oxygen delivery may be inadequate because of a decreased caliber of the coronary vessels, because of decreased oxygen-carrying capacity of the blood, or because of decreased blood flow resulting from increased viscosity of the blood. Myocardial oxygen demand is increased by an increased heart rate, by excess release of inotropic hormones (e.g., epinephrine), or by increased myocardial wall tension. An anginal attack is often preceded by a rise in systemic blood pressure and/or heart rate, resulting in increased oxygen demand by the heart muscle. Accordingly, the strategy for relieving anginal pain is aimed at reducing ischemia by increasing oxygen supply to the heart muscle or by decreasing cardiac work (oxygen demand).

Explanations for the physiological basis of anginal attacks have changed frequently over many years. In the past, treatment of angina pectoris involved administering drugs which dilate the coronary artery (the principal artery that supplies blood to the heart muscle), to increase blood supply to the heart. However, drugs were discovered which showed a marked ability to dilate the coronary arteries in animals and humans, but which had little or no ability to relieve pain in many angina victims. It was concluded that the pathophysiology of myocardial ischemia may be considerably more complex than mere coronary artery spasm. However, it has been demonstrated by X-ray contrast techniques that spasms of the coronary artery do indeed occur, and it is estimated that approximately one third of angina patients owe their pain to coronary artery spasm. In the remaining two thirds of angina victims the attacks are caused by other factors. The causes of the coronary artery spasms are not well understood. Regardless of their chemical/pharmacological category, antianginal drugs do not relieve pain by an analgesic effect. Rather, they correct the actual cause of the pain. Anginal pain is influenced by so many factors, particularly emotional ones, that evaluation of drug treatment is difficult, and it may be quite subjective.

Arterial spasms can result from a sudden influx of calcium cations into the cells of the blood vessel walls. Hence, calcium channel blockers have gained prominence. To understand these drugs better, some simplified biochemical aspects of the extremely complex process of muscle contraction will be described. The contractile mechanism is the same for all three types of muscle: striated, smooth, and cardiac. The fundamental contractile unit within the muscle fiber is the *sarcomere*. The contraction of the muscle fiber involves interaction between two linear proteins, *myosin* and *actin*. As illustrated in figure 15.1, when the actin filaments slide into channels between the myosin filaments, the overall length of the protein complex diminishes, thus contracting the muscle fiber.

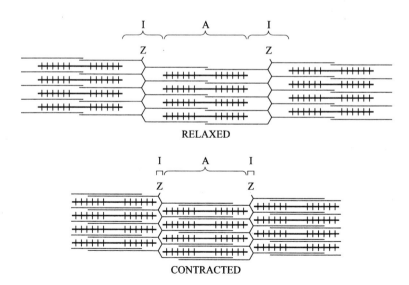

RELAXED

CONTRACTED

Figure 15.1 Muscle contraction. Relaxed and constricted states showing sliding of the actin filaments into the channels between myosin filaments. (Reproduced with permission from reference 1. Copyright 1996 Saunders)

The preliminary biochemical reaction in contraction involves stimulation by a neurotransmitter substance, usually acetylcholine. Calcium cations are present in the muscle fibers, chiefly bound to the A band, the region where the myosin and actin interact. In the activation stage (figure 15.2, diagram I), calcium cations enter the muscle cell (e.g., in a blood vessel wall) through calcium channels in the cellular membrane in response to depolarization of the membrane by acetylcholine.

These entering calcium cations trigger the release of additional Ca^{2+} which in the resting stage is stored in the cell (diagram II). By a series of steps involving cleavage of ATP into ADP (the source of energy in the process) the calcium ions promote interaction between the actin and myosin strands and the muscle fiber contracts. In the relaxation stage (diagram III) calcium ions are desorbed from the protein filaments. The filaments expand and the muscle fiber relaxes. A portion of the calcium cations leaves the cell through the calcium channels, and the remainder is stored in the muscle cell.

There are four different kinds of calcium channels in the human body, and they are classed according to their location and physiological function:

1. *L-type*, found in smooth, skeletal, and cardiac muscle
2. *T-type*, found in pacemaker cells (in the heart)
3. *N-type*, found in neurons and participate in release of neurotransmitter
4. *P-type*, found in Purkinje fibers in the heart; physiological role not understood

Calcium channel blocking drugs selectively interfere with the entry of Ca^{2+} into the smooth muscle cells of the blood vessel walls and thus the contraction of the muscle fiber is prevented. The blood vessel remains dilated. Calcium channel blocking drugs relax arterial smooth muscle but they have little effect on the walls of the veins. These drugs are effective in inhibiting contraction of the smooth muscle in the walls of the coronary arteries, thus increasing the supply of blood to the heart muscle. Because the smooth muscle of the coronary arteries is much less likely to contract in the presence of a calcium channel blocker, spasm of the coronary artery is prevented. This prevention is important in some angina patients. It is usually easier to prevent a spasm from occurring than to reverse it once it has begun. Additionally, elevated Ca^{2+} concentration inside the heart muscle fibers increases the contractile force of the myocardium (positive inotropic effect), which aggravates the anginal syndrome. Blockade of calcium channels prevents entry of a significant amount of Ca^{2+} into the heart muscle cells. As might be inferred from the preceding discussion, the effects of calcium channel blocking drugs in dilating arterioles combined with their negative inotropic effect establish their value in treating essential hypertension as well as myocardial ischemia.

15.2.2 Antianginal Therapy

Nifedipine **15.12**, *diltiazem* **15.13**, and *verapamil* **15.11** are typical calcium channel blockers utilized in treatment of myocardial ischemia. Historically, (aside from the calcium channel blockers) the most important antianginal drugs have been some organic nitrate esters and to a lesser extent, organic nitrite esters. These esters are powerful vasodilators. Glyceryl trinitrate **15.14** (widely known by the chemically incorrect name "nitroglycerin") is a short-acting antianginal drug. It is markedly lipophilic. A nitroglycerin tablet placed under the tongue permits rapid absorption across the mucous membrane, and relief of anginal pain is evident within 2–3 min. However, this

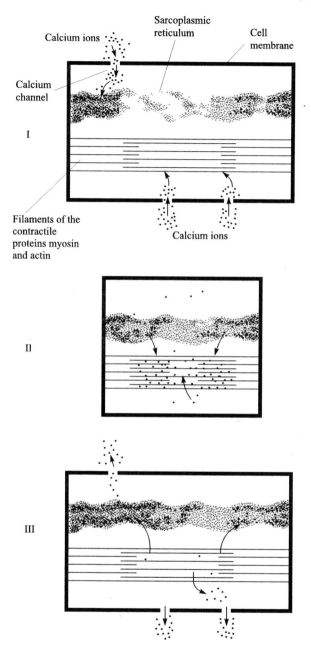

Figure 15.2 Diagrammatic representation of the role of calcium channels in the contraction of muscle fibers. (Reproduced from reference 2. Copyright 1982 American Chemical Society)

effect dissipates within a few more minutes and if the pain persists or returns, another tablet must be placed under the tongue.

The liver contains a high capacity organic nitrate reductase enzyme which, with participation of glutathione, removes the nitrate groups in a stepwise fashion (forming inorganic nitrite anion) and a more hydrophilic partially denitrated glycerin derivative. A by-product is the oxidized (R-S-S-R) form of glutathione. There is a major loss of pharmacological activity. Glyceryl trinitrate is rarely given orally (to be swallowed) because of extensive degradation in the gastrointestinal tract and first-pass metabolic inactivation.

The desired pharmacologic effects of organic nitrate and nitrite esters result from their enzyme-mediated release of the free radical form of nitric oxide in the vascular smooth muscle. The chemical details of this conversion are still unresolved. As cited previously (chapter 9) nitric oxide activates guanylate cyclase which results in increased production of a second messenger substance, guanosine 3',5'-monophosphate ("cyclic GMP"). In addition to its effect on intracellular Ca^{2+} levels, this second messenger stimulates a protein kinase system which eventually results in dephosphorylation (hydrolytic cleavage of phosphate ester groups) of one of the protein chains of myosin fibers. The phosphorylated form of this protein is believed to play an important role in the muscle contractile process. Thus, nitrate and nitrite esters diminish the ability of smooth muscle fibers of the vascular walls to constrict, leading to hypotension. In part, the antianginal effects of these drugs are directly caused by these vasodilating properties. The lowering of peripheral resistance also diminishes the contractile force of the heart required to eject blood, and the work of the heart and its oxygen demand are lessened.

The tolerance that develops to the hypotensive effect of glyceryl trinitrate has been explained by depletion of stores of the cofactor(s) required for enzymatic conversion of the drug into nitric oxide. However, this hypothesis has been challenged and a role for physiological counterregulatory hypertensive processes (such as activation of the renin-angiotensin system) has been suggested.

The lipophilicity of glyceryl trinitrate permits its efficient absorption across intact skin. A newer dosage form is an adhesive patch, impregnated with nitroglycerin (a "transdermal patch") which is attached to the skin of an arm or leg. A steady rate of release of drug from the patch and resulting steady rate of absorption provides a prolonged antianginal effect not possible with sublingual administration of a tablet.

$$H_2C-ONO_2$$
$$HC-ONO_2$$
$$H_2C-ONO_2$$

15.14 Glyceryl trinitrate

$$CH_3$$
$$CH-CH_2-CH_2-ONO$$
$$CH_3$$

15.15 Isoamyl nitrite

Isoamyl nitrite **15.15** is a rarely used short-acting antianginal drug. This lipophilic volatile liquid is administered by inhalation. It is inflammable and it has a pleasant, fruity odor. The drug is well absorbed across the mucosa of the respiratory tract, and presumably its mechanism of hypotensive action involves its metabolic conversion to nitric oxide. Isoamyl nitrite and a closely related ester, isobutyl nitrite, have been sold in "head shops" as sex stimulants under the names *Rush* or *Bolt*. They are said to enhance the sensation of orgasm. Adverse effects of these recreational drugs include postural hypotension and loss of consciousness.

Erythrityl tetranitrate **15.16** and *isosorbide dinitrate* **15.17,** unlike the preceding nitrate and nitrite esters, are solids. They are effective by sublingual administration. Their onset of action is somewhat slower, but the duration of their effect is decidedly longer. They have been used as prophylactic agents (taken chronically to prevent an anginal attack). In sufficiently large doses (to compensate for inactivation in the gastrointestinal tract and first pass hepatic metabolic inactivation) these two agents can be given by mouth as prophylactic agents. Erythrityl tetranitrate, like glyceryl trinitrate, is violently explosive.

$$\begin{array}{c} H_2C-ONO_2 \\ H_2C-ONO_2 \\ H_2C-ONO_2 \\ H_2C-ONO_2 \end{array}$$

15.16 Erythrityl tetranitrate

15.16 Isosorbide dinitrate

It will be recalled that sildenafil, the drug used to treat erectile dysfunction (chapter 9), exerts its effect by an indirect mechanism leading to protection of the second messenger, cyclic GMP, from metabolic destruction, thus leading to prolonged vasodilation. The action of the organic nitrates and nitrites in elevation of cyclic GMP levels in the smooth muscle cells of the blood vessels by release of nitric oxide, will contribute additively to the effect of sildenafil to result in a sometimes profound elevation of levels of cyclic GMP in the blood vessel walls, which can result in a severe, sometimes fatal, fall in blood pressure. Sildenafil should not be prescribed for patients receiving any form of a nitrate or nitrite ester.

Bibliography

1. Guyton, A. C. and Hall, J. E. *Textbook of Medical Physiology,* 9th ed. Saunders: Philadelphia, Pa., 1996; p. 76.
2. Sanders, H. J. New Drugs for Combating Heart Disease. *Chem. Eng. News* **1982,** *59,* 33.

Recommended Reading

1. Roden, D. M. Antiarrhythmic Drugs. In *Goodman and Gilman's The Pharmacological Basis of Therapeutics,* 10th ed. Hardman, J. G., Limbird, L. E. and Gilman, A. G., Eds. McGraw-Hill: New York, 2001; pp. 933–970. This reference includes a discussion of cardiac electrophysiology related to production of arrhythmias.
2. Thomas, R. E. Cardiac Drugs. In *Burger's Medicinal Chemistry and Drug Discovery,* 5th ed. Wolff, M. E., Ed. Wiley: New York, 1996; Vol. 2, pp. 152–261. This chapter begins with a discussion of cardiac physiology and transmembrane and intracellular signaling systems.
3. Guyton, A. C. and Hall, J. E. Contraction of Skeletal Muscle. In *Textbook of Medical Physiology,* 9th ed. Saunders: Philadelphia, Pa., 1996; pp. 73–85.
4. Kerins, D. M., Robertson, R. M. and Robertson, D. Drugs Used for the Treatment of Myocardial Ischemia. In Recommended Readings ref. 1, pp. 843–870.

5. Rang, H. P., Dale, M. M., Ritter, J. M. and Gardner, P. The Heart. In *Pharmacology*, 4th ed. Churchill-Livingstone: Philadelphia, Pa., 2001; pp. 250–277.
6. The Vascular System. In Recommended Readings ref. 5, pp. 278–300.
7. Keefer, L. K. Design of Nitric Oxide-Releasing Drugs. *Pharm. News* **2000**, *7*, 27–32.
8. Joshi, G. S., Burnett, J. C. and Abraham, D. J. Cardiac Drugs: Antianginal, Vasodilators, and Antiarrythmics. In *Burger's Medicinal Chemistry and Drug Discovery*, 6th ed. Abraham, D. J., Ed. Wiley: Hoboken, N.J., 2003; Vol. 3, pp. 1–53.
9. Fung, H.-L. Biochemical Mechanism of Nitroglycerin Action and Tolerance. Is This Old Mystery Solved? *Annu. Rep. Pharmacol. Toxicol.* **2004**, *44*, 67–86

16

The Cardiovascular System. III:
Congestive Heart Failure and Diuretics

Congestive Heart Failure

16.1.1 Etiology

Congestive heart failure results from an inability of the heart to pump sufficient blood to meet the body's needs. The efficiency of the heart as a pump is compromised. The primary causes of congestive heart failure are unknown, but biochemically there is an inability of the heart to translate ATP (the body's prime source of chemical energy) into mechanical work. When the heart begins to fail as a pump, compensatory mechanisms (many of which are mediated by the autonomic nervous system), which are aimed at maintaining cardiac output, become apparent. In the heart failure patient there is an increase in sympathetic nervous activity caused in part by a malfunction of the arterial baroreceptor reflex, which results in a decline in the body's ability to suppress the activity of CNS-directed sympathetic activity. Renal retention of sodium ions and water occurs, and this response is associated with increased blood volume. The actual size of the heart increases (cardiac hypertrophy). There is *tachycardia* (excessively rapid heart rate). Because of the inefficiency of the heart as a pump, the veins become engorged with blood and fluid escapes from the capillaries into the tissues. The result is edema of the lower legs, ankles, feet, and the lungs (pulmonary edema). Pulmonary edema occurs especially when there is a drastic failure in the efficiency of the left ventricle, which is responsible for sending blood to the periphery of the body. When the left ventricle fails, blood becomes static in the pulmonary vessels, and fluid leaks out into the lung tissue. The result is *dyspnea* (difficulty in breathing) which is common and characteristic in congestive heart failure patients. The principle of drug treatment of congestive heart failure is to improve cardiac function without undermining the compensatory mechanisms of the body.

16.1.2 Structure and Gross Pharmacological Effects of Cardiac Glycosides

The most widely used drugs for treating congestive heart failure are steroidal glycosides obtained from plants of the genus *Digitalis*. However, several other plant species (e.g., *Strophanthus*, *Convallaria*, *Urginea*) also produce cardioactive steroidal glycosides which are structurally similar to the digitalis glycosides. The most frequently and widely used of the cardiac glycosides is digoxin **16.1**. The biological activity of the cardiac glycosides is considered to reside in the *aglycone* (nonsugar) portion, but removal of the sugar moiety from the molecule results in loss of therapeutic value. The role of the sugar in part involves modifying the lipophilicity of the steroid portion, rendering the overall molecule more hydrophilic, to approach an optimal partition coefficient for efficient absorption and transport to appropriate sites of action in the body. In addition, the sugars participate in binding to the so-called digitalis receptor (vide infra). The mode of action of the cardiac glycosides is extremely complex and is not completely understood. It is likely that all of them, regardless of their specific chemical structures, have the same mechanism of action. Although the cardiac glycosides may differ somewhat from one another in the details of their effects in the body, they elicit the same qualitative response in the heart muscle: they increase the efficiency of the heart as a pump. The contractile force of the heart muscle is increased (*positive inotropic effect*), and the heart rate is slowed (*negative chronotropic effect*). In the congestive heart failure patient, these drugs bring about increased cardiac output, decreased heart size, lowered venous pressure, lowered blood volume, mobilization of edema fluid from the tissues, and increased urine output which results from the cardiac effects and mobilization of edema fluid. The cardiac glycosides themselves have no direct effect on the kidney, and they do not produce diuresis in the individual who has a normally functioning heart. Moreover, in the individual who has a normally functioning heart, the cardiac glycosides produce only a very moderate positive inotropic effect.

16.1

16.1.3 The Digitalis Receptor

Figure 16.1 is a proposed model for the binding of a digitalis glycoside to its in vivo receptor. Diagram I depicts a drug-induced conformational change in the receptor protein. Diagram II suggests some details of the agonist–receptor binding process. Significant ligand–receptor interactions include hydrogen bonding of the lactone carbonyl oxygen and electrostatic interaction of the carbon–carbon double bond of the lactone ring, hydrophobic interaction of components of the steroid nucleus, hydrophobic interaction of the lipophilic region of the desoxy sugar, and hydrogen bonding of the alcohol groups of the sugar(s). This diagram has been used to rationalize the experimentally observed necessity for the unsaturated lactone ring; the stringent requirements for

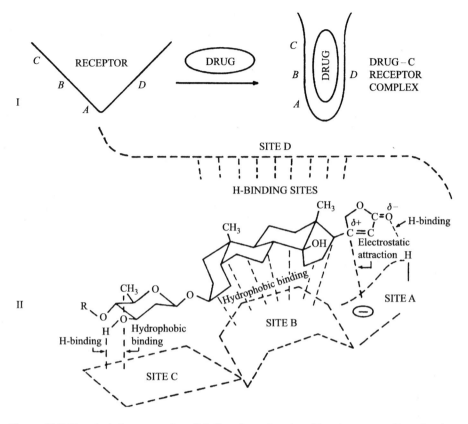

Figure 16.1 Hypothetical representation of binding of a cardiac glycoside to its receptor. (Reproduced with permission from reference 1. Copyright 1996 Wiley–Interscience)

the stereochemistry of ring fusion within the steroidal entity; and the pharmacological importance of the desoxy sugar.

16.1.4 Aspects of the Mechanism of Action of Cardiac Glycosides

These drugs inhibit *sodium- and potassium-dependent adenosine triphosphatase* ("transport ATPase"), the enzyme that catalyzes the conversion of adenosine triphosphate into adenosine diphosphate. Na–K ATPase has been described as the enzymatic equivalent of the cellular sodium pump. It is responsible for maintaining the unequal distribution of sodium and potassium ions across the membrane of heart muscle cells. It will be recalled that Na^+ is maintained at a higher concentration in the extracellular fluid, whereas K^+ is in higher concentration inside the cell. When a wave of depolarization passes through the heart, there is a change in the permeability of the heart cell membranes. Na^+ rapidly moves into the cell and K^+ moves out, both by passive diffusion.

After the heart beats, the direction of ion migration must be reversed: K^+ must be transported against its concentration gradient into the cell, and Na^+ must be transported against its concentration gradient out of the cell. The active transport systems for these ions (the sodium and potassium pumps) intimately involve the Na^+–K^+ ATPase system. The hydrolysis of ATP to ADP by the enzyme and the accompanying release of energy by the cleavage of the so-called high energy phosphate bond provide the energy needed to transport the sodium and potassium ions against their concentration gradients. In the digitalized heart, the inhibition of transport ATPase results in and maintains elevated intracellular Na^+ concentrations. This in turn enhances the function of the Na^+–Ca^{2+} exchange transport system that results in an increase in available intracellular Ca^{2+}, to produce the positive inotropic effect, the desired effect of cardiac glycoside therapy. A low level of K^+ in in the extracellular fluid compartment *enhances* and a high extracellular level of K^+ *antagonizes* the effects of the cardiac glycosides on the heart muscle. Thus, a common procedure for treating acute cardiac glycoside intoxication is the administration of a potassium salt to increase extracellular levels. These increased extracellular levels of K^+ stimulate the Na^+–K^+ ATPase pump system, and this promotes removal of excess Na^+ from inside the cells and subsequent lowering of intracellular Ca^{2+}. The elevated extracellular K^+ also apparently decreases the binding affinity of the cardiac glycoside molecule to the heart tissue. These factors operate in concert to antagonize the effect of the cardiac glycoside on the heart. The cardiac glycosides slow conduction through the AV node by a central nervous system action that stimulates the vagus nerve. It will be recalled that stimulation of the vagus nerve slows heart rate. The resultant slowing of the rate of ventricular contraction is a desirable effect.

Digoxin suppresses the renin–angiotensin system in congestive heart failure patients, and it also inhibits noradrenergic nervous activity. It was previously mentioned that increased noradrenergic activity is one of the events occurring in congestive heart failure. Some authorities propose that these factors contribute to the overall therapeutic utility of the cardiac glycosides. The cardiac glycosides seem to achieve their effects without a significant increase in oxygen consumption, and many authorities state that these drugs are the only cardiac stimulants suitable for long term use in congestive heart failure.

16.1.5 Pharmacologically Significant Physical Properties of Cardiac Glycosides

Despite the presence of hydrophilic sugar residues, the cardiac glycosides are poorly soluble in water. However, oral dosage forms are common. In general, those cardiac glycosides that have increased lipid solubility are the most efficiently absorbed. They also delay the onset of effect and they are the slowest to be excreted. Somewhat simplistically, it may be stated that the more hydroxyl groups a cardiac glycoside bears on its steroid nucleus, the more hydrophilic it is, and this is reflected in its pharmacological profile. Absorption of the cardiac glycosides from the gastrointestinal tract involves a passive diffusion process, dependent, at least in part, upon lipophilicity. An oral dose of digoxin demonstrates 70–85% bioavailability. The cardiac glycosides have no selective affinity for cardiac tissue. They are distributed over the entire body, and only a tiny percentage of a total dose manifests effects on the heart. Surprisingly the principal tissue reservoir for digoxin is skeletal muscle, not adipose tissue. Digoxin (and probably other cardiac glycosides) penetrate the placental barrier by passive diffusion.

16.1.6 Clinical Aspects of Cardiac Glycosides

It is difficult to designate a cardiac glycoside drug of choice. Digoxin is the most widely used in the United States, but it is difficult to state whether digoxin has any genuine pharmacological advantage over other commercially available glycosides. Among the criteria considered in selecting a specific cardiac glycoside for a specific patient are: (1) speed of onset of the drug's effect; and (2) duration of effect. All of the digitalis glycosides have about the same extremely small therapeutic index. A typical therapeutic dose is approximately 50–60% of the toxic dose level. All are potentially dangerous drugs.

In most individuals, digoxin is excreted in the urine virtually 100% chemically unaltered, a result of combined glomerular filtration and direct tubular secretion. However, some rare individuals metabolize digoxin extensively. The sugar portion is cleaved from the steroidal aglycone, and the resulting free hydroxyl groups on the steroid are conjugated with glucuronic or sulfuric acids.

16.2   Diuretics

16.2.1 Aspects of Renal Anatomy and Physiology: Urine Formation

In chapter 3 the role of the nephron, the fundamental physiological unit of the kidney, was described from the aspect of metabolism and drug excretion. Note in figure 3.1 the various regions of the renal tubule. In the *proximal convoluted tubule,* potassium and sodium ions, glucose, amino acids, phosphate and bicarbonate ions, and large amounts of water are reabsorbed into the general circulation. The fluid remaining in the tubule moves into the *loop of Henle,* first to the *descending limb* thence to the *ascending limb.* The loop of Henle is believed to be another important site for reabsorption of water. Next, the remaining fluid moves into the *distal convoluted tubule* where additional sodium is reabsorbed into the general circulation in exchange for some potassium ions and protons. This ion exchange in the distal tubule is stimulated by aldosterone. In the next region, the "collecting tubule", more water is reabsorbed, and this effect is regulated by a peptide released from the posterior pituitary gland, the *antidiuretic hormone* (arginine vasopressin) whose function is to promote increased reabsorption of water from the tubular fluid.

Thus, the body conserves ions that it utilizes in its normal physiology. These ion reabsorption processes (notably for Na^+, K^+, and Cl^-) involve active transport mechanisms. Waste materials of the body's metabolism, such as urea, are not reabsorbed but they remain in the tubular fluid and they are transported to the urinary bladder, whence they are excreted.

16.2.2 Diuretics and Methods of Producing Diuresis

Diuretics are substances that induce a net loss of fluid from the body in the form of urine. A true pharmacological diuretic drug exerts its effects on the kidney itself. By common usage, the term *diuresis* has two separate connotations: one refers to the increase in urine volume per se, and the other refers to a net loss of solute(s) and water. Probably the most important clinical use of diuretics involves mobilization of edema fluid: reduction of edema in congestive heart failure (adjunct to use of cardiac glycosides); relief of edema of pregnancy; and relief of premenstrual tension, based upon the hypothesis that this tension is caused in part by water and salt retention. The additional important use of certain diuretics in treating essential hypertension has been discussed previously (chapter 14).

Diuretics produce diuresis only in a kidney which is functional. If the kidney is nonfunctional or even severely hypofunctional, administration of a diuretic drug is unavailing and may be extremely dangerous.

Drugs used to produce diuresis may act by one or another of three different mechanisms increasing the osmotic pressure in renal tubules, altering the acid–base balance, or altering tubular transport mechanisms.

Increases in the osmotic pressure in the renal tubules

Neutral molecules can be administered intravenously for their diuretic effect. These molecules are removed from the blood by glomerular filtration, and they undergo only limited reabsorption across the tubular walls. They elevate the osmotic pressure in the tubules, thus impeding the reabsorption of water, largely from the loop of Henle and, to a lesser extent, from the proximal convoluted tubule. There is an accompanying modest increase in urinary excretion of almost all ionic species: Na^+, K^+, Ca^{2+}, Mg^{2+}, Cl^-, phosphate, and HCO_3^-. Most of the osmotic diuretics are sugars (e.g., glucose, sucrose) or polyols (e.g., mannitol, sorbitol). Urea is also classed as an osmotic diuretic. These osmotic pressure-elevating agents are not considered to be "general use" diuretic agents. They must be administered in very large doses and they are not very effective or dependable. However, there are some relatively rare clinical situations in which an osmotic diuretic may be indicated, such as lowering intraocular pressure in the eye and reducing cerebral edema before and after neurosurgery.

Alterations in the acid–base balance of the body

A drug may be employed which inhibits tubular reabsorption of sodium and bicarbonate ions, thus rendering the pH of the tubular fluid more alkaline. Elevation of the concentration of these ions in the tubular fluid exerts a strong osmotic effect, holding more water in the tubule and thus increasing the volume of urine output. When the tubular urine becomes alkaline, there is a *systemic acidosis*, elevated acidity in the blood and extracellular fluid. This category includes drugs such as *acetazolamide* **16.2**, whose mechanism of action involves inhibition of the enzyme *carbonic anhydrase*.

16.2 Acetazolamide

Carbonic anhydrase is found in red blood cells and in the walls of the renal tubules. It catalyzes and greatly speeds the attainment of equilibrium in the reaction of water with carbon dioxide to form carbonic acid (equation 16.1). Once formed, carbonic acid ionizes spontaneously to liberate a proton and a bicarbonate anion:

$$H_2O + CO_2 \xrightleftharpoons[\text{anhydrase}]{\text{Carbonic}} H_2CO_3 \xrightleftharpoons{H_2O} H_3O^+ + HCO_3^- \qquad (16.1)$$

In the absence of the enzyme, the entire set of equilibria is displaced far to the left. When a carbonic anhydrase (CA) inhibitor drug acts on the enzyme in the kidney, carbonic acid formation decreases, which results in decreased H_3O^+ in the walls of the tubules. Thus the normal physiological exchange of Na^+ for H_3O^+ in the tubular fluid is inhibited. The elevated concentration of Na^+ remaining in the tubular fluid exerts osmotic pressure to hold additional amounts of water in the tubules. The process of removal of HCO_3^- from the tubular fluid is also inhibited because an inadequate amount of H_3O^+ is present to react with the bicarbonate ion to form H_2CO_3 which immediately decomposes into CO_2 and H_2O. Therefore, the normally slightly acidic urine becomes alkaline. Inhibition of renal carbonic anhydrase also increases urinary excretion of K^+, which is usually undesirable.

Carbonic anhydrase is also found in the eye and in portions of the central nervous system. Inhibitors of this enzyme are instilled into the eye in treating certain types of glaucoma, to decrease the amount of fluid (*aqueous humor*) in the eyeball and thus to lower intraocular pressure.

The strategy of altering of acid–base balance to produce diuresis is considered to be outdated. The carbonic anhydrase inhibitors overall show a low level of diuretic efficacy.

Alterations in tubular active transport mechanisms

Tubular active transport mechanisms can be altered so that the ability of the tubular walls to reabsorb solutes and/or water is diminished. Retention of solutes in the tubular solution elevates the osmotic pressure of the fluid and promotes retention of water in the tubules, leading to increased urinary output.

Of the three possible pharmacological strategies for elevating urine output, only the alteration of tubular transport provides for extremely potent and dependable diuresis. Pharmacologists subdivide this category into several subtypes, as follows.

Inhibitors of the Na^+–Cl^- symport

This class includes thiazide diuretics, typified by *chlorothiazide* **16.3** and *hydrochlorothiazide* **16.4**. Pharmacological differences among the several thiazide diuretics are chiefly related to the size of the dose required to produce the maximal diuretic effect. All of them display approximately the same level of intrinsic activity. Hydrochlorothiazide is approximately 10 times as potent as chlorothiazide.

16.3 Chlorothiazide   **16.4** Hydrochlorothiazide   **16.5** Chlorthalidone

The normal active transport process for reabsorbing Na^+ from the lumen of the tubule across the tubular wall is coupled with a similar movement of Cl^- from the lumen across the tubular wall. The latter ion moves against a concentration gradient. This so-called *symporter* system (i.e., the transport of one ion coupled to that of another ion having the opposite charge, both transported across the membrane in the same direction) contrasts with an *antiport* system in which two ions of like charge are exchanged, one for the other, across a membrane. The thiazides gain access to the renal tubule mainly by direct secretion into the proximal tubule and to a lesser extent by glomerular filtration. Thiazides in the blood or in other body fluids do not seem to affect the sodium reabsorption process. Rather, the thiazide diuretics act directly in the tubule walls to block reabsorption of Na^+, Cl^-, and (to a some-what lesser extent) K^+. It has been speculated that the thiazide diuretics inhibit the Na^+–Cl^- symporter by competing for the Cl^- binding site. The resulting higher concentrations of Na^+, Cl^-, and K^+ in the tubular fluid exert a positive osmotic pressure, holding water, and the loss of water, sodium, and chloride is elevated. The thiazides exert their chief diuretic action in the distal convoluted tubule. However, the diagram of the nephron (chapter 3, figure 3.1) indicates that the great preponderance (at least 70%) of normal physiological reabsorption of sodium occurs, not in this distal tubular region, but rather in the proximal tubule where the thiazide diuretics have a relatively minor effect. Thus, the primary site of action of the thiazide diuretics is in a tubular region where sodium reabsorption is relatively small and the magnitude of diuretic effect of these drugs is self-limiting. Nevertheless, the ability of the thiazides to increase sodium excretion is pronounced, and they produce effective, depend-able, and long-lasting ($t_{1/2}$=2–7 h) diuresis. It was mentioned previously that the thiazides' long duration of effect is a factor in their utility in treatment of essential hypertension.

The thiazide diuretic's disruption of physiological processes for tubular K^+ transport can result, with chronic use, in a urinary loss of potassium sufficiently severe to require dietary supplementation. Hypokalemia can cause dangerous cardiac arrhythmias. Chlorothiazide and hydrochlorothiazide are excreted in the urine virtually 100% unchanged. However, some of the newer thiazide diuretics are extensively metabolized.

Newer agents, related pharmacologically but not chemically to the thiazides, are typified by *chlorthalidone* **16.5**. This drug exhibits an onset of action within 12 h, and it has a prolonged duration of action ($t_{1/2} \approx 47$ h). It can be given on alternate days. Most authorities describe the mechanism of diuretic effect of chlorthalidone as "thiazide-like".

Inhibitors of the Na^+–K^+–$2Cl^-$ symporter

Drugs of this category are frequently termed "loop diuretics" or "high-ceiling diuretics". There has been a need in the therapeutic armamentarium for diuretics with very high intrinsic activity, capable of producing powerful diuresis not attainable with thiazide drugs. This is the role of the high-ceiling agents. Their principal site of action is the ascending loop of Henle. In this region, the Na^+–K^+–Cl^- symporter concurrently trans-ports the three ions from the tubular lumen into the wall. It has been demonstrated experimentally that, in the absence of Na^+, the Cl^- uptake is inoperative. Typical of the heterogeneous chemical structures in this class of drugs are *ethacrynic acid* **16.6**, *furosemide* **16.7** and *bumetanide* **16.8**. These agents are capable of causing 15–25% of the Na^+ in the glomerular filtrate to be excreted. As with the thiazide diuretics, the loop diuretics can produce potentially dangerous hypokalemia. The mechanism by which the

16.6 Ethacrynic acid   **16.7** Furosemide

16.8 Bumetanide

high ceiling diuretics block the Na^+–K^+–Cl^- symporter is not well understood. It has been speculated that they attach to the Cl^- binding site. Ethacrynic acid, furosemide, and some other high-ceiling diuretics are extensively bound to plasma proteins and thus they are not expeditiously filtered through the glomerulus. However, there is an efficient physiological system by which they are secreted directly into the lumen of the proximal convoluted tubule, and thereby they gain access to the Na^+–K^+–Cl^- symport. Ethacrynic acid bonds covalently to in vivo sulfhydryl groups via Michael addition of SH across the, α, β-unsaturated ketone moiety. This sulfhydryl adduct is said to be the active form of the drug. All of the loop diuretics can cause ototoxicity (tinnitus, deafness, vertigo) which effects are usually but not always reversible. Ethacrynic acid appears to produce ototoxicity more often than the other loop diuretics.

The time of onset of diuresis and the duration of effect for the loop diuretics are considerably shorter than for the thiazide diuretics. The half-lives of the loop diuretics are typically 0.3–1.5 h. They undergo some degree of metabolic inactivation which in most cases is secondary to their elimination in the urine chemically unchanged. A complicating factor is that the intensity of diuresis produced by these agents calls into action physiological compensatory mechanisms which counteract the drugs' diuretic effect. These agents find their major use in the relief of acute pulmonary edema, secondary to congestive heart failure.

Aldosterone antagonists

The steroid hormone aldosterone (a mineralocorticoid) acts on nether regions of the proximal convoluted tubule and on the collecting tubule. Aldosterone receptors, when activated, initiate a sequence of biochemical events leading to stimulation of sodium–proton exchange such that Na^+ is reabsorbed and H_3O^+ is released into the tubule. The aldosterone receptors are representative of a third pharmacological category of receptors (in addition to membrane-bound G-protein-linked receptors and ion channels, previously described). These steroid receptors are not membrane-bound, but are soluble proteins, found in the cytoplasm of the tubular wall cells. As illustrated in figure 16.2, aldosterone is transported in the bloodstream to the cell membrane surface where (because of their strong lipophilicity) aldosterone molecules passively diffuse into the cell. In the cytoplasm, aldosterone interacts with its protein receptor and the resulting aldosterone–receptor

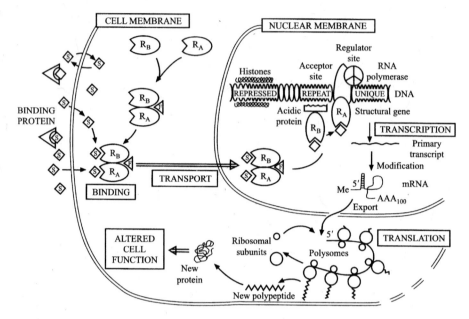

Figure 16.2 A general model for soluble steroid receptors. (Reproduced with permission from reference 2. Copyright 1978 Marcel Dekker)

complex migrates into the cell nucleus where it binds to a specific *nuclear acceptor*, a portion of a DNA molecule. The result of this nuclear reaction is the expression of multiple gene products, *aldosterone-induced proteins*, which in turn activate Na^+ channels and Na^+ pumps and promote reabsorption of Na^+ from the lumen of the tubule. There are also membrane-bound aldosterone receptors, activation of which stimulates Na^+ reabsorption from the tubules via a $Na^+-H_3O^+$ exchange process.

Aldosterone also promotes potassium excretion. When a drug blocks aldosterone receptors, there is an increase in the sodium content of the urine and a concomitant retention of potassium. Drugs which block aldosterone's renal effects are sometimes termed *potassium sparing diuretics*. *Hyperkalemia* (excess levels of K^+ in the blood) may occur in individuals who chronically take relatively large doses of an aldosterone antagonist and whose diet is consistently high in potassium-rich foods such as bananas or oranges. Probably the most serious effect of hyperkalemia is on the heart. High potassium levels disrupt the normal ion flux which maintains heart muscle contraction, and hyperkalemic arrhythmias can result, which can be life-threatening.

Spironolactone **16.9** has some aldosterone antagonist activity, but it is believed that the majority of its pharmacological effect derives from *canrenone* **16.10**, a Δ^6 metabolite resulting from elimination of the elements of thiolacetic acid from the spironolactone molecule. Orally administered spironolactone undergoes extensive ($\approx 35\%$) first-pass metabolic inactivation; it undergoes enterohepatic recirculation, and it has a short half-life (≈ 1.5 h). However, canrenone has a half-life of approximately 16 h, which prolongs the pharmacologic effects of a dose of spironolactone. Canrenone is employed clinically, but it is not available in the United States. The therapeutic utility of spironolactone and canrenone is somewhat limited by their low water solubility. *Potassium canrenoate* **16.11**, the salt of the lactone hydrolysis product of canrenone, is sufficiently soluble that it can be formulated in a solution for parenteral administration. In vivo, the lactone ring recloses and canrenone is formed. Potassium canrenoate also is not available in the United States.

16.9 Spironolactone   16.10 Canrenone

16.11 Potassium canrenoate

Spironolactone and canrenone competitively inhibit the binding of aldosterone to the cytoplasmic mineralocorticoid receptor (cf. figure 16.2). The receptor–canrenone complex does not attach to the nuclear DNA acceptor. Thus, expression of the proteins necessary for activating Na^+ transport mechanisms is inhibited. Earlier literature suggested that the cytoplasmic mineralocorticoid receptor exists in two conformations: active and inactive. It was proposed that spironolactone binds to the receptor and prevents it from assuming the active conformation.

Miscellaneous potassium sparing diuretics

Two drugs in this category, *triamterine* **16.12** and *amiloride* **16.13**, probably have similar mechanisms of action. They block sodium channels in the nether regions of the distal convoluted tubule and in the collecting duct. An indirect result of this inhibition of Na^+ reabsorption is a decreased excretion rate of K^+. Because the regions affected by triamterine and amiloride physiologically have a limited capacity to reabsorb sodium ion, these diuretics are capable of producing only a very modest increase in sodium ion excretion and hence only a modest degree of diuresis. Their major therapeutic utility is in combination with such as the thiazides and the loop diuretics, to counteract the potassium ion depletion that they cause. These potassium-sparing drugs are used more often than is spironolactone because they are better tolerated by most patients.

16.12 Triamterene   16.13 Amiloride

Xanthines

Caffeine **9.6** and *theophylline* **9.7** have diuretic effects, but their clinical utility is limited by their low potency, development of tolerance, and side effects such as cardiac stimulation and psychomotor effects. The mechanism of diuretic effect of caffeine and theophylline has not been clearly established. It has been suggested that, grossly, caffeine-induced diuresis results from vasodilation of the afferent glomerular arterioles, causing an increased glomerular filtration rate. There may be involvement of blockade of adenosine receptors (see chapter 9). Adenosine produces antidiuretic effects in several animal species, which are competitively antagonized by theophylline.

Bibliography

1. Thomas, R. E. Cardiac Drugs. In *Burger's Medicinal Chemistry and Drug Discovery*, 5th ed. Wolff, M. E., Ed. Wiley–Interscience: New York, 1996; Vol. 2, p. 241.
2. Schrader, W. T., Kuhn, R. W., Buller, R. E., Schwartz, R. J. and O'Malley, B. W. In *Receptors in Pharmacology*. Smythies, J. R. and Bradley, R. J., Eds. Marcel Dekker: New York, 1978; pp. 67–95.

Recommended Reading

1. Ooi, H., Colucci, W. Pharmacological Treatment of Heart Failure. *In Goodman and Gilman's The Pharmacological Basis of Therapeutics,* 10th ed. Hardman, J. G., Limbird, L. E. and Gilman, A. G., Eds. McGraw-Hill: New York, 2001; pp. 901–932.
2. Repke, K. R. H., Weiland, J., Megges, R. and Schön, R. Approach to the Chemotopography of the Digitalis Recognition Matrix in the Na^+-K^+-Transporting ATPase as a Step in the Rational Design of New Inotropic Steroids. In *Progress in Medicinal Chemistry.* Ellis, G. P. and Luscombe, D. K. Eds. Elsevier: Amsterdam, 1993; Vol. 30, pp. 135–202.
3. Joshi, G. S., Burnett, J. C. and Abraham, D. J. Cardiac Drugs: Antianginal, Vasodilators, and Antiarrhythmics. In *Burger's Medicinal Chemistry and Drug Discovery,* 6th ed., Abraham, D. J., Ed. Wiley: Hoboken, N.J., 2003; Vol. 3, pp. 1–53.
4. Thomas, R. E. Cardiac Drugs: Transmembrane and Intracellular Signaling Systems. In *Burger's Medicinal Chemistry and Drug Discovery,* 5th ed. Wolff, M. E., Ed. Wiley: New York, 1996; Vol. 2, pp. 165–189.
5. Thomas, R. E. Congestive Heart Failure. In Recommended Reading ref. 4, pp. 223–261.
6. Jackson, E. K. Diuretics. In Recommended Reading ref. 1, pp. 757–788.
7. Rang, H. P., Dale, M. M., Ritter, J. M. and Gardner, P. The Kidney. In *Pharmacology,* 4th ed. Churchill-Livingstone: New York, 2001; pp. 351–369.
8. Fink, C. A., McKenna, J. M. and Werner, L. H. Diuretic and Uricosuric Agents. In Recommended Reading ref. 3, Vol. 3, pp. 55–154.
9. Guyton, A. C. and Hall, J. E. Urine Formation in the Kidneys: 1. Glomerular Filtration, Renal Blood Flow, and their Control. 2. Tubular Processing of the Glomerular Filtrate. In *Textbook of Human Physiology.* W. B. Saunders: Philadelphia, Pa., 2000; pp. 279–312.

17

Pharmacology of Some Histamine-Implicated Diseases

Allergy

17.1.1 The Immune Response

The immune reaction and the related allergic response in the body are chemically mediated. Several endogenous chemical entities have been identified as participants in the triggering and propagation of these responses, but it is extremely difficult to describe how all of these agents act in concert in the physiology of the body. The following discussion is based on a compilation of the results of experimental observations, and it does not present a comprehensive or cohesive exposition of immune reactions, allergy, and/or asthma. The physiological processes involved are exquisitely complex and they are biochemically interactive.

Lymphocytes are a type of white blood cell formed in part in the lymph tissue. They are transported in the blood to their sites of use in the body. *T cells*, a type of lymphocyte, originate in the bone marrow. On exposure to appropriate antigens, large numbers of activated T cells are released, and they are distributed through the body, passing through capillary walls into the tissue spaces. There are several kinds of T cells:

- *Helper T cells*, which regulate virtually all immune functions by forming a series of protein mediators, lymphokines.
- *Cytotoxic T cells*, which kill microorganisms and, at times, even the body's own cells. They kill cells by attaching to them and secreting a protein, *perforin*, which literally punches large round holes in the membrane of the attached cell, thus destroying the integrity of the membrane.
- *Suppressor T cells*, which suppress the functions of the other two types of T cells.

An important side effect of immunity is the development, under some conditions, of allergy, of which there is more than one type. Allergy induced by activated T cells can cause skin eruptions (rash) in response to drugs or chemicals such as cosmetics or household chemicals to which the skin is exposed. Poison ivy is a classic example of a delayed-reaction allergy which is caused by activated T cells and not by antibodies.

The toxin of poison ivy (urushiol) in itself does not cause much harm to the tissues. However, upon repeated exposure it causes formation of activated helper and cytotoxic T cells. Then, after subsequent exposure to poison ivy toxin, within a day the activated T cells diffuse from the blood into the skin to respond to the poison ivy toxin and produce a cell-mediated immune reaction. This type of immune response releases many toxic substances from the activated T cells, and the eventual result of some delayed-reaction allergies can be serious tissue damage.

17.1.2 Histamine-Derived Allergic Reactions

Some individuals have an allergic tendency. Such a condition is termed *atopic allergy* because it is caused by a nonordinary response of the immune system. Chemical allergy (hypersensitivity) is an adverse phenomenon that results from previous sensitization to a particular chemical and it involves reactions mediated by the immune system. This immune response involves the production of antibodies in an animal injected with an antigen and the subsequent combination of the antigen with the antibody. Antigens may be proteins, polysaccharide–protein complexes, or whole cells or cell constituents. When a single administration of an antigen causes the production of antibodies, the reaction is termed a *primary response*. When the antigen is subsequently administered, the level of circulating antibodies is greatly increased; this is the *secondary response*. Only proteins or polysaccharides are known to elicit formation of antibodies, although smaller molecules, called *haptens*, if coupled to a protein, stimulate production of antibodies specific for the particular hapten molecule attached to the antigen. This antibody generally does not react with uncoupled protein (that is, protein lacking the hapten molecule). Allergic (hypersensitivity) reactions can result in acute or chronic conditions such a sneezing, runny eyes and nose, skin rash, and urticaria. *Anaphylactic shock* may develop within seconds in a presensitized individual who comes in contact with an appropriate antigen. Anaphylaxis is an intense systemic allergic reaction that can be fatal. Dose–response relationships are usually not apparent for triggering an allergic reaction.

As a part of the allergic response to an antigen, specific antibodies ("IgE", immunoglobulin E, also known as reagins or sensitizing antibodies) are generated, and these bind to receptors on the membrane of *mast cells* which are found in connective tissue. As many as a half million IgE antibodies can bind to a single mast cell. An *allergen* is an antigen that reacts with a specific type of IgE reagin antibody. When an allergen enters the body, the allergen–reagin reaction takes place and the allergic response occurs. The mast cell membrane surface also contains receptors for members of the *complement system*, an enzyme cascade with nine major components. The complement subtypes of significance in allergy are C3a and C5a. The mast cells can be activated to secrete mediators through these complement receptors. These mediators cause activation of the enzyme phospholipase C, leading to generation of second messengers (diacyl glycerol and inositol-1,4,5-tris -[dihydrogen phosphate]) which cause an elevation of intracellular Ca^{2+}, which in turn triggers release of a number of chemical substances from the mast cells. One of the principal substances released is histamine, which is stored inside the mast cells complexed with an acidic protein and a high molecular weight heparin. There are other agents present in mast cells (serotonin, acetylcholine, bradykinin, and leukotriene), and it is likely that these also contribute to allergic and anaphylactic responses. Substance P and some other tachykinins are found in small diameter visceral sensory nerve fibers and in enteric sensory neurons. The release of tachykinins from the peripheral ends of these neurons may play a role in neurogenic inflammatory responses to local injury by triggering histamine release from the mast cells and by acting directly on blood vessels to produce vasodilation and leakage of fluid into the tissues (edema).

17.1.3 Histamine H_1 Receptors in Allergy

H_1 receptors, which are primarily involved in the allergic response, are located peripherally in the smooth muscle of the bronchi of the lungs, intestines, blood vessels, and in the pain and itch nerve endings in the skin. H_1 receptors are linked to G-protein transduction systems which increase intracellular Ca^{2+}, leading to activation of protein kinases and phospholipase in the cell, which are responsible for the characteristic observable responses. Stimulation of peripheral H_1 receptors dilates small blood vessels which results in flushing of the blush areas of the face and neck and a fall in blood pressure. Vascular H_1 receptor stimulation also increases the permeability of the small blood vessels, resulting in outward passage of plasma proteins and fluid into the intracellular spaces, and the development of edema. Stimulation of H_1 receptors in the lung causes constriction of the smooth muscle of the bronchi (bronchospasm). Very small amounts of histamine evoke intense bronchoconstriction in asthmatics. By subcutaneous injection, histamine elicits a so-called "triple response". First, a localized red spot which extends a few millimeters around the site of injection. This appears within a few seconds and reaches a maximum in about 1 min. Second, a brighter red flush extends approximately 1 cm beyond the original red spot and appears more slowly. Third, a wheal that can be seen in 1–2 min and occupies the same area as the original red spot. The red spot results from the direct vasodilation action of histamine. The flush is caused by an indirect nervous stimulation by histamine, and the wheal demonstrates histamine's ability to produce edema. The triple response is manifested in the clinical condition of urticaria. In hay fever, the allergen–reagin reaction occurs in the nose. Histamine released in response to the reaction causes local vasodilation and increased capillary permeability, causing fluid leakage, and the nasal linings become swollen and secretory. Other products of the allergen–reagin reaction cause irritation of the nose, producing sneezing.

17.1.4 Anaphylaxis

This condition is usually caused by the release of histamine into the general circulation, which produces bodywide vasodilation. Death can follow within minutes unless epinephrine is administered to counteract the hypotensive effect of the histamine. H_1 blocking agents are not sufficient to prevent death. In anaphylaxis, leukotrienes are also liberated, which cause bronchoconstriction, and the individual may die of the resulting suffocation. Epinephrine also reverses this leukotriene effect. Some instances of systemic anaphylaxis are mediated by serotonin rather than by histamine.

17.1.5 H₁ Receptor Antagonists in Therapy

There is no currently recognized therapeutic use for peripheral H_1 receptor agonists. However, H_1 antagonists have an established and valued place in the symptomatic treatment of various hypersensitivity reactions. H_1 antagonists compete with histamine for the receptor sites and they are useful in acute types of allergy: rhinitis, urticaria, dermititis, conjunctivitis. However, beneficial effects are limited to those symptoms related to histamine release by the antigen–antibody reaction. H_1 blockers have little value in treating bronchial asthma, and they are of no value in treating the type of systemic anaphylaxis in which chemical mediators other than histamine are involved.

There is a large number of H_1 antagonists available. The older H_1 antagonists, typified by diphenhydramine **17.1**, produce marked CNS depression: somnolence, diminished alertness, and slowed reaction times. It will be recalled that histamine is a neurotransmitter in the CNS. However, H_1 antagonists vary in the severity of their CNS depressant actions, and individuals vary in their susceptibility to this side effect of these drugs. Diphenhydramine is also marketed as an over-the-counter sedative under a variety of brand names. This drug, like several other H_1 antagonists, displays prominent muscarinic receptor blocking actions, and it is occasionally used to reverse the extrapyramidal side effects of phenothiazine-derived antipsychotic agents. Prior to the introduction of levodopa, diphenhydramine was employed in relief of the symptoms of Parkinson's disease. The occasionally beneficial effects of the drug are not understood. Another antimuscarinic side effect of these H_1 blockers is dryness of the mouth and respiratory passages. This effect was the basis for the early use of H_1 antagonists in relieving the symptoms of the common cold. But, contrary to persistent past and contemporary belief, H_1 antagonists are without value in treating the common cold. Diphenhydramine and several other H_1 antagonists have significant activity in preventing motion sickness. Based on the fact that scopolamine, a muscarinic receptor blocking agent, has a powerful antiemetic effect, it was speculated that the antimotion sickness action of H_1 receptor blocking drugs is referrable to their antimuscarinic effects, which do not directly involve the chemoreceptor trigger zone in the medulla oblongata. Some H_1 antagonists are employed as prophylactics against nausea and vomiting, but their utility is frequently compromised by their prominent sedative action.

17.1 Diphenhydramine

Structures **17.2–17.6** illustrate other contemporary H_1 antagonists. *Terfenidine* **17.2** does not readily penetrate the blood–brain barrier to reach central histamine receptors. Hence, therapeutic doses of this drug produce little or no sedation. Terfenidine was found, in rare instances, to induce a potentially fatal cardiac arrhythmia if its hepatic metabolism was impaired, and it was withdrawn from the market. *Fexofenidine* **17.3** is a primary metabolite of terfenidine. It too is a selective peripheral H_1 receptor blocker and it produces no sedative or other CNS effects. This drug lacks the tendency of terfenidine to cause cardiac arrhythmias.

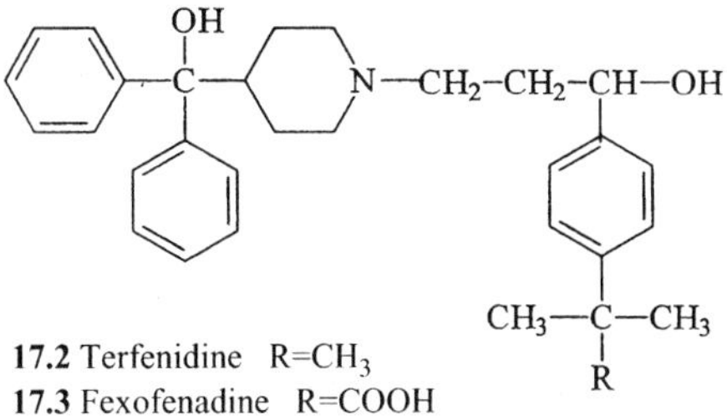

17.2 Terfenidine R=CH$_3$
17.3 Fexofenadine R=COOH

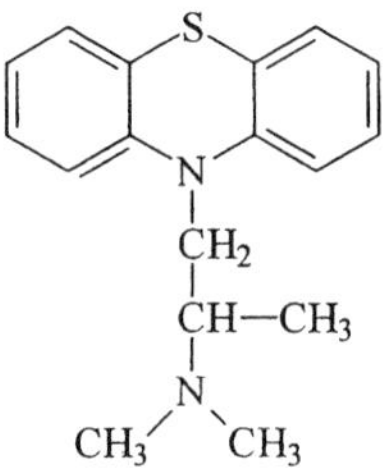

17.4 Peomethazine

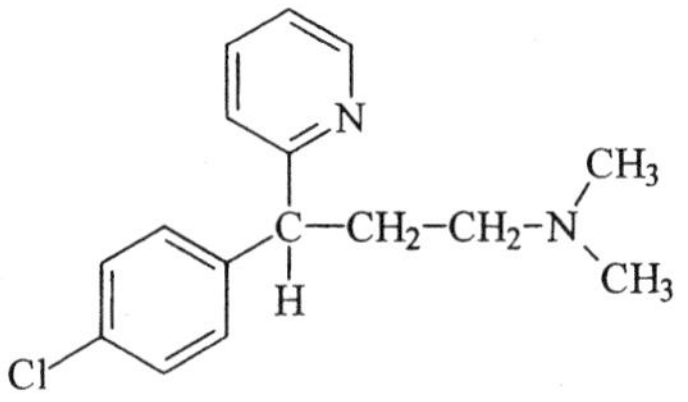

17.5 Chlorpheniramine

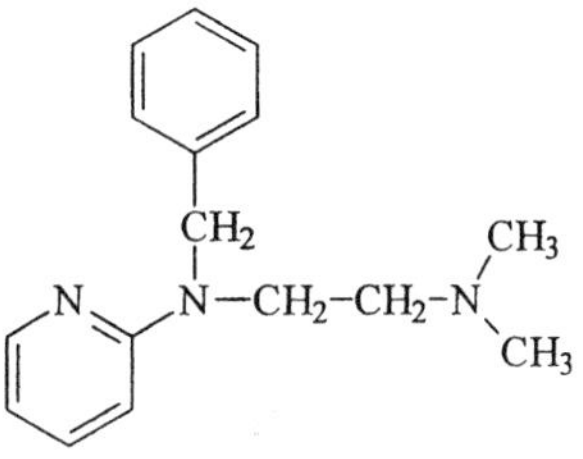

17.6 Tripelennamine

Bronchial Asthma

17.2.1 Physiology

Asthma is a syndrome in which there is a recurrent obstruction of the airways in response to stimuli that are not in themselves noxious and which do not affect nonasthmatic individuals. The asthmatic patient suffers intermittant attacks of wheezing, cough, and *dyspnea* (difficulty in breathing). In *status asthmaticus* (acute severe asthma) the dyspnea can last for days, and in some cases it can be fatal. Asthma was formerly thought to be a hypersensitivity response involving an allergen–IgE interaction on mast cells, leading to the release of histamine and other substances, which produce the bronchoconstriction. In asthmatics, inhalation of even small doses of histamine produces an intense bronchospasm. However, in healthy, nonasthmatic individuals, inhalation of small doses of histamine does not elicit this response. It is now recognized that not all cases of asthma are caused by allergies. Nevertheless, allergic asthma has been used as a model for the general study of asthma, partly because the condition can be induced at will by exposure to appropriate allergens. The development of the asthmatic syndrome probably involves both genetic and environmental factors. In allergic asthma, exposure of the individual to an allergen such as pollen or fecal proteins of the house dust mite causes sensitization that involves activation of T cells which in turn leads to IgE production. When the IgE binds to its receptors on the mast cells, the system is now in a condition such that subsequent re-exposure to the allergen triggers an asthmatic attack. The initial response occurs abruptly and is caused by spasm of the bronchial smooth muscle. This effect is caused by mast cells which release histamine as well as by other cells (white blood cells, blood platelets) which release a series of prostaglandins and leukotrienes (see figure 11.3). These agents are probably more significant spasmogens in bronchial asthma than is histamine. Adenosine triggers the release of histamine in the lung. Inhalation of adenosine by asthmatics produces bronchoconstriction which is related to release of histamine and leukotriene.

It has been speculated that nonallergic asthma (including exercise-induced asthma) arises from the stimulation of "irritant receptors" which trigger the release of neuropeptide mediators from sensory nerve fibers. These neuropeptides activate the mast cells.

Bronchospasm is considered to be only the first stage in the asthma syndrome. A later phase is a progressing inflammatory reaction, differing somewhat from other inflammatory syndromes, and mediated by a variety of peptides and other agents which are released by several types of the body's migratory cells which were attracted to the bronchial area. Leukotrienes (chapter 11, figure 11.3a) are directly involved in mediating bronchial inflammation. Asthma should be categorized as an inflammatory disease.

17.2.2 Drug Therapy of Asthma

The two phases of the asthma syndrome (bronchospasm and inflammation) require different therapeutic approaches. Bronchospasm is treated with β_2-adrenoceptor agonists, whose pharmacology was described in chapter 7. Based on the premise that asthmatic symptoms reflect an imbalance between the bronchoconstrictor action of the cholinergic nervous system and the bronchodilator effect of the β-noradrenergic system, an antimuscarinic agent, *ipratropium* **17.7** has been employed. A solution of this drug, frequently combined with a β_2-adrenoceptor agonist, is administered as an aerosol into the airway via the mouth, frequently by means of a metered dose inhaler. Because of its quaternary ammonium moiety, ipratropium is poorly absorbed and transported. It produces little or no systemic effects when given by aerosol and it does not penetrate the blood–brain barrier. Its actions are restricted to the bronchi. The beneficial effects of this combination therapy are slightly greater and more prolonged than with either drug alone.

17.7 Ipratropium

Theophylline **9.7** has been widely employed as a bronchodilator in asthmatics, although its use is declining. The mechanism of theophylline's bronchodilating action is as yet not well understood. In part it may involve inhibition of phosphodiesterases, the enzymes which metabolically inactivate cyclic AMP. The resulting increase in intracellular levels of this second messenger leads to a prolonged bronchodilation (a β_2-adrenoceptor effect). Theophylline, like caffeine, is a competitive antagonist at adenosine receptors, whose role in leukotriene and histamine release has been described previously. Theophylline is believed to block A_1, A_2, and A_3 adenosine receptors, and this effect may contribute to theophylline's bronchodilator effect. However, many authorities do not accept either the phosphodiesterase theory or the adenosine receptor blockade theory. The therapeutic dose range for theophylline is very close to its toxic dose range. Fatal intoxication with theophylline has been recorded. Death is caused by cardiac arrhythmias induced by the drug, especially following rapid intravenous administration. Even therapeutic doses of theophylline frequently produce unacceptable side effects: diarrhea, headache, insomnia, nausea, vomiting. The metabolic products of theophylline, partially N-demethylated xanthines (but not uric acid) are excreted in the urine.

The inflammatory phase of asthma is frequently treated with one or another of the semisynthetic steroidal *glucocorticoids*, so named because of their physiological role in glucose uptake and utilization, and typified by *triamcinolone acetonide* **17.8**, *prednisolone* **17.9**, and *beclomethasone* **17.10**. These drugs are not bronchodilators, and they are ineffective in the early stages of asthma. But in chronic asthma where inflammation of the bronchi is a major problem, their efficacy is impressive. They can be life-saving in acute severe asthma. The antiinflammatory effects of the glucocorticoids result from a multiplicity of actions, among which are inhibition of the production of prostaglandins and leukotrienes in a variety of sites involved in bronchial asthma. The glucocorticoids inhibit the expression of the COX-2 enzyme by inhibiting transcription of the relevant gene. Production of *interleukins* (an extensive family of peptides also involved in the inflammatory response) is blocked. IgE-dependent release of histamine is also inhibited. The antiasthmatic glucocorticoids are most frequently given by inhalation, but prednisone is given orally. Unwanted side effects of these agents reflect their many physiological actions. There can be suppression of the response to infection or injury, and wound healing may be impaired. There may be metabolic effects related to these drugs' role in maintaining water–electrolyte balance. Oral *thrush* (oropharyngeal candidiasis: fungal infection caused by members of the genus *Candida*) frequently occurs when the glucocorticoids are inhaled.

17.8 Triamcinolone acetonide

17.9 Prednisolone

17.10 Beclomethasone

The likely role of leukotrienes in asthma has been cited previously. On this basis, leukotriene receptor antagonists and leukotriene synthesis inhibitors have been introduced. *Zafirlukast* **17.11** is an example of leukotriene receptor antagonists. *Zileuton* **17.12** is an inhibitor of leukotriene synthesis, by virtue of its inhibition of 5-lipoxygenase, a critical enzyme in leukotriene synthesis. These agents have been proven to be effective prophylactic treatment for mild asthma. However, the scope of their therapeutic value in the asthmatic syndrome has not yet been established.

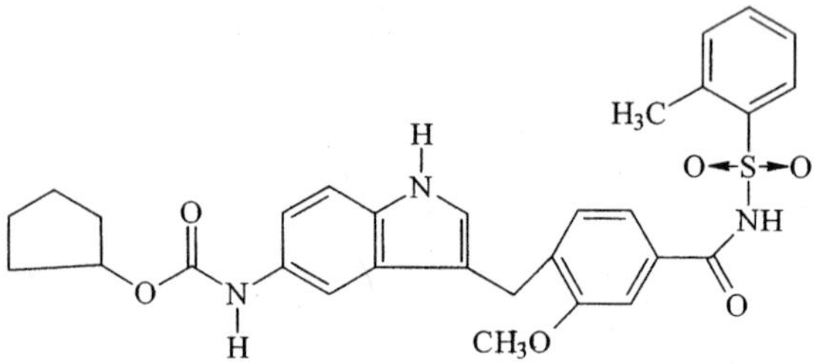

17.11 Zafirlukast 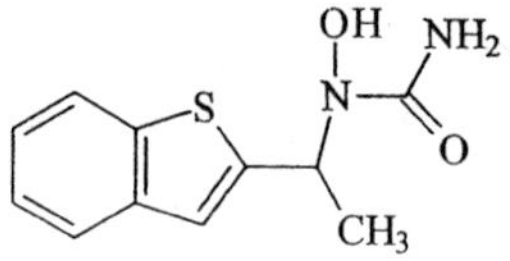   **17.12** Zileuton

Cromolyn sodium **17.13** and a related drug, *nedocromil sodium* **17.14**, are not bronchodilators, nor do they have any effect on smooth muscle. They do not inhibit the actions of any of the known smooth muscle stimulants. If given prophylactically, they reduce both the immediate and the later stage asthmatic responses. They are effective against antigen-induced, exercise-induced, and irritant-induced asthma, although not all asthmatic patients respond favorably to the drug.

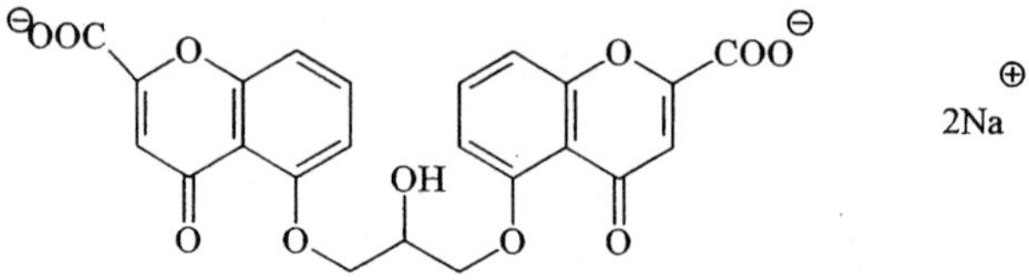

17.13 Cromolyn sodium

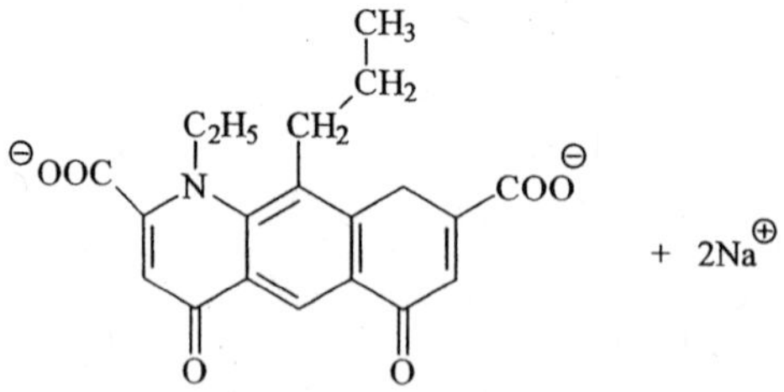

17.14 Nedocromil sodium

The mechanism of action of these drugs is not understood. Both drugs are extremely poorly absorbed from the gastrointestinal tract, and they are given by inhalation. Because they are so poorly absorbed from any site of administration, cromolyn and nedocromil produce only minor systemic side effects. Cromolyn is said to be less efficaceous than the glucocorticoids in controlling asthma, and the use of cromolyn in combination with inhaled glucocorticoids yields no additional benefit. Because of their limited potency, the use of these two drugs is said to be decreasing.

17.3 Gastric Hypersecretion

17.3.1 Some Aspects of Gastrointestinal Physiology

In addition to its principal physiological role in digestion and absorption of food, the gastrointestinal tract is one of the major endocrine systems of the body. It also has its own integrative neural network, the enteric nervous system. The components of the gastrointestinal system under neuronal and hormonal control are the smooth muscle, the blood vessels, and the glands. The stomach secretes approximately 2.5 L of gastric juice per 24 h period. The principal exocrine secretions in gastric juice are hydrochloric acid, *pepsinogen* (a zymogen: an inactive precursor to the proteolytic enzyme pepsin which is activated by the hydrochloric acid of the stomach), and *intrinsic factor*, a protein necessary for the absorption of vitamin B_{12}. Mucus is secreted by special cells on the inner surface of the stomach. Bicarbonate ions are also secreted and they are trapped in the mucus, creating a pH gradient of 1–2 in the lumen of the stomach to 6–7 at the mucosal surface. The mucus–bicarbonate mixture forms a gel-like layer which protects the stomach wall from the adverse effects of the components of the gastric juice. Autonomic nervous control of the gastrointestinal system has been addressed previously.

Gastrin, a peptide hormone synthesized in the stomach and duodenum, stimulates secretion of hydrochloric acid by the parietal cells of the stomach. Release of gastrin is inhibited when the pH of the stomach contents falls below 2.5. Other agents which participate in hydrochloric acid secretion are acetylcholine and histamine. Chloride ions, accompanied by K^+, are actively transported by a symport carrier from the parietal cells into small canals which connect directly with the lumen of the stomach. The K^+ ions are substrates for a K^+/H_3O^+ antiport system which carries the K^+ back into the parietal cells and delivers H_3O^+ into the canals, thence to the lumen of the stomach, completing the delivery of the ionic components of HCl into the stomach. Inside the parietal cells, carbon dioxide (from cellular metabolism) and water combine (under catalysis by carbonic anhydrase) to form carbonic acid which dissociates to provide the store of protons required for the potassium–proton exchange process, and HCO_3^- which exchanges across the cell membrane to bring more Cl^- from the plasma into the cell.

Stimulation of a branch of the vagus nerve (a cholinergic nerve) and the release of gastrin stimulate the release of histamine which in turn activates H_2 receptors in the parietal cells. These receptors are G protein-linked to adenylate cyclase, and the resulting elevated levels of cyclic AMP stimulate the $H^+–K^+$ ATPase proton pump system. This proton pump is also stimulated by a calcium-dependent pathway which is activated by muscarinic (M-2 and M-3) receptors. In addition to stimulating the liberation of gastric mucus, prostaglandins produced and liberated in the stomach wall inhibit histamine-stimulated adenylate cyclase activity, thus reducing acid secretion. In these gastric actions, histamine is acting physiologically as a "local hormone", and not as a neurotransmitter.

Xanthine alkaloids (caffeine and theophylline) stimulate secretion of gastric acid and digestive enzymes, and this is the basis for withholding coffee from the diet of peptic ulcer patients. However, even decaffeinated coffee has a potent stimulant effect on gastric hydrochloric acid secretion.

17.3.2 Peptic Ulcer and Reflux Esophagitis

Peptic ulcer arises from an imbalance between the activity of acid secretory mechanisms (excessive) and of the mucosal protective factors (deficient). Ulcer therapy aims
to restore a proper balance. Some types of ulcers are associated with normal HCl secretion but with deficient gastric mucus production, which exposes an unprotected area
of the stomach wall to the irritant effects of 0.1 M HCl and to the proteolytic action of
the stomach's digestive enzyme, pepsin. Some ulcers are associated with hypersecretion
of HCl. The causes of peptic ulcer include, bacterial infection (*Helicobacter pylori*),
chronic use of nonsteroidal antiinflammatory agents, and malignancy.

The esophagus, the tube that carries food from the mouth to the stomach, contains
(close to the entrance to the stomach) a *sphincter valve* which relaxes and opens to
permit swallowed food to enter the stomach. Once the bolus of food has passed into
the stomach, this valve closes. However, in some individuals the sphincter does not
close completely and gastric juice can enter the lower portion of the esophagus.
The hydrochloric acid of the gastric juice can irritate the wall of the esophagus, causing pain ("heartburn"). If this condition becomes chronic, the acid can seriously damage
the esophageal wall and this condition is called *reflux esophagatis*. The condition can,
in some instances, become life-threatening.

17.3.3 Drug Therapy of Peptic Ulcer and/or Reflux Esophagitis

In the past, ulcer therapy was based upon use of orally administered antacids and
muscarinic blocking agents, which were intended to lower gastric acidity, thereby
relieving symptoms and promoting healing. Neither therapy was completely satisfactory, nor was a combined therapy adequate. Orally administered antacids have a short
duration of effect, and the muscarinic receptor blockers produce a host of unacceptable
side effects.

Bismuth-containing drugs

Ulcers of bacterial origin are sometimes treatable with antibiotics, often combined with
a bismuth subsalicylate product (*Pepto-Bismol®*). This bismuth-containing substance
does not neutralize gastric acid. Possible components of its beneficial action include
enhancing mucus secretion, inhibiting pepsin activity in the stomach, and mechanically
coating the ulcer crater, but these effects are secondary to its antibacterial action against
the *H. pylori* organism.

H_2 receptor blockers

Histamine H_1 receptor antagonists do not inhibit gastric acid secretion, and they are
worthless in treating peptic ulcer. However, specific H_2 receptor competitive antagonists are powerful suppressants of hydrochloric acid secretion; they have no effect at
H_1 receptors. The H_2 antagonists inhibit fasting and nocturnal HCl secretion. There is a
reduction in the volume of acid secreted and the proton concentration is also lowered.
The output of pepsin usually falls parallel to the fall in HCl. *Cimetidine* **17.15** and

ranitidine **17.16** are representative H$_2$ receptor antagonists. Cimetidine inhibits the activity of cytochrome P-450 enzyme systems, thereby slowing the metabolism of many drugs that are substrates for these hepatic enzymes. Cimetidine prolongs the half-lives of such drugs as phenobarbital, quinidine, theophylline, propranolol, and some benzodiazepines. Other H$_2$ receptor antagonists have less effect on the P-450 systems. Cimetidine binds to androgen receptors and antiandrogenic effects, such as gynecomastia in men and galactorrhea in women, have been noted.

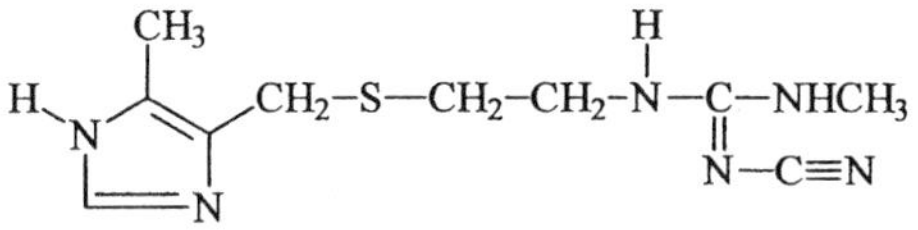

17.15 Cimetidine

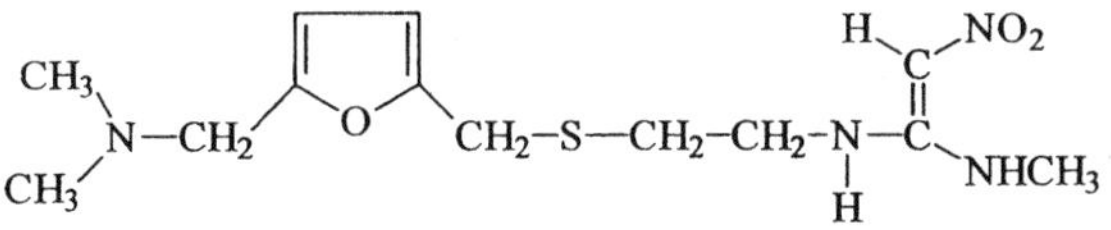

17.16 Ranitidine

Inhibitors of H$^+$–K$^+$ ATPase

As was previously described, the action of the H$^+$–K$^+$ ATPase system (a proton pump) is the final step in gastric hydrochloric acid secretion. Inhibition of this pump would be an alternative therapeutic strategy to inhibiting histamine H$_2$ receptors. Proton pump inhibitors are useful in individuals suffering from reflux esophagitis and in those whose peptic ulcer disease is not well controlled by H$_2$ antagonists. Characteristically, the proton pump inhibitors contain a sulfoxide group. At neutral pH, these drugs are chemically stable, lipid-soluble weak bases that are devoid of inhibitory activity. *Omeprazole* **17.17** is representative of this category. This drug undergoes a series of proton-mediated rearrangements in the acidic environment of the parietal cells of the stomach wall, to form a sulfenamide **17.18**. This chemically reactive compound covalently bonds to sulfhydryl groups of cysteine residues from the extracellular domain (especially cysteine 813) of H$^+$–K$^+$ATPase (structure **17.19**), thus inactivating the enzyme and inhibiting acid production. Complete inhibition of the enzyme requires

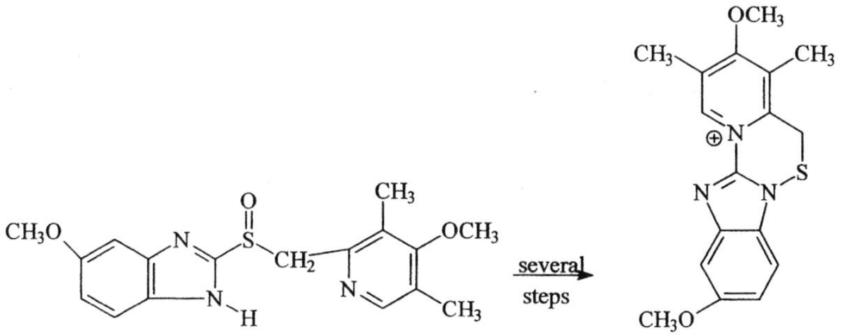

17.17 Omeprazole

several steps

17.18

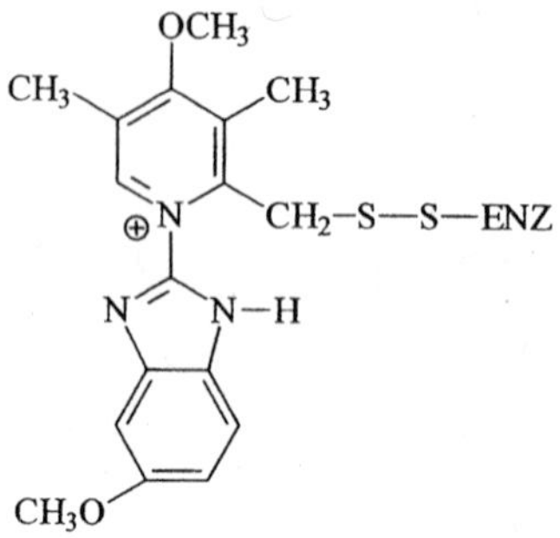

17.19

binding two molecules of the inhibitor per molecule of the enzyme. Because these drugs are irreversible inhibitors of H^+-K^+ATPase, their pharmacological effect persists after the drug has disappeared from the plasma. Inhibition of gastric acid production from a single dose of omeprazole lasts 3–4 days. Like cimetidine, omeprazole inhibits liver cytochrome P-450 metabolic reactions.

Because omeprazole undergoes acid-catalyzed rearrangement, oral administration of a simple tablet or aqueous suspension results in conversion of the omeprazole in the acidic environment of the stomach lumen into products that are not well absorbed nor transported to the appropriate locations of Na^+-K^+ ATPase in the stomach wall, and hence the drug has little or no beneficial action. Effectual oral dosage requires a microencapsulated, delayed release solid form of the drug in which the coating of the solid drug particles is stable in the acidic environment of the stomach, but disintegrates in the higher pH of the intestine, releasing the drug to be absorbed and transported to the parietal cells in the stomach where the acidic environment induces the rearrangement in the locale of the target enzyme. Because of its unique mechanism of action, inhibition of the final enzymatic step in HCl formation, omeprazole causes an inhibition of gastric acid secretion which is independent of the kind of physiological stimulation of HCl secretion that is involved.

A synthetic prostaglandin (PGE_1) analog, *misoprostol* **17.20**, inhibits gastric secretion. This drug exhibits both antisecretory and cytoprotectant effects of the natural prostaglandins. It is effective orally and after absorption it undergoes enzyme-mediated ester cleavage to generate the free carboxylic acid which is the pharmacologically active form. Its only approved use is in the prevention of mucosal injury caused by the non-steroidal antiinflammatory agents (e.g., salicylates).

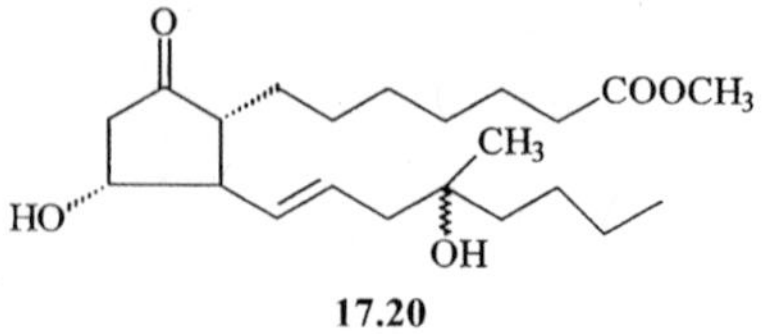

17.20

Gastric wall coating agent

The integrity of the bicarbonate-buffered mucus layer of the stomach wall is severely compromised in peptic ulcer. The observation that sulfated polysaccharides inhibit pepsin-mediated protein hydrolysis led to the development of protective agents that mimic this effect.

Sucralfate is the product of interaction between sucrose octasulfate and aluminum hydroxide. It is represented as $C_{12}H_6O_{11}[SO_3^- Al_2 (OH)_5]_8 \cdot nH_2O$. This material, a white powder, is insoluble in water. When it is exposed to the acidity of the stomach, it polymerizes and cross-links to form a sticky, viscid gel that adheres to the epithelial cells and to the crater of an ulcer, where it remains for longer than 6 h. Antacids and foods do not affect the integrity of the adherent gel. Thus, a strong protective coating, a replacement for missing mucus, is provided which promotes ulcer healing. For therapeutic success, the drug must be taken on an empty stomach 1 h before each meal and at bedtime. Sucralfate is said to be as efficacious as H_2-blocking agents, but the required daily multiple dose regimen is not appealing to many patients. Sucralfate polymer adsorbs and reduces the bioavailability of several drugs: tetracycline, digoxin, cimetidine, and phenytoin among others. This effect is minimized by administering other medication 2 h before administering sucralfate.

Recommended Reading

1. Roberts, S. and McDonald, I. M. Inhibitors of Gastric Acid Secretion. In *Burger's Medicinal Chemistry and Drug Discovery*, 6th ed. Abraham, D. J., Ed. Wiley: Hoboken, N.J., 2003; Vol. 4, pp. 85–127.
2. Hoogerwerf, W. R. and Pasricha, P. J. Agents for Control of Gastric Acidity and Treatment of Peptic Ulcers. In *Goodman and Gilman's The Pharmacological Basis of Therapeutics*, 10th ed. Hardman, J. G. Limbird, L. E. and Gilman, A. G. Eds. McGraw-Hill: New York, 2001; pp. 1005–1020.
3. Brown, N. J. and Roberts III, L. J. Histamine, Bradykinin, and Their Antagonists. In Recommended Reading ref. 2, pp. 645–668.
4. Rang, H. P. Dale, M. M. Ritter, J. M. and Gardner, P. Local Hormones, Inflammation, and Allergy. In *Pharmacology*, 4th ed. Churchill-Livingstone: New York, 2001; pp. 198–228.
5. Guyton, A. C. and Hall, J. E. *Textbook of Medical Physiology*, 9th ed. W.B. Saunders, New York, 1996; pp. 445–455.
6. Undem, B. J. and Lichtenstein, L. M. Drugs Used in Treatment of Asthma. In Recommended Reading ref. 2, pp. 733–754.
7. Tortora, G. J. The Digestive System. In *Principles of Human Anatomy*, 8th ed. Addison Wesley Longman: Menlo Park, Calif., 1999; pp. 720–725.

Glossary

acidosis Accumulation of acid or loss of base in the body.

active transport Movement of a chemical species across a membrane, which requires the expenditure of chemical energy. This process permits transport against a concentration gradient.

adjuvant A substance which aids another, such as an auxiliary remedy.

ADP Adenosine diphosphate.

afferent nerve A nerve carrying impulses toward the central nervous system; a sensory nerve.

affinity Magnitude of the attraction of an organic molecule for its receptor.

agonist A molecule that interacts with a receptor and produces a pharmacological response.

alkalosis Accumulation of base or loss of acid in the body.

allergen A substance which is capable of inducing allergy or a specific susceptibility.

allosteric site Some pharmacologically significant region on a biopolymer molecule that is physically separate from a drug (or other ligand) receptor site.

alveoli (sing., **alveolus**) Small sac-like dilatations in the lung through whose walls the gaseous exchange takes place.

anaphylactic shock (anaphylaxis) An exaggerated or unusual response of an individual to foreign protein or other substance. In **shock** there is a sudden and drastic fall in blood pressure.

anesthesia Reduction of pain perception. **Local anesthesia** is reduction of pain perception confined to one part of the body.

antiport system A physiological process or mechanism by which an ion is exchanged for another across a membrane.

antipyretic A drug capable of lowering body temperature in case of fever.

antispasmodic A drug which counteracts spasms of muscle fibers.

antitussive A drug which depresses the cough reflex.

apolipoprotein A protein associated with lipids in lipoprotein.

apoptosis Refers to cellular death.

arrythmia A variation from the normal rhythm of the heart beat.

arteriole A tiny arterial branch, especially one just proximal to a capillary.

ataractic A drug which induces mental calmness.

atheroma Arteriosclerosis with marked degenerative changes.

atherosclerosis A lesion in large arteries with deposits of yellowish plaques containing cholesterol and other lipoid material.

ATP Adenosine triphosphate.

atrioventricular [AV] node A collection of specialized cardiac muscle fibers located between the right auricle and the right ventricle. An impulse from the SA node, directing the ventricles to contract, is delayed slightly at the AV node, thus permitting the auricles to contract before ventricular contraction.

atrophy A wasting away or diminution in size of a body cell, tissue, organ or part.

autoimmune disease A condition resulting when then body's system is sensitized by endogenous proteins that are recognized as "foreign" antigens. This results in the formation of antibodies which react with these antigens to produce destructive changes in the tissues.

autoreceptor Presynaptic receptor that influences the release of neurotransmitter from the nerve terminal. Usually this is a negative feedback phenomenon. The terms autoreceptor and presynaptic receptor are frequently used synonymously.

baroreceptor reflex Sensory nerve endings in certain arteries detect changes in arterial blood pressure and transmit afferent impulses to the CNS, where they are translated into efferent impulses which lead to correction of the arterial pressure change.

bioavailability The fraction of drug that is absorbed unchanged and then reaches its site of action or the systemic circulation, following administration by any route.

bolus administration (of a drug) Usually refers to intravenous administration of the entire drug dose at once, as contrasted with intravenous drip administration.

bradycardia Abnormal slowness of heart rate.

brain stem All of the brain except the cerebrum, the cerebellum, and some ancillary connected tissue. The brain stem includes motor and sensory tracts.

cardiac output Volume of blood pumped from the heart per unit of time.

catabolism Any destructive process by which complex substances are converted by living cells into simpler compounds; destructive metabolism.

ceiling dose That dose of a drug above which no further increase in magnitude of response is noted.

chemoreceptor trigger zone (CTZ) That region of the medulla oblongata where the emetic response originates.

chronotropic Refers to the rate of contraction of a muscle, especially cardiac muscle.

chylomicron Triglyceride and cholesterol-rich lipoprotein, found in the plasma.

circadian rhythm Rhythmic repetition of certain phenomena in living organisms at about the same time each 24-h period.

clearance The volume of blood (or plasma or serum) completely cleared of total drug or of unbound drug per unit of time (e.g., ml/min^{-1}).

compartment Hypothetical spaces in the body into which a drug is assumed to be uniformly distributed.

confidence limits Numerical boundaries which are expected to contain the "true" value of a statistic at some selected level of probability.

conjunctiva The membrane that lines the eyelid and covers the exposed surface of the eyeball.

connective tissue The tissue which binds together and is the support of the various structures of the body.

cytoplasm The colloidal material within the cell membrane, exclusive of the nucleus.

cycloplegia Loss of the ability of the pupil of the eye to change size to accomodate for changes in light intensity.

dementia A general term for mental deterioration.

depolarization A displacement of the electrical potential across a nerve membrane to a less negative value. The opposite of hyperpolarization.

dermatitis Inflammation of the skin.

diastole (adj. diastolic) The period of dilatation (relaxation) of the heart muscle.

diuretic An agent which increases the output of urine by a direct effect on the kidney.

down-regulation Lowering of the sensitivity of a receptor to its specific agonist substance. Also termed **desensitization** or **refractoriness**.

drug action A drug's biological mechanism; indicates where and how the drug produces its pharmacological response.

drug effect An observed drug-induced alteration of the normal function of an existing biological process, for example: fall in blood pressure or change in heart rate.

duodenum The upper portion of the small intestine, nearest to the stomach.

dyspepsia Impairment of the power or function of digestion.

dysphoria Malaise; depression.

dyspnea Difficult or labored breathing.

EC$_{50}$ (effective concentration 50) The concentration of agonist in an assay producing 50% of the maximal response. Determined from the agonist dose–response curve.

eclampsia Convulsions and coma occurring in pregnant or puerperal women.

edema Collection of abnormally large amounts of water in the intercellular spaces in the body.

efferent nerve A nerve that carries impulses from the central nervous system to the periphery; a motor nerve.

efficacy Measure of the magnitude of response produced by a drug.

embolism A free-floating clot in an artery or vein, which travels from a larger vessel to a smaller one and obstructs it.

endocrine Secreting internally; applied to organs which secrete into the blood or lymph a substance that has a specific effect on some other organ or body part.

endogenous Developing or originating within an organism.

endoplasmic reticulum A network of protoplasmic material located in the central portion of the cytoplasm of the cell.

endothelial cell The layer of epithelial cells that lines the cavities of the heart and of the blood and lymph vessels.

endothelin A series of 21-residue peptides, some of which have been thought to be involved in hypertension and ischemic heart disease.

enteric Pertaining to the intestines.

enterohepatic circulation Absorption of a substance from the intestine and its re-excretion by the liver into the lumen of the intestine as a component of bile.

epithelial cells Cells that cover the external and internal surfaces of the body.

etiology The cause of a disease condition.

euphoria An exaggerated sense of well-being.

exocrine Secreting outwardly (as in the case of a gland); the opposite of endocrine.

exogenous Developing or originating outside the organism.

extrapyramidal effects Occurring in the brain, but outside the pyramidal nerve tracts there.

facilitated diffusion A carrier-mediated process that does not involve the input of energy. This process cannot occur against a concentration gradient.

fenestration A window-like opening in an anatomic structure.

fibroblast A connective tissue cell.

first-order kinetics A rate (e.g., of drug absorption or transport) whose change is exponential; maximum at the instant that transport begins, then constantly decreasing. Characteristic of passive diffusion.

flaccid paralysis Paralysis in which the muscle is limp and without tone.

flickering cluster Proposed model for the structure of liquid water, suggesting a highly dynamic state in which localized regions of crystal lattices are constantly forming and collapsing.

fluid mosaic Hypothesis that biological membranes have many of the properties of a fluid and that membrane-associated proteins are in a dynamic state, constantly moving to different positions in (on) the membrane matrix.

foam cell Cells with a peculiar vacuolated appearance due to the presence of complex lipids.

formed elements (of the blood) The red and white blood cells and the blood platelets.

galactorrhea Excessive or spontaneous flow of milk.

ganglion (autonomic) A collection of synapses between nerve fibers originating in the central nervous system and those terminating in an effector organ.

glial cell Cells associated with and surrounding neurons. Neuroglia support, nurture, and protect the neurons and maintain the chemical balance of the fluid that surrounds the neurons.

gynecomastia Excessive development of the male mammary glands.

hapten That portion of an antigenic molecule or antigenic complex that determines its immunological specificity.

hemisubstrate A compound that is chemically changed by the action of an enzyme, but the product of the catalysis is not desorbed from the enzyme catalytic surface, thus rendering the enzyme inert.

hyperkalemia Abnormally high potassium ion content in the blood.

hyperpolarization A shift toward a more negative value of the electrical potential across a nerve membrane. The opposite of depolarization.

hypertension Abnormally high blood pressure.

hypertrichosis An abnormal, excessive growth of hair.

hypertrophy Excess growth of an organ or a part caused by an increase in the size of its constituent cells.

hypokalemia Abnormally low potassium ion content in the body.

hypotension Abnormally low blood pressure.

IC_{50} (inhibitory concentration 50) In enzyme inhibitor assays, the concentration of inhibitor causing 50% of the maximal inhibition. Determined from the inhibitor dose–response curve.

iceberg The stable lattice of ordered, hydrogen bonded water molecules that cover the surfaces of a cell membrane.

idiopathic Of unknown cause.

inflammatory syndrome Chronic pathological condition, especially in the joints. Characterized by redness, swelling, and pain.

inotropic Refers to the contractile force of a muscle.

intrinsic activity The ability of a molecule which has interacted with a receptor subsequently to produce a pharmacological response.

in vitro Latin, "in glass"; in the test tube.

in vivo Latin, "in life"; in the living body.

ischemia Deficiency of blood delivered to some body part, due to constriction or obstruction of a blood vessel.

keratinocyte Cell in the outer layer of the skin which synthesize keratin.

latentiated drug A pharmacologically active compound that is converted into a simple derivative that exhibits superior chemical stability in vitro, or superior absorption or transport ability in vivo, and is readily metabolically converted to the originally pharmacologically active molecule.

Lewy bodies Intracellular inclusions found in the substantia nigral region of the brains of Parkinsonian syndrome victims. Presence is diagnostic of the disease.

ligand A chemical substance that interacts with a receptor.

ligand-gated channel An ion channel in a membrane whose opening and closing are mediated by the interaction of some endogenous chemical substance with a receptor associated with the channel.

limbic system One of the regions of the brain where, inter alia, pain, mood, and emotion are mediated and controlled.

lumen A cavity or channel within a tube.

lupus A destructive, characteristically chronic type of skin condition involving a local degeneration and ulceration.

macrophage A large mononuclear wandering phagocyte cell that originates in the tissues.

methemoglobinemia A condition in which the divalent (ferrous) iron in hemoglobin has been oxidized to the trivalent (ferric) state.

microsome Fragments of cells or of cell structural components formed in vitro by mechanical fragmentation.

mineralocorticoid A steroid hormone of the adrenal cortex involved physiological processes of sodium retention and potassium loss.

miosis Contraction of the pupil of the eye.

mitochondrion (pl. **mitochondria**) An anatomic component of the cytoplasm of cells. The principal site of oxidative reactions by which the energy in foodstuffs is made available to the cell for its normal physiological functions.

mucopolysaccharide Polysaccharides which contain hexosamine units and may or may not be linked to protein material. When dispersed in water, mucopolysaccharides form many of the mucins, the chief constituents of mucus.

mydriasis Dilation of the pupil of the eye.

myocardial infarction The formation of a coagulation necrosis in a blood vessel in the heart muscle, resulting from interruption of blood flow.

myocardium The heart muscle.

narcotic A term formerly designating addicting morphine-like analgesics. The term is still employed in legal terminology, but it is considered to be obsolete for pharmacological use.

negative feedback A condition in which a negative response is evoked by a stimulus.

nephron The microscopic-sized functional unit of urine formation in the kidney.

neuroleptic Formerly used synonymously with antipsychotic. Contemporary usage of the term is restricted to emphasis on the neurological aspects of a psychotropic drug.

neuropathic pain Spontaneous burning pain caused by injury to a nerve. Also, by definition, includes pain of cancer, diabetes, AIDS, and herpes zoster.

nociception The first phase of the pain phenomenon, involving the perception of the unpleasant sensation.

node of Ranvier Regularly spaced constrictions on myelinated nerve fibers where the myelin sheath is absent. Involved in saltatory conduction phenomena.

nigrostriatal pathway Nerve pathways in the brain involved in regulating the tone in certain groups of voluntary muscles.

opiate A drug which is an analog, derivative, or congener of an opium alkaloid.

opioid Any drug, natural product or synthetic, which has morphine-like analgesic and addicting properties.

OTC "Over the counter". A drug available without a prescription.

ototoxicity Having a deleterious effect upon the nerves and/or organs of hearing and balance.

parenteral Introduced into the body via some route other than the gastrointestinal tract. In common usage, refers to subcutaneous, intramuscular, or intravenous injection.

partition coefficient A numerical expression of the hydrophilicity–lipophilicity of an organic molecule. The ratio of its solubility in a nonpolar solvent to its solubility in water or an aqueous buffer.

peripheral resistance Resistance to the flow of blood through the arteries.

phagocyte Any cell that ingests microorganisms or other cells or foreign particles.

pharmacodynamics The study of the pharmacological responses to drugs.

pharmacokinetics The study of movement of drugs within the body: absorption, distribution, metabolism, elimination, especially the rates at which these processes occur.

pinocytosis Formation of incuppings (invagination) in cell membranes, which entrap protein molecules, leading eventually to their transport across the membrane.

placebo A pharmacologically inactive substance; a dosage form containing no medicinal agent. Used in controlled studies to assess the efficacy of medicinal substances.

plasma The fluid portion of the blood from which the formed elements (red and white cells, platelets) have been removed.

pleural cavity That membrane-lined space in the chest that contains the lungs.

portal vein Venous system which, inter alia, carries absorbed materials from the gastrointestinal tract to the liver.

postganglionic fiber That segment of an autonomic (motor) nerve extending from the synapse to the effector organ.

postural (orthostatic) hypotension Lowered blood pressure occurring when an individual suddenly changes posture, as from supine to erect position or standing up from sitting position or bending over from an erect posture.

preganglionic fiber That segment of an autonomic (motor) nerve extending from the spinal cord to the synapse.

pressor A substance that causes a rise in blood pressure.

prodrug A pharmacologically inactive compound that is converted in vivo into a pharmacologically active compound.

prognosis A forecast as to the probable result of an attack of disease. The prospect for recovery.

prophylaxis The prevention of disease; preventive treatment.

prostanoid Members of the families of prostaglandins and thromboxanes.

prothrombin time The experimentally determined coagulation time of a tube of oxalic acid-treated plasma to which graduated amounts of calcium chloride have been added.

pruritis Itching.

psychotropic drug A chemical substance that affects psychic functions.

Purkinje fibers Muscle fibers found in the ventricles of the heart. Thought to be involved in conduction of stimuli from the atria to the ventricles.

pyramidal tracts Pathways in the brain that are involved in voluntary movements of skeletal muscles.

quantal response An "all-or-none" pharmacological reponse, for example sleep or death.

quantitative response A graded pharmacological response, for example, change in heart rate or change in blood pressure.

receptor A cellular macromolecule with which an agonist molecule can interact and produce a pharmacological/physiological response.

recognition site Any region(s) on a macromolecule of the body to which an endogenous or exogenous molecule can bind.

reflex arc The nervous route employed in a reflex (involuntary) response to a stimulus.

REM (rapid eye movement) sleep That phase of physiological sleep in which the eyeballs move rapidly under the closed eyelids; associated with dreaming.

reticular activating system (RAS) Pathways in the brain from lower centers to the cerebrum, stimulation of which leads to excitation of the cerebral cortex.

rhinitis Inflammation of the mucus membranes of the nose.

saltatory conduction Nerve impulse conduction in myelinated nerve fibers characterized by "jumping" of the region of depolarization from one node of Ranvier to the next.

saluresis The excretion of sodium and chloride ions in the urine.

serum The clear fluid portion of the blood that lacks the formed elements and the clotting components.

sigma (σ) receptors Once called σ opioid receptors. They are not true opioid receptors, but are believed to be the site of action of certain psychotomimetic drugs, at which some opioids also interact. Their physiological role(s) is/are not well understood.

sinoatrial (SA) node A bundle of atypical cardiac muscle fibers located in the right atrium of the heart, where impulses initiating the rhythm of cardiac contraction originate. The pacemaker of the heart.

smooth muscle Involuntary muscle fibers which lack the surface striational markings characteristic of voluntary muscle.

somatic pain Pain emanating from muscle and bone, for example, toothache or headache.

spasm Sudden, violent, involuntary contraction of a muscle or a group of muscles.

sphincter A ringlike band of muscle fibers that constricts a passage or closes a natural orifice.

steady state concentration That concentration of a drug in the plasma when the rate of entry into the plasma equals the rate of elimination from the plasma.

striated muscle Voluntary muscle; muscle fibers bear surface striational markings.

sublingual Beneath the tongue.

symporter system The transport of an ion across a membrane, coupled to that of some other (oppositely charged) ion, both being transported in the same direction.

symptom Any visible evidence of disease or of a patient's condition.

synapse The region of contact/communication between two adjacent neurons, where the nerve impulse is transmitted from one neuron to the next. Also the contact/communication between a neuron and an effector organ, where an efferent nerve impulse is translated into a stimulus for the effector.

synaptosome A nerve ending that is isolated from homogenized nerve tissue.

syncope A sudden loss of strength or a temporary suspension of consciousness.

syndrome A set of symptoms that occur together.

synovial fluid An aqueous fluid found, inter alia, in joint cavities and tendon sheaths.

systole (adj. systolic) Period of contraction of the heart muscle.

tachycardia Abnormally rapid heart rate.

tachykinin A family of peptides that occurs mainly in the nervous system and performs a variety of specific biochemical tasks. Some are believed to be neurotransmitters.

tachyphylaxis Decreasing magnitude of pharmacological responses following consecutive doses of a drug given at short intervals.

tardive dyskinesia Impairment of voluntary movement that occurs late in the course of drug therapy.

threshold dose The dose level below which a drug does not produce any discernible response.

tinnitus A noise in the ears, such as ringing, buzzing or clicking.

tolerance Progressive diminution of susceptibility to the effects of a drug, resulting from its continued administration.

tone The normal degree of vigor and tension of a muscle.

trauma A wound or injury.

ulcerative colitis Chronic ulceration of the lower portion of the large intestine (colon).

urticaria A vascular reacion of the skin marked by the transient appearance of elevated patches that are redder or paler than the surrounding skin. Often accompanied by intense itching.

vasoconstriction Constriction of the blood vessels.

vasodilation Relaxation or stretching of the blood vessel walls.

vasomotor tone The normal degree of tone of the smooth muscle of the blood vessel walls.

vertigo A hallucination of movement; a sensation as if the external world were revolving around the individual or as if the individual were revolving in space. The term is sometimes used erroneously as a synonym for dizziness.

villus (pl. villi) A small vascular protrusion from a membrane surface.

visceral pain Pain involving nonskeletal parts of the body, for example gastric pain or intestinal cramp.

voltage-gated channel An ion channel in a membrane whose opening and closing are mediated by the passage of a nerve impulse.

wheal A smooth, slightly elevated area on a body surface which is redder or paler than the surrounding area. The typical lesion of urticaria.

xenobiotic A substance unnatural to a living organism or animal.

zero-order kinetics A rate (e.g., of drug absorption or transport) that is linear and constant. Characteristic of active transport.

zymogen An inactive precursor that is converted into an active enzyme by the action of an acid, another enzyme, or some other means.

Index

(Tables in *Italics*, Figures in **Bold**)

acetaldehyde, 45–46
acetaminophen, 16, *27,* 204, **204–205**
acetanilide, 42, **43,** 204, **204**
acetazolamide, 272, **272**
acetic acid, 45–46
acetophenetidine, 44, **44, 204,** 204–205
acetorphan, 218, **218**
acetylators, 53
acetyl β-methylcholine, 135, **136**
acetylcholine
   allergic responses and, 281
   blood-brain barrier and, 133
   hydrolysis, 131, 138–139, **139**
   -inactivating enzymes, 131 (*see also*
      acetylcholinesterase)
   and nerve activity, 102, 105
   pain and, 194
   receptors, 102, 105
   release, 104, 132, 141
   stereospecificity, 63
   synaptic uptake, 132–133, **133**
acetylcholinesterase, 131, 135,
      138–142, 148
acetylcoenzyme A, 133
acetylsalicylic acid, 12
   as analgesic, 195
   anion, 12

   excretion, 195
   gastric distress and, 201
   half-life, *27*
   mechanism, 198
   metabolism, 195
   tinnitus and, 195
   warfarin and, 16
acid-base balance, 272–273
actin, 261, **261**
active transport, 10, 17, 23, 110
addiction, 210, 213
adenosine, 70, **71,** 160
   diphosphate (ADP), 14, 100, 194
   kinase, 221
   phosphate esters, 160
   receptors, 160
   triphosphatase, 269, 289, **290**
   triphosphate (ATP), 14, 100, 105, 117,
      160, 194, 221, 267
adenylate cyclase, 117–118
ADME, 21
adrenal glands, 110
adriamycin, 58, **60**
alcohol dehydrogenase, 45
alcoholism, 46
aldehyde dehydrogenase, 45, 154
aldosterone, 234, 236, 275–276

allergic reactions, 120, 280–281
allergy, 120, 280–281
    atopic, 280
allodynia, 191
α-methyldopa, 238, **238**
α-methyldopamine, 238, **238**
α-methyl norepinephrine, 238, **238**
α-synuclein, 128
alprazolam, 27
Alzheimer's disease, 138, 147–149
amantadine, 27
amiloride, 277, **277**
amino acid transport, 10, 17–18, 33
amiodarone, 259, **259**
amitryptyline, 27
ammonium chloride, 34
amobarbital, 188, **188**
amoxicillin, 27
amphetamine, 18, **119**, 119–121
analgesia. *See* analgesic(s)
analgesic(s), 191
    agonists, 217–219
    assessment, 192–193
    coal tar, **204**, 204–205
    drug therapy novel approaches, 221
    endogenous, 220
    non-opioid, 195–204
    opioid, 208–221
anandamide, 173, **173**
anaphylactic shock, 280–281
anesthesia, 224. *See also* anesthetics
anesthetics
    action, 97
    general, 222–226
        body fat partitioning in, 19
    inhalation, 63, 222–225, **225**
        minimum alveolar concentration
            (MAC), 223
    intravenous, 225–226, **226**
    local, 227–228
angina. *See* myocardial ischemia
angiotensin, **235**, 235–236, 239, 242–243
    converting enzyme (ACE) inhibitors,
        242–243
angiotensinogen, 235, **235**
aniline, 44, **44**, 49, **49**
animal testing, 75, 78, 85, 167–169, 192–193

anterior pituitary gland, 125. *See also* brain
antianxiety agents, 175–177
anticoagulant(s), 25, 39
    clotting time and, 87
    heparin, 250, **251**
    -metabolizing enzymes, 37–39
    protamine, 251
    prothrombin time and, 252
    vitamin K antagonists, 251, **252**
    warfarin, 15, 25, **251**, 251–252
antidepressants, **169**, 169–71
antidiuretic hormone (arginine
        vasopressin), 271
antigen-antibody reaction, 282. *See also*
        immune response
antihistaminics, 176
antimalarials, 58, 256. *See also* quinidine
antipsychotics, 177–179
antispasmodics, 145
antithrombin III, 250
apnea, 145
apolipoproteins, 243–244
apoptosis, 128
appetite suppressants, 120
arachidonic acid, **199**, 253
area under the curve (AUC), 22
arecoline, 136, **136**, 148
arene oxide, 43, **43**
arginine vasopressin (antidiuretic
        hormone), 271
aromatic gorge, 148
aspartic acid, 157
aspirin. *See* acetylsalicylic acid
assays
    affinity (binding), 74–76
    analytical dilution, 77
    bioassay designs, 78
    biological variation and, 77
    cell-based, 76
    comparative, 78
    dose-response curve, **64–65**, 76–77, **78**,
        79, 82–84, 86, **86**
    high throughput, 75
    median effective concentration and, 81
    parallel line, 79
    quantal, **83**, 83–85
        cumulative frequency and, 83

assays (*cont.*)
    quantification, 76–79
    radioimmunoassay, 75
    slope ratio, 79
    therapeutic index (ratio), 85–87
asthma, 202, 239, 279, 283–286
ataractic agent, 152
atheroma, 243
atherosclerosis, 243–249
atomoxetine (strattera), 121, **122**
atorvastatin, 248, **248**
Atropa belladonna, 146
atropine, 48, **48**, 146–147, **147**
attention deficit disorder (ADD), 121
autonomic nervous system. *See* nervous
    system: autonomic
azathioprine, 27

baldness, 241
barbital, 188, **188**
barbiturate(s), 11, 40, 176, 179, 187–189, **188**
    anticoagulants and, 252
    categories, 188–189
    duration of action, 188
    hangover, 187
    intoxication, 189
    oxygen, 188
    therapeutic uses, 189
benzene metabolic fate, **43**
benzocaine, 227, **227**
benzodiazepines, 176, 186–187
benzyl alcohol, 42, **42**
β-endorphin, 217
bethanechol, 135, **136**, 142
bioavailability. *See* pharmacokinetics:
    bioavailability
biological membrane, 3–8
    anesthetics and, 227
    cholesterol in, 7, **7**
    drug action and, 4
    ion channels (*see* ion channels)
    lipid bilayer, 4, **6–7**, 223
    penetration by organic molecules, 9–10
    protein molecule mobility in, 7
    structure, 3–4, 7–8
        fluid mosaic model, 7
    water permeability, 8

biological screens, 76–77
bipolar disorder, 174
birth defects, 247
bismuth subsalicylate, 288
black widow spider venom, 104
blood
    albumin, 15
    anticoagulants, 15, **15**, 250–253 (*see also*
        anticoagulant(s); *specific*
        anticoagulant)
    clot, 249–250 (*see also* blood:
        anticoagulants)
    clotting, 249–250, **250** (*see also* blood:
        anticoagulants)
    formed elements, 32 (*see also* platelet(s);
        red blood cells; white blood cells)
    platelets (*see* platelet(s))
    proteins, 15–16, *16*
    serum, 250
blood-brain barrier, 16–18, 120
    anesthetics and, 222
    catecholamines and, 111
    drug enhancement of, 18
    drug penetration by, 17, 111, 133,
        135–136, 146–148, 156, 210
    enzymatic, 17
    molecular weight and, 17
    physiology, 16–17
    proteins and, 17–18
    small organic molecules and, 17
    stress and, 18
blood pressure, 123
    baroreceptor reflex and, 233
    elevated (*see* hypertension)
    postural hypotension, 171
    regulation, 110–111, 123, 233–234
bonds
    covalent, 60
    hydrogen, 8, **9**, 14
    hydrophobic, 4, 60–62, **62**
    ionic, 62
botulinum toxin, 102
bradykinin, 194, 242, 281
brain, **93**, 93–94
    cerebellum, 94
    cerebrum, 93–94, 147
    diencephalon, 93

brain (*cont.*)
  extrapyramidal tracts, 166, **167**, 179
  functions, 166–167
  hypothalamus, 94, 125, 155, 197
  limbic system, 166, **166**, 178
  medulla oblongata, 94, 125, 128–129,
      137, 147, 209–210
    chemoreceptor trigger zone of (CTZ),
      125, 128, 147, 210, 282
  microcapillaries, 17
  neurotransmitters, 152–165
  nigrostriatal pathway, 125, 128
  pineal gland, 155
  pons, 94
  pyramidal tract, 166, **167**
  respiratory center, 94, 137, 189
  reticular activating system (RAS), 167, **168**
  thalamus, 93, 154, 157
  See also hemorrhage: intracranial; nervous
      system: central
bretylium, 259, **259**
bronchial
  asthma (*see* asthma)
  dilation, 123
bumetanide, 274, **275**
buspirone, 177, **177**

caffeine, 160, **160**, 278, 287. *See also*
      xanthines
calmodulin, 117
cannabis, 172–174
Cannabis sativa, 172
cannabinoid receptors, 173
canrenone, 276–277, **277**
capsaicin, 205, **206**
captopril, 242, **242**
carbachol, 135, **136**
carbamazepine, 46, **46**
carbidopa, 128, **129**
carbinolamine, 44, **45**
carbocation, 55, **55**
carbohydrates, 7, 62
carbonic acid, 34
carbonic anhydrase, 272
  inhibitors, 273
cardiac. *See* heart
cardiac glycosides, 267–270, **269**

catecholamines
  biosynthesis, **112** (*see also specific
      compound*)
  metabolic fate, **113**
catechol-*O*-methyltransferase (COMT),
      111, 113–114, 127, 129
Celebrex®, 203, **203**
celecoxib, 203, **203**
cellular respiration, 104
central nervous system. *See* nervous
      system: central
cerebrospinal fluid (CSF), 94
cheese syndrome, 40
chemoreceptors, 94
  trigger zone (CTZ), 125, 128, 147,
      210, 282
chloral hydrate, 47, **47**, 55, 63, 185, **185**
chlorothiazide, 273, **273**
chlorpromazine, 24, **178**, 179
chlorthalidone, 237, **237**, **273**, 274
cholesterol, 7, **7**, **246**. *See also*
      hyperlipidemia
cholestyramine, 247, **247**
choline acetyltransferase, 133
cholinergic
  agents, 138–141
  nerve terminal, **132**, 132–134
  receptor stimulants, 141–142
cholinesterase inhibitors, 138–141
chylomicrons, 245
cimetidine, 288, **288**
circadian rhythms, 155
cis-aconitic acid, 35, **36**
citric acid, 35, **36**
clearance. *See* pharmacokinetics:
      clearance
clonidine, 122, **122**, 219, 238
clozapine, 179, **179**
coagulation. *See* blood: clotting
cocaine, 171–172, **172**, 227
codeine, 45, **45**, **209**, 211
coformycin, 70, **71**
colon, 12
combinatorial chemistry, 75
compartments, body, 22–23, 25
complement system, 280
computational chemists, 14

confidence (fiducial) limits, 81
conjugation reactions, 49–50
constipation, 209
cotinine, 137, **137**
cough center, 209
COX enzymes, 201, 204
crack cocaine, 172. *See also* cocaine
cromolyn sodium, 286, **286**
cyclic AMP, 117, **117**, 126, 155, 160
cyclic GMP, 264–265
cyclic nucleotide phosphodiesterase, 117
cyclooxygenase, 201, 204
cyclophosphamide, 55, **55**
cyclopropane, 223
cytochrome oxidase P-450, 39, 41, **41–42**

daidzin, 154
date rape drugs, 157, 186
DDT, 18
dementia praecox, 151
dependence, 220. *See also* addiction
depression, 40, 169–71. *See also*
        antidepressants
desflurane, 225, **225**
desoxyephedrine, 44, **44**, 121, **122**
dextromethorphan, 27, 211
diabetes, 244. *See also* insulin
diacyl glycerol, 156, **156**
diarrhea, 146, 211
diazepam, 16
diethyl ether, 223, **223**, 224
diffusion passaue, 9–10, 34, 134
diffusion, facilitated, 134
diflunisal, 195, **195**
digitalis, 267–271
digoxin, 16, 25, 35, **268**, 270
dihydroxyphenylacetic acid, 125, **125**
dihydroxyphenylalanine
        (DOPA, levodopa), 127
diisopropylfluorophosphate (DFP),
        141, **141**
diltiazem, 260, **260**, 262
dimethyl sulfide, 46, **46**
dimethyl sulfone, 46, **46**
dimethyl sulfoxide, 46, **46**
diphenhydramine, 282, **282**
diphenoxylate, 209, **210**

disulfiram (antabuse), 45–46, **46**
diuresis, 271
diuretics, 79, 237, 271–278
donepezil, 148, **149**
DOPA (dihydroxyphenylalanine,
        levadopa), 127–128
dopamine, 63, 121, 125–126
DNA, 58, **59–60**, 225
drug
    absorption, 11–16, 28–29, 38
    acidity, 11–12
    action, 3
    administration, 14
    adrenergic drugs, 119–122 (*see also*
        epinephrine; isoproterenol;
        norepinephrine)
        blockers, 122–124
        direct-indirect, 118–119
        imidazoline derived, 122
        mixed acting, 119
    agonists, 64–67, **68** (*see also specific*
        agonist drug)
    antagonists, 64–66, **68**, 82, **82**
    basicity, 11–12
    bioavailability (*see* pharmacokinetics:
        bioavailability; *specific drug*)
    blood levels, 15, 24
    carcinostatic, 58
    clinical usefulness, 87
    cumulation, **80**, 80–81
    disposition tolerance, 39–40
    dissolution, 8–9
    distribution, 15
    dose
        ceiling, 77
        effective, 81–82, 85–87, **86**
        lethal, 81, 85–87, **86**
        loading, 29
        numerical expression, 87
        -response curve, **64–5**, 76–77,
            82–84, 86, **86**
        therapeutic index, 85
        threshold, 83
    effects, 3 (*see also specific drug*)
    efficacies, 65, 67
    elimination, 28–29
    -enzyme interactions, 69

drug (*cont.*)
excretion, 34–35
half-life, 27–28
inhibitors
competitive, 70
irreversible, 70
noncompetitive, 70
reversible, 70
suicide substrate, 71–72, **72**
transition state analogs, 70
See also specific drug
intravenous administration, 21, 38
latentiated, 54–55
metabolism
age and gender differences, 54
human genetic variation in, 52
interspecies differences in, 50, **53**
in vivo, 37–40
Phase I, 41–49
Phase II, 49–50, **51–52**
of xenobiotics, 31
metabolites storage, 18–19
occupancy theory, 64–66
passive diffusion of, 9–10
potencies, 65–66
protein binding and, 25, 33
-receptor interactions, 58–62
receptors (*see* receptor(s))
solubility, 11
specificity, 63–64
steady state concentrations, 23
storage sites, 18–19
therapeutic range, 23–24
threshold concentration, 26, 77
tolerance, 79–80, 238
toxicity, 87
transport, 11–16
See also pharmacokinetics; *specific drug*
duodenum, 12
dynorphins, 165, 217, **217**
dyspnea, 267

echothiophate, **141**, 142,
eclampsia, 234
ecstasy (MDMA), 122, **122**
edema, 267, 271, 281
edrophonium, 140, **140**

eicosanoids, 198
electroencephalogram, 183
electroshock therapy, 145
embolism, 249
entacapone, 129, **129**
enalapril, 242, **242**
enalaprilate, 242, **242**
enantomers, 36–37, 62
endorphin(s), 17, 165, 208, 217, **217**, 219, 220
endothelins, 234
enflurane, 225, **225**
enkephalins, 165
blood-brain barrier and, 17
enolic acids, 203
enzyme(s)
-drug interactions, 69–70
induction, 39
inhibition, 40, 67–69
kinetics, 69–70
as receptors, 67–72
–substrate complex, 69
epibatidine, 138, **138**, 219
ephedra, **119**, 119–120
ephedrine, **119**, 119–120
epinephrine, **110**, 281
epoxidation, 46
erectile dysfunction, 163, 265
erythrityl tetranitrate, 265, **265**
estrogens, 39
ethacrynic acid, 237, 274–275, **275**
ethanol
abuse, 46
anticoagulants and, 252
antipsychotics and, 179
barbiturates and, 189
craving, 154
depressant effect of, 184–185
dopamine production and, 126
excretion, 35
liver and, 40
metabolism, 39, 45
placental penetration of, 18
structural nonspecificity, 63
ethchlorvynol, 185, **185**
etomidate, 226, **226**
euphoria, 146, 171, 178, 209–210, 214, 216

eye
    aqueous humor, 142, 273
    canal of Schlemm, 142
    drug administration, 14
    mydriasis, 146
    REM in, 153, 157, 183–184 (*see also*
        sleep: rapid eye movement)

fat, blood supply, 19
fentanyl, 212, **212**, 226
fetus, 18
fexofenidine, 282, **283**
fibrin, 249
    monomer, 249
    stabilizing factor, 249
fibrinogen, 249
fibrinolysis, 250
fight or flight, 105, 153. *See also* nervous
        system: autonomic: physiology
first-dose phenomenon, 240
first-pass effect, 21, 38–39, 42, 123–124,
        208, 211, 216, 248, 258, 264, 276
5–hydroxyindoleacetic acid, 154
5–hydroxytryptamine. *See* serotonin
5'-AMP, **117**
flaccid paralysis, 144–145
flunitrazepam (rohypnol), 186, **186**
fluoroacetic acid, 35, **36**
fluoroacetylcoenzyme-A, 35, **36**
fluorocitric acid, 35, **36**
fluoxetine, 170, **170**
flurazepam, 186, **186**
formaldehyde, 46
formic acid, 46
furosemide, 237, 274–275

galvanic skin response, 176
gamma aminobutyric acid (GABA),
        148, 156–157, 176
gamma aminobutyric acid-transaminase,
        156–157
gamma-hydroxybutyric acid (GHB), 157
gastric hypersecretion, 287
gastrin, 287
gastrointestinal tract, 9, 11–13, 38
Gaussian distribution curve, 83
gemfibrozil, **247**, 247–248,

glial (Schwann) cells, 96, 100
glaucoma, 141–142
glutamate. *See* glutamic acid
glutamic acid, 157–159
    receptors, 157–158, **159**, 166, 194, 211, 221
glyceryl trinitrate, 48, **48**, 262, 264
glycine, 159
glycoproteins, 16
guanabenz, 239, **239**
guanfacine, 239, **239**

hallucinogens, 180
haloperidol, **178**, 179
halothane, **47**, 224, **225**
hapten, 75, 280
headache, 40, 45, 163
heart
    anatomy, 231–233, **232**
    arrest, 145
    arrhythmias, 145, 255–60, 282
        atrial, 255
        causes, 255–6
        Class I drugs for, 256–258
        Class II drugs for, 258–259
        Class III drugs for, 259
        Class IV drugs for, 259–260
        premature contractions, 255
        ventricular fibrillation, 255
    cardiac output, 19, 233, 236
    congestive failure, 267–271
    hypertension and (*see* hypertension)
    infarction (*see* myocardial infarction)
    ischemia (*see* myocardial ischemia)
    pacemaker, 232
    Purkinje fibers, 259
    tachycardia, 120, 123, 174, 210–211,
        259, 267
Helicobacter pylori, 288. *See also* peptic ulcer
hemiaminal, 44, **44**
hemorrhage, 16,
    intracranial, 40, 122
heroin, 55, 126, 210, 211, **212**
hippuric acid, 50, **50**
histamine
    as chemical messenger, 161
    formation, 160–161
    metabolic fate, **162**

histamine (*cont.*)
  pain and, 194
  receptor, 281–282, 288–289
    antagonists, 146, 161
    types, 161
  release, 145, 281
HMG-CoA, 248, **248**
homovanillic acid, 125, **125**
Huntington's chorea, 158, 167
hydralazine, 241, **241**
hydrazide nitrogen, 52, **53**
hydrocholorothiazide, 237, **237**, 273, **273**
hydrolytic cleavage, 48–49
hydrophilicity, 11
hyoscine, 146, **147**
hyoscyamine, 146, **147**
hyperalgesia, 191
hyperforin, 39
hyperkalemia, 145, 276
hyperkinesia (ADD), 121
hyperlipidemia, 243–249. *See also*
      specific lipids
hyperlipoproteinemia, 243
hypertension,
  acute, 40
  drug-induced, 120
  primary (essential), 235, 272
  pulmonary, 236
  renal, 235–236
  secondary, 235
  treatment, 236–237, 242
    drugs, 237–243
    sympatholytic agents, 238–242
hypertrichosis, 241
hypertriglyceridemias, 248
hypnotics, 11, 183, 185–186
hypothalamus, 121

ibuprofen, **202**, 203
ice, 8
immune response, 279–280
indomethacin, 202, **202**
inflammation, and prostaglandins, 196–197.
      See also immune response
inflammatory syndrome, 196–201
inhibitors. *See* drug: inhibitors
inosine, 70, **71**

inositol-1,4,5-triphosphate, 156, **156**
inositol phosphate metabolic pathway, 174, **175**
insulin, 10
ion channels, 4–6, 58
  antiport, 274, 287
  calcium, 262, **263**
  ligand-gated, 6
  sodium, 4–5, **6**, 273
  symport, 274
  transmitter-gated, 6
  voltage-gated, 6, 221
insulin, assay for, 77
intercalation, 58, **60**
intrinsic factor, 287
isamyl nitrite, 264, **264**
isobutyl nitrite, 264
isoflurane, 223–225, **225**
isoniazid, 52
isoproterenol, **110**, 115–116
isosorbide dinitrate, 265, **265**
isozymes, 42

kallidin, 194
ketamine, 226, **226**
ketone bodies, 114
kidney
  acid-base balance and, 272–273
  Bowman's capsule, 31, 32
  drug elimination by, 15, 31, 33–34
  filtrate, 32
  functional microanatomy, 31–33
  glomerular filtration, 31–34, 36, 137, 271
  glomerulus, 33
  lipophilic drugs and, 33
  Loop of Henle, 271, 274
  nephron, 31, **32**
  plasma proteins and, 32
  tubular function, 33–34, 271–273
  tubules, 32–33, 271–272, 274
  urine production, 31–33
  See also diuretics
kinetics
  exponential (first order), 10, 25–27, **26**
    single compartment model, 25–26
    two-compartment drug model, 26
  zero-order, 10
Krebs cycle, 35, 45

labetalol, 124, **124**, 239
lansoprazole, *27*
leprosy, 37
lethal synthesis, 35–36
leu-enkephalin, **217**, 217–218
leukotrienes, 198, **200**, 281
levallorphan, 215, **215**
levorphanol, 211, **212**
Lewy bodies, 128
lidocaine, 48, **49**, 227–228, **227**, 258, **258**
ligand
    radiolabelled, 74
    sites of loss and, 74
Lineweaver-Burk plot, 70
linopirdine, 141, **141**
lipophilicity, 11
lipoprotein
    chemistry, 244, **244**
    high-density (HDL), 244–245
    intermediate-density (IDL), 244–245
    low-density (LDL), 243, 246
    receptors, 248
    very low-density (VLDL), 243, 248
    See also hyperlipidemia
lisinopril, 242, **242**
lithium cation, 16, 35, 174–175
liver
    disease, 24
    drug metabolism, 29
    endoplasmic reticulum, 37, **38**
    enzymes, 39
    microsome, 37–38
        age and, 54
        enzymes, 37, 41
        metabolism, 37
    reductase systems, 47
lobeline, 138, **138**
losartan, 242, **243**
lovastatin, 248, **248**
lymphocytes, 279. *See also* white blood cells
lysergic acid diethylamide (LSD), 180, **180**

magnetic resonance imaging, 193
malathion, 51–52
malignant hyperthermia, 225
marijuana, 172–174
mast cells, 280

master biological clock, 155
MDMA (ecstasy), 122, **122**
mean residence time in central compartment
    (MRCT), 27
mecamylamine, 143, **143**
mefenamic acid, 202, **202**
mefloquine, *27*
melatonin, 154–155
meperidine, 48, **48**
meprobamate, 43, **43**
mescaline, 180, **181**
metabolic inactivation, first pass, 21, 38–39
metabolism, drug. *See* drug: metabolism
metabolite solubility, 36
metanephrine, 111
met-enkephalin, 217–218, **217**
methadone, 212–213, **212**
methamphetamine, 121–122, **122**
methanol, 46, 63
methemoglobinemia, 205
methylphenidate (Ritalin), 121, **122**
methyl salicylate, 195, **195**
metoclopramide, 12
metoprolol, 123, **124**, 239, **240**
mevalonate, 248, **248**
mexiletene, 258, **258**
Michaelis-Menten equation, 69–70
minoxidil, 55, **55**, 240–241, **240**
misoprostol, 290, **290**
mitochondrion, 104, **104**
molindone, **178**, 179
monoamine oxidase (MAO), **72**, 154
    -A, 114
    -B, 114, 129, 214
    catecholamine inactivation by, 40, 111,
        113, 125, 127
    distribution, 114
    inhibition, 148–149
mood-stabilizing agents, 174–175
morphine
    age-based effects of, 54
    antagonists, 214–215
    bioavailability, 208
    clearance, 24
    esterification, 55
    first-pass metabolic inactivation, 38, 208
    half-life, *27*

morphine (*cont.*)
 inhalation anesthetics and, 226
 metabolism, 49, **49**
 structure, **209**
MPTP, 213, **214**
muscarine, 134, **134**
muscarinic agonists, 135–136
muscarinic blockers, 145–147
muscle
 contraction, **261**, **263**
 involuntary, 97
 sarcomere, 261, **261**
 smooth, 97, **98**, 106
  multiunit, 97, **98**
  visceral, 97, **98**
 striated, **98**, 106
 voluntary, **97**
mushrooms, 134, 180
myasthenia gravis, 142
myocardial infarction, 244, 249–253
myocardial ischemia
 angina pectoris, 244, 260–265
 arterial spasms and, 261–262
 classifications, 260
 pathology of, 260–262
 therapy for, 262–265
myoinositol, 174
myosin, 261, **261**

nadolol, 239, **239**
N-acetylprocaineamide, **257**, 257–258
N-allylnormorphine (nalline, nalorphine),
  214, **214**
naloxone, 215, **215**
naltrexone, 215, **215**
naproxen, **202**, 203
narcolepsy, 120–121
nausea, 45, 127–128, 147, 157, 202,
  214, 282
N-dealkylation, 44–45, **44–45**
nedocromil sodium, 286, **286**
neostigmine, 140, **140**, 142
nerve
 action potential, 99, **99**
  saltatory conduction of, 100, **101**
  transmission, 102–105
  velocity, 100

adrenergic, 107
autonomic, 97 (*see also* nervous system:
  autonomic)
cell, 91, **92**, 93, **101**, **103**
 axon tree, 102
 axoplasm, 100
 dendrite, 102
 fiber, 98
 neurotransmitter (*see* neurotransmitters)
 vesicles, 102
cholinergic, 108
depolarization, 99
ending, 97
impulses (*see* nerve: action potential)
motor (efferent), 91, 96
 autonomic, 96
 voluntary, 96, 102
myelin sheath, 100, **101**, 115
myoneural junction, 97
 blocking agents for, **143**, 143–145
neuromuscular junction, 97
nodes of Ranvier, 100, **101**
noradrenergic, 107
parasympathetic, 96, 107, **108**
postganglionic, 106
preganglionic, 106
recovery process, 99–100
resting potential, 99
sensory (afferent), 91
sympathetic, 96, 107
synapse, 91, **92**, **103**, 106
 chemical mediation of, 102–105
 postsynaptic membrane, 102
synaptic cleft, 102, 106, 110
synaptic knobs, 102, **104**
trunk, 91, **92**
See also glial (Schwann) cells
nerve gas, 141
nervous system
 adrenergic, 107, **109**
 autonomic
  nomenclature, 107–108
  physiology, *105*, 105–106
 central
  depressants, 63, 282
  neurotransmitters, 152, **153**, 154–166
   (*see also specific neurotransmitter*)

nervous system (*cont.*)
  central (*cont.*)
    psychotropic agents in, 151–182
    receptors, 74
    stimulation, 120
  cholinergic, 131–150
  dopaminergic, 125–129
  functional organization, 97
  ganglia, 106, 143
  heroin and, 211
  noradrenergic, 107, 109–125
  parasympathetic, 107
  sympathetic, 107
  See also brain; nerve; spinal cord
N-ethylglycine, 258, **258**
neuroleptic, 152
neuromodulator, 160
neuropeptide(s)
  hormones, *165*
  hypothalamic-releasing, *165*
  pituitary, *165*
  opioid, *165*
neurosis, 151
neurotransmitters
  depletion, 124–125
    enzymatic inactivation, 40, 111–114
      of epinephrine (*see also* monoamine
        oxidase)
      metabolic fate and norepinephrine, **113**
      re-uptake of norepinephrine, 112
  epinephrine, **110**, 281
  nitric oxide (*see* nitric oxide)
  norepinephrine, **110**, 110–111, **112**
  peptide, 164–165, *165*
  physiology, 165–166
  psychic role of, 153
  release, 106, **111**
nicotine, 126, 134, **134**
  addiction, 137
  agonists, 136–138
  pharmacology, 136–137
  receptors, 134, **135**
nicotinic acid, 245–247
nifedipine, 260, **260**, 262
nitric oxide, 161–164
  effects, 241
    cytotoxic, 163

gastrointestinal, 163
release, 162, **162**
synthesis, 161, **162**, **220**
nitric oxide synthase, 161
nitrobenzene, 47, **47**
nitroglycerin, 262
nitrosobenzene, 44, **44**, 47, **47**
nitrous oxide, 225, **225**
nociception, 191–192, **194**, 219
norepinephrine, **110**, 110–113, **112**
normetanephrine, 111
N-oxides, 44, **44**
N-phthaloylglutamine, 37

omeprazole, **289**, 289–290
opiates. *See specific drug*
opioid antagonists, 214–215
opium, 208–209
oral cavity, transmembrane transport in, 11
oral contraceptives, 39
orphanin FQ, 219
osmosis, 8
oxidation, 41–46
oxycodone, 211, **212**
oxycontin, 211, **212**

p-acetylaminophenol, 44, **44**
pain, 191
  acute, 191
  chronic, 191
  neuropathic, 191
  neurotransmitters in, 194
  receptors, 194
  response, 193
  sensation, 194
  sensory nerve fibers and, 194
  somatic, 191, 206
  threshold, 191
  visceral, 191, 206
paracellular absorption, 10
paraldehyde, 185, **185**
pargyline, 40, **40**, **72**
parkinsonian syndrome, 126–129, 138, 147,
    179, 213–214, 282
  DOPA side effects in, 127–128
  symptoms, 126
  treatment, 127–129

partition coefficient, 11, 13, 17, 31,
    120, 222–223
pentazocine, 216, **216**
pentobarbital, 188, **188**
pepsinogen, 287
peptic ulcer, 288, 291
peptide
    kinin, 194
    synthesis, 75
pharmacodynamics, 3
pharmacodynamic tolerance, 189
pharmacogenomics, 54
pharmacokinetics, 3, 21–30
    bioavailability, 21–22
        absolute, 22
        incomplete, 21
        oral administration and, 38
        relative, 22
    blood-to-plasma ratio and, 24
    clearance (CL), 23–24, 26
    combined absorption and elimination,
        28, **28**
        K_m, 29, 70
        t_{max}, 28
        Y_{max}, 28
    definition, 21
    half-life, 27–28 (*see also specific drug*)
    kinetics and (*see* kinetics)
    loading dose, 29
    mean residence time in central
        compartment (MRCT), 27
    nonlinear, 28–29
    volume
        of distribution (V_d) and, 24-25
        at steady state (V_{ss}), 25
pharmacology
    computer use in, 84
    definition, 3, 21
    See also pharmacokinetics
phenacetin, **204**, 205
phencyclidine (PCP, angel dust),
    159, **159**, 226
phenobarbital, 16, 39, 188, **188**, 289
phenol, 49, **49, 201**
phenoxybenzamine, 123, **123**
phenothiazine, **178,** 178–179
phentolamine, 123, **123**

phenylacetone, 44, **45**
phenylalanine, 110, **112**. *See also*
    epinephrine
phenylhydroxylamine, 44, **44**
phenylketonuria, 114–115
phenylpyruvic acid, 114
phenytoin, 16
phocomelia, 37
phosphatidyl inositol-4,5-bisphosphate
    (PIP), 174
phosphodiesterases, 160
phospholipids, 5, 7
physostigmine, 140, **140,** 142, 144, 148
pilocarpine, 136, **136**
piloerection, 211
pineal gland, 154–155
pinocytosis, 10, 17
pirenzepine, 146, **147**
piroxicam, 202, **202**
pK_a, 11, 17, 25
placebo, 193
placenta, drug transport across, 18.
    See also fetus
plasmin, 253
plasminogen, 253
platelet(s)
    aggregation, 201
    antiplatelet drugs and, 253
    kidney filtration and, 31–33
    plug, 249
poison ivy, 279
polysaccharides, 58
positron emission tomography, 193
potassium canrenoate, 276, **277**
prazosin, 240, **240**
probits, 83
procaine, 48, **48**, 227, **227**, 257
procaineamide, 257, **257**
prodrugs, 54–55
propofol, 226, **226**
propranolol, 123, **123**, 289
    as an antiarrhythmic, 258–259
    as β-adrenoceptor blocker, 123, 239
    clearance, 24
    racemates, 36
propanthaline, 12
prostaglandins, 194, 196, **197**, 198, 236

prostanoids, 198
proteases, 149
protein
    blood, 15–16
    displacement reactions, 16
    plasma, 32
    transporters, 14
prothrombin activator, 249
proton pump inhibitors, 146, 289
prazosin, 123, **123**
pseudoephedrine, 120
psilocin, 180, **181**
psilocybin, 180, **181**
psychedelic drugs, 180
psychoactive drugs
    classification, 152
    partitioning in body fat, 19
    See also specific drugs
psychological performance tests, 187
psychosis, 151
psychotomimetics, 180–181
psychotropic agents, 151–182
pyramipexole, 129, **129**
pyridostigmine, 140, **140**, 142
pyridoxine, 127
pyrogallol, 114

QSAR, 14
quinidine, 256, **256**, 289

radioimmunoassay, 75
ranitidine, 289, **289**
receptor(s)
    acetylcholine, 102, 134–135, **135**
    activation, **68**
    adenosine, 160
    adrenergic
        α, 116, **116**, 120, 123, 161, 240
        β, 115–116, **116**, 120, 234
        stimulants, 120–122
    affinity, 65–66
    analgesic, 215–216
    benzodiazepine (BZD), 176–177
    cannabinoid, 173
    cardiac glycosides, **269**
    central nervous system, 93–95
    chemical nature of, 58

    chemo-, 94
    cholinergic, 134–135, **135**, 141, 146
    cloning, 58
    definition, 57
    dopamine, 126
    -drug receptor interactions, 58–62, **62**
    efficacy, 65–66
    enzymes as, 67–72 (*see also* enzyme(s))
    GABA, 148, 156–157, 176, **176**, 188,
        219, 226
    G-linked protein, 117, **117**, 281, 287
    glutamic acid, 157–159, **159**, 166, 194,
        211, 221
    histamine, 160, 161, 281–282,
        288–289
    induced fit and, 66–67
    isolation, 58
    lipoprotein, 248
    nicotinic, 58, 142
        blockers, 142–145
    noradrenergic, 115–118
    norepinephrine, 118
    opioid, 219
    pain, 194
    presynaptic, 106–107, **107**, 171
    postsynaptic, 106–107, **107**
    purinergic, 160
    quantitative binding of, 75
    recognition site, 57
    sites of loss, 57
    steroid, 275–276
    three-point attachment hypothesis and,
        62, 62–63
red blood cells, 32, 131, 272–273
reduction reactions, 47
reflex arc, 94–95
reflux esophagitis, 288–289
renal. *See* kidney
renin, 235, 240. *See also* renin-angiotensin
        system
renin-angiotensin system, 234, **235**, 236,
        270. *See also* angiotensin
reserpine, **124**, 124–125
Reye's syndrome, 201
rheumatoid arthritis, 196–201
rhinitis, 146, 282
righting reflex, 84

rofecoxib, 203, **203**
ropirinole, 129, **129**
rule of five, 14

St. John's wort, 39
salicylamide, 195, **195**
salicylates, **195**, 195–196, 198–199
salicylism, 204
sarin, 141, **141**
schizophrenia, 138, 151, 179
scopolamine, 146, 282
secobarbital, 188, **188**
second messenger, 117–118
secretases, 149
sedatives, 11, 183, 185–187
selegiline, 129, **129**
semiquinones, 43, **43**
serotonin (5-HT)
    action, 152, 281
    biosynthesis, 153, **154**
    as CNS neurotransmitter, 152–154
    pain and, 194
    receptor
        stimulants, 12
        subtypes, 155–156
    release, 145
    re-uptake, 122, 169
shock, 46, 126
side effects, 57
sildenafil, 163, **164**, 236, 265
skin transmembrane transport, 11
sleep
    natural, 183
    paradoxical, 184
    rapid eye movement, 183–184, **184**, 186
        (*see also* eye: REM in)
    restful, 184
sodium bicarbonate, 34, 36
sodium nitroprusside, 241–242
sodium/potassium pump, 100
spasm, 144–145
speed, 121
spinal cord, 94–95, **95**, 96
    dorsal horns, 94
    gray matter, 94, 159, 194
    peripheral nerves and, 95, **96**
    ventral horns, 94

white matter, 95
    See also nervous system: central
spironolactone, 276–277, **277**
statins, 248–249. *See also* hyperlipidemia
stomach
    acid secretion, 146
    aspirin absorption, 12
    emptying time, 12
    pepsin secretion, 146
    villi, 12–13, **13**
streptokinase, 253
stroke, 147, 158, 244. *See also* hemorrhage:
        intracranial
strychnine, 159
substance P, 164–165, 194, 206, 281
succinyl dicholine, 48, **48**, 52, **144**, 144–145
succinylsulfathiazole, 48, **49**
sucralfate, 291
suicide, 94
sulfanilamide, 36, **36**
sulfisoxazole, *27*
sulfonamides, 36
sympatholytic agents, 124

tabun, 141, **141**
tachycardia, 120, 123, 174, 210–211, 259,
        267. *See also* heart
tachykinins, 164, 194, 281
tachyphylaxis, 79–80, 137. *See also* drug:
        tolerance
tacrine, **148**, 148–149
tardive dyskinesias, 179
T cells, 279. *See also* white blood cells
teeth, 19
terazosin, 240, **240**
terfenidine, 282, **283**
tetanus toxin, 159
tetracaine, 227, **227**
tetracycline antibiotics, 13, 19
tetrahydrocannibinol, 172, **173**
tetramethylammonium cation, 140
thalamus gland, 155
thalidomide, 36–37, **37**
theobromine, 160
theophylline, 160, **160**, 278, 287, 289
thiopental, 188, **188**
thiorphan, 218, **218**

3'-hydroxycotinine, 137, **137**
threshold drug concentration, 26
thrombolytic drugs, 253
thromboxanes, 198, **199**, 201
thrombus, 243
thyroid hormone receptors, 259
thyrotropin-releasing hormone, 17
thyroxine, 16
tinnitus, 195
tissue plasminogen activator, 253
tolazoline, 123, **123**
tolcapone, 129, **129**
toluene, 43, **43**
Tourette's syndrome, 138
tranquilizers, 152
transdermal patch, 264
transducer protein complex, 117–118
triamterine, 277, **277**
triazolam, 186, **186**
trichloroethanol, 47, **47**, 55
trifluoroacetic acid, 225
triglycerides, 245
trimethaphan, 143, **143**
trimethylamine, 44, **44**
tripelennamine, 216, **283**
tropolone, 114
tubocurarine, *27*, 140, **144**, 144–145
2-arachidonylglycerol, 173, **173**
2'-hydroxycotinine, 137, **137**
2,6-xylidine, 258, **258**
tyramine, 40, **40**
tyrosine, 110. *See also* epinephrine

ulcer, 288, 291
urine formation, 31–33. *See also* kidney

urinary bladder, 33, 142
urushiol, 280

valproic acid, *27*
van der Waals interactions, 4, 7, 15
vasoconstriction, *116*, 122, 126, 172,
    227, 234, 239
vasodilation, 123, 163, 234, 238, 240–241,
    247, 265, 278, 281
verapamil, 260, **260**, 262
Vioxx®, 203, **203**
volume
    of distribution (V_d), 24–25
    at steady state (V_{ss}), 25
vomiting, 45, 127–128, 147, 156–157, 202,
    211, 214, 282

warfarin, 15, **15**, 25, **251**, 251–252
water
    body, 23
    flickering cluster model of, 8
    pharmacological significance, 8
    structure, 8, **9**
white blood cells, 32, 279
wintergreen oil of, 195, **195**

xanthines, 278, 287
xenobiotics, 31, 50

zafirlukast, 286, **286**
zileuton, 286, **286**
zolpidem, 186, **187**
zymogen, 250